HAZARDOUS MATERIALS OPERATIONS

Chris Weber

PEARSON

Boston   Columbus   Indianapolis   New York   San Francisco   Upper Saddle River   Amsterdam
Cape Town   Dubai   London   Madrid   Milan   Munich   Paris   Montreal   Toronto   Delhi
Mexico City   São Paulo   Sydney   Hong Kong   Seoul   Singapore   Taipei   Tokyo

Publisher: Julie Levin Alexander
Publisher's Assistant: Regina Bruno
Senior Acquisitions Editor: Stephen Smith
Associate Editor: Monica Moosang
Development Editor: Eileen M. Clawson
Editorial Assistant: Samantha Sheehan
Director of Marketing: David Gesell
Marketing Manager: Brian Hoehl
Marketing Specialist: Michael Sirinides
Marketing Assistant: Crystal Gonzalez
Managing Production Editor: Patrick Walsh
Production Liaison: Julie Boddorf

Production Editor: Lisa S. Garboski, bookworks editorial services
Senior Media Editor: Amy Peltier
Media Project Manager: Lorena Cerisano
Manufacturing Manager: Alan Fischer
Creative Director: Jayne Conte
Cover Designer: Karen Salzbach
Cover Photo: Chris Weber
Composition: S4Carlisle Publishing Services
Printing and Binding: R.R. Donnelley/Willard
Cover Printer: Lehigh-Phoenix Color/Hagerstown

Credits and acknowledgments borrowed from other sources and reproduced, with permission, in this textbook appear on appropriate pages within the text. Unless otherwise stated, all photos have been provided by the author.

The information contained in this book is believed to be accurate at the time of publication. No warranty, express or implied, is given to the accuracy or applicability of the contents of this book to a given hazardous materials or weapons of mass destruction incident. Multiple reference sources should be used at all hazardous materials incidents. In order to safely and effectively operate at a hazmat incident the reader must be a member of an organized hazardous materials response team that trains regularly and adheres to all applicable portions of 29 CFR 1910.120 and NFPA parts 471, 472, and 473. Hazardous materials response entails risk that must be effectively managed through proper training and preincident planning.

Many of the designations by manufacturers and sellers to distinguish their products are claimed as trademarks. Where those designations appear in this book and the publisher was aware of a trademark claim, the designations have been printed in initial caps or all caps.

Library of Congress Cataloging-in-Publication Data
Weber, Chris, Ph.D.
   Hazardous materials operations / Chris Weber.
      p. cm.
   Includes bibliographical references.
   ISBN 0-13-219027-3
1. Hazardous substances—Fires and fire prevention. 2. Hazardous substances—Accidents. 3. Emergency management—Safety measures. I. Title.
   TH9446.H38W43 2012
   628.9'2—dc23

                                                    2011018689

Pearson® is a registered trademark of Pearson PLC.

10 9 8 7 6 5 4 3 2 1

ISBN 10:   0-13-219027-3
ISBN 13: 978-0-13-219027-5

DEDICATION

I would like to dedicate this book to four firefighters that have played a tremendous role in the Washtenaw County fire and hazmat community and lost their lives much too early: Brad Patton, Matt Tuttle, Amy Schnearle-Pennywitt, and Rick LaPensee. You will not be forgotten.

CONTENTS

Chapter 11 Operations Level Responders Mission-Specific Competencies: Air Monitoring and Sampling 286

Glossary

Chapter 11 Operations Level Responders Mission-Specific Competencies: Air Monitoring and Sampling 286

Glossary

Hazardous Materials Operations is designed for firefighters, law enforcement officers, military personnel, emergency medical services (EMS) personnel, and industry employees that may be faced with hazardous materials emergencies. Although emergency responders that have been trained to the hazardous materials awareness and operations levels have historically been firefighters, the law enforcement and EMS communities have been increasingly embracing hazardous materials training.

Law enforcement has been faced with clandestine drug labs that it must characterize, render safe, and collect and process evidence from. Unfortunately, a lack of hazardous materials training has led to significant injuries to law enforcement personnel. Likewise, EMS workers are called upon to decontaminate and treat victims in the warm zone as well as respond to industrial accidents involving chemical releases.

Numerous "cleanup contractors" operate at the scene of emergencies, many times before the incident is downgraded to a nonemergency situation. This is often due to the lack of training or inexperience of the first responders on the scene of the hazardous materials emergency. These industry employees should be trained to paragraph Q (emergency response operations) of 29 CFR 1910.120 (HAZWOPER), not just to paragraph E (uncontrolled hazardous waste sites) intended for waste-site cleanup workers.

This text closely follows the 2008 NFPA 472 competencies and is organized in a sequence that follows the logical progression of a hazardous materials response. Chapter 1 lays the legal and conceptual foundations of hazardous materials response. Chapter 2 covers the awareness level competencies found in Chapter 4 of NFPA 472 (2008). Chapters 3, 4, and 5 cover the core operations level competencies found in Chapter 5 of NFPA 472 (2008). Chapters 6 through 13 cover the mission-specific competencies found in Chapter 6 of NFPA 472 (2008).

The addition of mission-specific competencies to NFPA 472 (2008) marks a significant philosophical shift from past practices. Their inclusion effectively brings what were historically technician level competencies to the operations level. This allows personnel with significantly less training than a hazardous materials technician to perform in the hot zone. With the inclusion of mission-specific competencies comes the increased need for hands-on training with the equipment and tools the authority having jurisdiction (AHJ) provides to the employees. The information in this book is complemented with hands-on activities designed to reinforce key response concepts. The online component, offered through Resource Central on www.bradybooks.com, is available to both instructors and students, and contains suggested hands-on activities and exercises for each chapter. Resource Central also contains step-by-step visual guides for hands-on activities such as decontamination and donning and doffing personal protective equipment. Any operations level class is not complete without hands-on skills stations and exercises pertaining to the actual equipment students will be using in the field.

Any authority having jurisdiction that chooses to train its personnel to one or more mission-specific competencies must ensure that the appropriate equipment and refresher training are available to its personnel. This textbook provides technical information that can be used for initial training and periodic recurrent training. The Occupational Safety and Health Administration's (OSHA's) HAZWOPER law requires that emergency response personnel operating under paragraph Q receive annual refresher training. This training must be "of sufficient content and duration to maintain their competency." OSHA did not assign a length of time to refresher training. Remember that the more complicated the equipment and procedures that personnel are required to carry out, the more periodic training will be required to maintain proficiency. Stay safe!

Chris Weber

ACKNOWLEDGMENTS

This book has been the work of the better part of 5 years as I have obsessed about what to include and what not to include; welcomed the new NFPA 472 standard that made my life much easier; and spent many days trying to ensure that what I was trying to explain was actually getting explained. We all stand on the shoulders of giants, and I have had a lot of help and support from many friends and colleagues over the years, not to mention the accumulated work of the legends in the field of hazardous materials response.

I would like to thank the Washtenaw County Hazardous Materials Response Team for making me a hazardous materials responder. Many lessons were learned the easy way, during classes, training, and exercises. Over the years, we as a team honed our skills through education, training, and much practice. Some lessons were learned the hard way, but we always strived to learn the lessons well and not repeat our mistakes. I have learned so much and received so much support, especially from past and present team members such as the following: Chief Jim Roberts, Chief Bill Steele, Chief Bill Wagner, Chief Mike Skrypec, Chief Tom Edman, Chief Darwin Loyer, Chief Mark Nicholai, Captain Dan Cain, Captain Max Anthouard, Lieutenant Scotty Maddison, Supervisors Chuck Rork and Dean Lloyd, Captain Craig Liggett, Brad Tanner, Don Dettling, Tim Andrews, Doug Armstrong, Ernie Close, Jim Rachwal, Jeff French, Lieutenant Shaun Bach, Fred Anstead, Mike Loria, Andy Box, Chief Chris Bishop, and all of the other hardworking and dedicated team members.

The Michigan State Police Emergency Management and Homeland Security Training Center asked me to become an adjunct instructor in 2003, and what an eye-opening experience that was. There I had the opportunity to polish my instructing skills with sage advice from program managers and mentors Bob Cook; Tony Garcia; and Ed Halcomb. The experience and advice of the following center instructors have proved invaluable in framing my understanding of the hazardous materials world: Brian Stults, Dean Blauser, Gary Brannock, Dennis Herbert, Chuck Doherty, Don Tillery, Pat Nael, Jeremy Connell, Tim James, Don Owsley, Larry Kosmalski, Chad Tackett, Phil Salinas, Gregg Ginebaugh, Dennis Reilly (now a program manger himself), Sergeant Don Tillery, Sergeant Emmitt McGowan, and Matt Ratliff. The center wouldn't run like a well-oiled machine without the tireless efforts of Sergeant Dennis Harris, Lieutenant Mike Johnson, Stacy Theis, Renee Osborne, Rudy Walsdorf, Wendy Galbreath, Lieutenant Dave Wood, and Jim Porcello.

Over the years, I have had the honor and privilege of training thousands of students nationally and internationally. The most valuable feedback for an instructor and an author comes from the students themselves. The questions they ask steer the class and make it better. Students, please ask questions! This gives your instructor valuable information regarding your background, your interests, your temporary gaps in knowledge, and consequently what information needs to be addressed. Questions also guide the discussions and spur the instructor to add information that may be missing from the syllabus, or information and anecdotes that he is reminded of through the question. I would like to thank every student that asked a question during one of my classes, that asked questions during break that I could address immediately after break, and that honestly filled out the evaluation at the end of class. You have played a large role in shaping not only this text but also this instructor.

The Longmont Fire Department, Longmont Police Department Special Enforcement Unit, Michigan State Police Emergency Management and Homeland Security Training Center, Colorado WMD-CST, Las Vegas Metro Police ARMOR, Oakland County Regional Response Team, Pittsfield Township Fire Department and Ypsilanti City Fire Department were kind enough to provide me with opportunities to take photographs at exercises and training scenarios.

Some individuals have helped by discussing aspects of the book, others have starred in photos or provided photos and case histories, and still others have read and edited chapters to make sure I didn't leave anything out and to make sure the copy wasn't gibberish. My co-instructors Bob Cook and John Meyers have given me tremendous feedback on many of the chapters; thank you. I would also like to thank the following colleagues for their generous help during this project: Mike Becker, Jeff Peterson, Martin McFarland, Michele Goldman, Todd Feaster, Craig Paull, Jeff Moll, Brian Jackson, Ethan Unwin, and Jared Mannering (Longmont FD); SSG Jay Lemons (8th WMD-CST); Brian Wagner (National Institutes of Health Fire Department); Major Dennis Fivecoat (92nd WMD-CST); Doug Huffmaster (Las Vegas Metro PD/ARMOR); Steve Curry (Columbia, SC FD); Dean DeMark (Fort Bragg); Jesse Taylor (JTL); Tony Saucedo (Michigan State Police Methamphetamine Interdiction Team); Tony Garcia, Dennis Harris, and Chad Tackett (Michigan State Police Emergency Management and Homeland Security Training Center); Jay Daum (U.S. Customs and Border Protection); Alan D'Agostino, Sean Gleason, Jeff Foulke, and Dave Dorian (Pittsfield Township Fire Department); Jeff French (Superior Township Fire Department); Don Ostrowski (Troy Police Department); Chris Waier (Lansing Fire Department); Gary Sharp (Independence Township Fire Department); MSgt William Bennett (51st WMD-CST); and Chris Wrenn (Environics USA).

I would like to thank the following individuals for putting forth the time and effort to review this book. Their comments and suggestions were invaluable throughout the process.

John P. Alexander
Senior Adjunct Instructor
Connecticut Fire Academy
Windsor Locks, CT

Maryam Azarbayjani, Ph.D.
California State University, Los Angeles
Los Angeles, CA

Shankara Babu Ph.D., CHMM, CPEA, CEM
Professor
ESH Program
College of Southern Nevada
Henderson, NV

David Kidder
Fire Chief
East Texas Regional Airport Fire Department
Longview, TX

Amber L. Reamy, MA
Instructional Designer
Institute Development Section
Maryland Fire and Rescue Institute
University of Maryland
College Park, MD

Douglas Rohn
Lieutenant
Madison Fire Department, Station 6
Madison, WI

Derek Silva, EMT-P
CBRN/Medical Specialist
New York, NY

Hezedean A. Smith, BS, MA, EMT-P
Lieutenant
Orlando Fire Department
Orlando, FL

Jeff Travers
Director of Public Safety
Great Oaks Institute
Cincinnati, OH

David Williams, EMT-P, HMT, CSM
Training Officer
Omak Fire Department
Omak, WA
Corporate HSE/Risk Manager
VECA
Seattle, WA

Gray Young
Chief of Fire Training
South Bossier Fire
Elm Grove, LA
Adjunct Instructor
Fire and Emergency Training Institute
Louisiana State University
Baton Rouge, LA

I would also like to thank David Heskett, an aspiring artist and graphic designer, for much of the artwork in this book. He has done a tremendous job. His other work can be viewed at the website www.DHdesigns.org.

I would like to thank my family for their patience and support during all of my writing projects, including this book. Long writing projects are difficult not only for authors but also for their families. Sometimes the project appears all consuming to everyone involved. I deeply appreciate their understanding and help through the duration!

Chris Weber teaches hazardous materials topics internationally, actively develops and instructs numerous hazardous materials and counterterrorism courses, and runs full-scale training exercises through his training and consulting firm, Dr. Hazmat, Inc. (www.drhazmat.com). Chris regularly instructs illicit laboratory response classes for the U.S. military; tactical chemistry classes for hazmat technicians, 24-hour Hazmat Medic classes for EMS personnel; such regulatory classes as HAZWOPER, RCRA, and DOT courses and refreshers for industry; and numerous other courses for hazardous materials responders including:

Hazardous Materials Operations Train-the-Trainer Course
Hazardous Materials Technician Train-the-Trainer Course
24-, 40-, 80-, 160-, and 240-hour Hazardous Materials
    Technician Courses
Air Monitoring and Sample Identification Course
Tactical Chemistry Course
Illicit WMD and Drug Laboratory Recognition and Response Course

Chris is currently a subject matter expert (SME) with the Longmont (Colorado) Fire Department's Hazardous Materials Response Team. Chris's past experience includes serving on the Washtenaw County (Michigan) Hazardous Materials Response Team for over a decade in positions that include hazmat technician, training officer, and deputy director. He has been a firefighter for over 20 years and is an adjunct instructor with the Michigan State Police Emergency Management and Homeland Security Training Center in Lansing, Michigan. He has extensive experience involving hazardous materials chemistry, including a Ph.D. in cellular and molecular biology and biological chemistry from the University of Michigan, Ann Arbor, and 15 years' research experience in the biomedical sciences. He can be reached at cweber@drhazmat.com.

His other publications include:

Pocket Reference for Hazardous Materials Response

Introduction

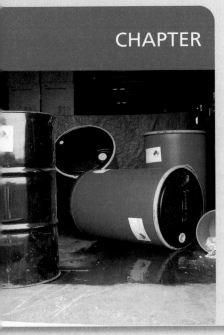

OBJECTIVES

- Describe the requirements of HAZWOPER in regard to hazardous materials emergency response and hazardous materials remediation work.
- List the five levels of hazardous materials training.
- Describe the training requirements at the hazardous materials awareness level.

- Describe the training requirements at the hazardous materials operations level.
- Summarize the requirements of the Resource Conservation and Recovery Act (RCRA).
- Summarize the requirements of CERCLA and EPCRA.
- Summarize the requirements of the Hazardous Materials Regulations (HMR).
- Describe the differences between the HAZWOPER regulation and the NFPA 472 (2008) standard.

Resource Central

For additional review and practice tests, visit **www.bradybooks.com** and click on Resource Central to access book-specific resources for this text! To access Resource Central, follow directions on the Student Access Card provided with this text. If there is no card, go to **www.bradybooks.com** and follow the Resource Central link to Buy Access from there.

You Are at Work! What Do You Do?

Your Town, North America

It is a late June summer evening. A parent calls 911 stating that his twin 9-year-old boys are feeling ill. They are having difficulty breathing and have severe rashes on their face, legs, hands, and arms. While resources are being dispatched to the location, two other 911 calls are received from parents indicating their children have the same signs and symptoms. All three families live within a half mile of each other. The dispatcher finds this suspicious and asks all three callers questions that would indicate whether the children had a common history. It turns out that all five children had played at an abandoned industrial facility that afternoon. Due to the quick thinking of the dispatcher, all responding units are notified that this medical emergency may involve a hazardous materials exposure.

- How would your EMS agency handle this incident?
- How would your fire department handle this incident?
- How would your law enforcement agency handle this incident?
- Do your response personnel have the appropriate training? What is the appropriate training?
- Do your response personnel have the appropriate equipment and resources to handle this medical emergency?
- Would patient decontamination be necessary? Who would perform it? How, and with what equipment? Is personal protective equipment (PPE) available?
- Who has the responsibility to secure the abandoned industrial facility? How would that be accomplished?

Let's try to answer some of these questions in this chapter and the ones that follow.

hazardous material ■ A substance that, because of its quantity, concentration, corrosiveness, flammability, reactivity, toxicity, infectiousness, or radioactivity, constitutes a threat to human health, public safety, or the environment.

corrosive ■ A liquid or solid that causes destruction of human skin at the site of contact or reacts with steel or aluminum at a specified rate.

radioactive material (RAM) ■ A material containing an isotope that spontaneously emits ionizing radiation.

Hazardous materials are chemicals that pose an unreasonable risk to people, property, or the environment. Where would you encounter hazardous materials? They are all around us. Hazardous materials are in your home in the form of gasoline, the **corrosives** in drain cleaners, the potent toxins in pesticides, and even the **radioactive material (RAM)** in your smoke detector. When these hazardous materials in the home are well managed, they do not pose an unreasonable risk. However, when they are mismanaged or get into the hands of small children, they can be very dangerous. Of course, it is not the hazardous materials in your own home that are covered in this book, but rather the hazardous materials that are transported, produced in factories, and used in industry on a daily basis throughout North America. The biggest differences between home use and industrial use of hazardous materials are the quantities and concentrations that are involved. When hazardous materials are shipped and used in large, concentrated quantities, any accidents tend to become significant incidents or even emergencies. The better we understand how hazardous materials behave; how hazardous materials can be dangerous to us; and what we can do to protect ourselves, our coworkers, and the public, the better off we all will be.

Awareness Level Personnel Versus Operations Level Responders

You may have asked yourself one or more of the following questions since you found out you were going to take this class: Why do I have to take an awareness level or operations level hazardous materials class? Will I ever encounter hazardous materials while doing my job? What is so bad about hazardous materials anyway? Let's explore some of the scenarios you may encounter. The following are a mixture of hypothetical situations and actual incidents. How would you react to them?

CANCER FORCES EARLY RETIREMENT FOR HIGHLY DECORATED FIREFIGHTER

Fort Lauderdale, Florida
February 21, 1993, Sun-Sentinel by Tao Woolfe, staff reporter
 The firefighters who have worked with Herbie Blabon wish his retirement was just another one of his practical jokes.
 They wish the cancer forcing him to retire would simply go away. But Blabon, the most decorated firefighter in the city, was prepared to walk out of the fire house this morning at 8 a.m. after working his last 24-hour shift.
 It wasn't easy.
 "I'll always be a fireman," he said on Saturday night, his voice almost breaking. "I'll still get excited when I see a truck go by."
 Blabon was one of 100 firefighters who fought the January 13, 1969, Everglades Fertilizer Plant fire that burned for three days and spewed chemical-laced smoke that peeled the paint off trucks [Figure 1-1]. The firefighters battled the blaze with no breathing masks. Blabon is the 29th of those firefighters who has been diagnosed with cancer. Nineteen have died.
 Blabon, 49, who has been with the department for 27 years, said his illness, which has caused the tumor to grow in his neck, is forcing him to retire a year early. Although the cancer is in remission, he must go through chemotherapy treatments that will render him too weak to work.
 "It hit everyone around here like a baseball bat. The guys said: 'How could it be him? How could be have cancer?'" Battalion Chief Stephen McInerny said. "The guys couldn't look each other in the eyes.
 Blabon had always been there for everyone else. And his practical jokes were legendary.
 "You could be having a bad day and reach up to get a bowl out of the cabinet and have it be half filled with water, and it would make you laugh," said firefighter Time Habig, 33, whose bunk was next to Blabon's. "He was like my surrogate father to me. I'll miss having someone to lean on."

— *Reprinted with permission from the Sun Sentinel.*

FIGURE 1-1 A burning factory or store can release toxic gases and smoke that can sicken first responders and the public, as well as cause contaminated water from firefighting operations to run into rivers, streams, and the groundwater table. *Art by David Heskett*

Awareness level personnel are any workers in almost any industry that could discover either a hazardous materials release or be present during a **weapons of mass destruction (WMD)** incident. This includes just about every worker in the United States. Some workers are more likely to encounter these incidents than others. Dispatchers, utility workers, teachers, public works employees, employees of colleges and universities, and chemical industry workers are among the people that should be trained to the hazardous materials awareness level. You might have noticed that I left out firefighters, police officers, and emergency medical service personnel. This is because those people are considered first responders and should be trained to a minimum of the hazardous materials operations level per the **National Fire Protection Association (NFPA)** 472 standard (2008).

RESIDENTIAL MERCURY SPILL

Staff from the local public health department recently responded to a small-volume mercury release in an apartment. Two adults and two boys (ages 2 and 11) resided in the apartment. A wall-mounted thermostat was determined to be the likely source of mercury **contamination**. The residents reported that an electrician had recently replaced the living room thermostat. The mercury was discovered a day later when the parents observed the 2-year-old boy lying down on a chair and pushing a bead of mercury around with his tongue. Closer inspection revealed many small mercury beads on the chair and surrounding carpet. The tenants immediately took both children to the hospital. The hospital collected biological samples from both children (blood and urine). Blood samples collected from the 2-year-old showed 10 micrograms per liter of mercury. The 2-year-old boy was subsequently admitted to the hospital for two nights. X-rays revealed three "dense metallic" forms in his gastrointestinal tract. Three days later, the mother noted that the 2-year-old became extremely violent—an abnormal behavior for the child. Subsequent X-rays revealed two mercury beads had passed out of the child, but one bead, approximately 0.125 inch in diameter, remained lodged in his gastrointestinal tract. On November 13, 2001, the parents took the 2-year-old boy to the Detroit Children's Hospital where the child was physically manipulated (suspending upside down, rotating, etc.), which worked to free the remaining mercury and allowed it to pass from the child.

— Excerpted and adapted from Agency for Toxic Substances & Disease Registry (ATSDR), *Health Consultation: Mercury Spill Assist in Watervliet, Michigan* (http://www.atsdr.cdc.gov/hac/pha/watervliet/msa_p1.html)

CLANDESTINE METHAMPHETAMINE LABORATORY RESPONSE

Hazmat Times, Yourtown, USA

After receiving a tip today, police responded to a residence on Main Street that contained a **clandestine laboratory** in the basement (Figure 1-2). Officers entered the residence after obtaining a search warrant and proceeded to investigate. During the investigation they observed numerous chemical bottles and what appeared to be laboratory glassware containing reacting chemicals. Approximately 30 minutes into the investigation, one officer reportedly started to feel ill. Before being overcome by fumes, Officer Smith stated he noticed a strong, pungent odor coming from a pot on a makeshift stove. Sergeant Jones helped Officer Smith out of the residence, and paramedics treated and transported both officers to the hospital. Sergeant Jones was treated and released, and Officer Smith is listed in stable condition at Memorial Hospital. The state occupational safety and health administration announced today that it will investigate the police department's response to this incident.—CHW

FREIGHT TRAIN DERAILMENT

At 5:07 A.M. on August 7, 1981, the residents of Bridgman, Michigan, awoke to the unmistakable screeching of grinding and bending steel. Fourteen cars of a northbound C&O train with a total of 79 cars had derailed at the 60th car. This car belonged to

---

weapon of mass destruction (WMD) ■ Any weapon composed of explosives, chemical agents, biological agents, or radioactive material designed to inflict large numbers of casualties and/or large amounts of property damage.

National Fire Protection Association (NFPA) ■ The voluntary membership organization that promotes and improves fire protection and prevention, establishes safeguards against loss of life and property by fire, and promulgates many safety standards.

contamination ■ The unintentional transfer of a spilled hazardous material to personnel and equipment. Contamination must be removed through decontamination to prevent cross contamination of other personnel and equipment.

clandestine laboratory ■ A location where illegal drugs, chemical agents, biological agents, or explosives are made using precursor chemicals, reagents, and solvents. Also known as a clan lab, illicit laboratory, or illicit lab.

FIGURE 1-2 A clandestine methamphetamine laboratory.

Dow Chemical out of Midland, Michigan, and contained 9655 gallons of fluorosulfonic acid. A large white plume began escaping from the overturned car almost immediately. A total of 3200 gallons of liquid would escape from the tank that day before the leak was eventually plugged.

Various police, fire, and EMS agencies responded to the accident. **Evacuations** were started, a command post was set up, the scene was secured, and eventually mitigation efforts were started. A law enforcement officer positioned at the inner perimeter of the release scene, approximately 100 yards from the overturned railcar, was responsible for keeping motorists, the media, and curious onlookers away from the leaking tank car. He spent a total of 13 hours in close proximity to the white plume. Several other law enforcement officers also complained of breathing discomfort and skin irritation. Several officers even returned to the station with tarnished weapons from the corrosive plume. On the morning of August 29, 1981, after a coughing fit, the 37-year-old law enforcement officer died at his home of a massive heart attack. This death was subsequently ruled as a line-of-duty death caused by one component of the noxious fumes, hydrogen fluoride.

evacuation ■ The removal of endangered people from a hazard zone.

The goal of this book and this training program is to prepare you to be able to handle exactly these types of situations in a safe, competent, and professional manner. The more quickly a dangerous situation can be faced and handled in a calm and efficient manner, the fewer injuries will result and the less damage will be done.

Legal Foundation of Hazardous Materials Response

One reason your employer is putting you through this training program, along with being concerned for your safety, is that several laws and federal regulations require it. State and federal governments regulate hazardous materials response due to the danger these types of incidents pose to responders and the impact hazardous materials incidents have on the public, property, and the environment. State regulations may be stricter than federal regulations; therefore, you should also be familiar with your state's laws and regulations. A good place to start checking is your state's Department of the Environment or Department of Environmental Quality.

Three agencies are primarily responsible for hazardous materials regulation at the federal level: the Environmental Protection Agency (EPA), the Occupational Safety and Health Administration (OSHA), and the Department of Transportation (DOT). Congress authorized these agencies to promulgate regulations by passing legislation mandating that the agencies regulate how hazardous materials are disposed of and affect the environment, how hazardous materials are handled in the workplace, and how hazardous materials are transported.

Two significant hazardous materials regulations are contained in the Superfund Amendments and Reauthorization Act (SARA). SARA Title I is commonly known as HAZWOPER, which stands for **Hazardous Waste Operations and Emergency Response**. HAZWOPER is the federal statute that regulates public sector and private sector response to hazardous materials incidents; **hazardous waste** sites; and the operation of treatment, storage, and disposal facilities. SARA Title III is commonly known as EPCRA, which stands for **Emergency Planning and Community Right-to-Know Act**. Let's explore each of these in turn because together they lay the legal foundation for hazardous materials response in the United States.

HAZARDOUS WASTE OPERATIONS AND EMERGENCY RESPONSE (HAZWOPER) 29 CFR 1910.120

HAZWOPER (Hazardous Waste Operations and Emergency Response) is a comprehensive hazardous materials worker regulation. The Environmental Protection Agency (EPA) has also adopted a version of HAZWOPER, which is found in 40 CFR 311. Personnel that may not be covered by OSHA's regulation are covered by EPA's. Your actions as awareness level personnel or as an operations level responder will primarily be governed by the HAZWOPER regulation. The following explores this regulation in detail.

Paragraph A: Scope

Unless the employer can demonstrate that the operation does not involve or have the reasonable possibility to involve employee exposure to safety or health hazards, HAZWOPER applies to:

1. Cleanup operations required by a governmental body at uncontrolled hazardous waste sites
2. Corrective actions involving cleanup at **Resource Conservation and Recovery Act (RCRA)** sites
3. Voluntary cleanup operations at governmentally recognized uncontrolled hazardous waste sites
4. Hazardous waste operations at treatment, storage, and disposal (TSD) facilities
5. *Emergency response operations for releases of, or substantial threats of releases of, hazardous substances without regard to the location of the hazard*

Paragraphs B Through O: Waste-Site Worker Regulations

These regulations differ substantially in certain areas with respect to the regulations for emergency response workers, which are described in paragraph Q (see below).

Paragraph F: Medical Surveillance

Emergency response employers and employees must comply with this regulation, including public safety agencies operated by local and state governments. Covered employees are:

1. Employees who are, or may become, exposed to hazardous substances at or above the permissible exposure limit (PEL) for 30 or more days per year
2. Employees who wear **respirators** for 30 or more days per year

3. Employees who are injured, become ill, or develop signs and symptoms due to a job-related exposure
4. *Members of hazardous materials response teams*

Medical exams shall include a medical and work history with emphasis on hazardous substance exposure and fitness to wear **personal protective equipment (PPE)**. All exams must be performed by or under the supervision of a licensed physician and shall be provided without cost to the employee. Certain information must be provided to the physician by the employer. This includes:

1. A copy of the HAZWOPER standard
2. A description of the employee's duties
3. The employee's anticipated (or actual) exposure levels
4. A description of the PPE to be used by the employee
5. Information pertaining to previous medical exams that are unavailable to the physician
6. Any other information required by 29 CFR 1910.134 (Respiratory Protection regulation)

HAZWOPER also stipulates when a medical exam must be performed, and the frequency of follow-up medical exams and consultations:

1. Prior to assignment
2. At least once every 12 months, unless the attending physician believes a longer interval (not to exceed 24 months) is appropriate
3. At termination of employment or assignment (if there has not been an exam within the preceding 6 months)
4. As soon as possible if the employee has been exposed above the PEL or has developed signs and symptoms consistent with a job-related exposure to hazardous substances
5. More frequently if the examining physician believes it medically necessary

The employer must give the employee a copy of the physician's written opinion, and the physician must inform the employee of the findings of the medical exam. The physician may not inform the employer of specific findings or diagnoses unrelated to occupational exposure. The employer must keep the medical records for the length of time specified in 29 CFR 1910.120. The records shall contain the name and social security number of the employee, the physician's written opinion(s), employee medical complaints, and a copy of the information supplied to the physician by the employer.

Paragraph G: Engineering Controls, Work Practices, and PPE

Engineering controls, work practices, personal protective equipment (PPE), or a combination of these shall be implemented to protect employees from exposure to hazardous substances and safety and health hazards. Whenever feasible, engineering controls and work practices need to be instituted to reduce and maintain employee exposure below the **permissible exposure limit (PEL)**. Examples of engineering controls include pressurized cabs or control booths on equipment and the use of remotely operated material handling equipment. Examples of work practices are staying upwind of the release, keeping nonessential personnel out of the **hot zone**, and wetting down dusty operations.

When engineering controls and changes in work practices are insufficient to eliminate the health and safety hazards, PPE shall be selected and used that will protect employees from the hazards and potential hazards they are likely to encounter (as identified during the **site characterization** and analysis). PPE selection shall be based upon hazard- and site-specific performance criteria. Furthermore, supplied air breathing apparatus must be used in **immediately dangerous to life or health (IDLH)** conditions. Level A (gas and vapor tight) protection must be used if a skin absorption hazard is present that may

personal protective equipment (PPE) ▪ A combination of chemical protective clothing (CPC) and respiratory protection designed to prevent hazardous materials from contacting and harming the wearer.

engineering control ▪ Mechanical or structural devices or equipment that reduces the risk or extent of injury or exposure to a hazardous material or atmosphere.

permissible exposure limit (PEL) ▪ The OSHA enforced time-weighted average threshold limit value to which workers can be exposed continuously during an 8-hour work shift and a 40-hour work week without suffering ill effects.

hot zone ▪ The exclusion zone in which hazardous materials are located, could reasonably be expected to migrate towards, or the area effects may be felt upon catastrophic container failure (such as from a boiling liquid expanding vapor explosion, or BLEVE).

site characterization ▪ The process of determining the dangers to responders and the public at a hazmat materials release.

immediately dangerous to life or health (IDLH) ▪ The atmospheric concentration of any substance that poses an immediate threat to life, causes an irreversible or delayed adverse health effect, or interferes with an individual's ability to escape during a 30-minute period. IDLH atmospheres may not be entered unless supplied air respiratory protection is used.

OSHA considers personal protective equipment (PPE) the last line of defense against hazardous materials, not the first! Always attempt to mitigate the effects of hazardous materials using engineering controls and safe work practices first; then use PPE as a final precautionary measure.

cause an IDLH exposure. The level of PPE must be increased if and when additional information or on site condition indicates an exposure above the permissible exposure limit (PEL) is possible. Conversely, the level of PPE may be decreased if site conditions improve and an exposure above the PEL will not result from downgrading to a lower level of PPE. Totally encapsulating chemical protective suits (level A) must protect employees from the hazards, be capable of maintaining positive air pressure, and prevent inward test gas leakage of more than 0.5%. And last but not least, a written PPE program must be part of the employer's safety and health program and contain the following elements:

1. PPE selection criteria that are based upon site hazards
2. How to use the PPE and limitations of the equipment
3. Work mission duration
4. PPE maintenance and storage
5. PPE **decontamination** and disposal
6. PPE training and proper fitting
7. PPE donning and doffing procedures
8. PPE inspection prior to, during, and after use
9. Evaluation of the effectiveness of the program
10. Limitations during temperature extremes, heat stress, and other appropriate medical considerations

decontamination ▪ The process of removing unwanted contaminants from personnel and equipment.

Paragraph P: Treatment, Storage and Disposal Facility Regulations (RCRA)

These regulations differ substantially in certain areas with respect to the regulations for emergency response workers, which are described in paragraph Q (see below).

Paragraph Q: Emergency Response to Hazardous Substance Releases

Paragraph Q specifies how public and private emergency response agencies must handle hazardous materials incidents. First it stipulates that an emergency response plan must be developed to address anticipated emergencies (transportation, pipeline, fixed site facilities, etc.). The emergency response plan should address the following key points if not addressed elsewhere, such as in standard operating guidelines, the local or state emergency response plan, or any applicable SARA Title III facility plans:

1. Pre-emergency planning and coordination with outside agencies
2. Personnel roles, lines of authority, training, and communication
3. Emergency recognition and prevention
4. Safe distances and places of refuge
5. Site security and control
6. Evacuation routes and procedures
7. Decontamination
8. Emergency medical treatment and first aid
9. Emergency alerting and response procedures
10. Critique of any hazardous materials incident responses and follow-up actions
11. Personal protective equipment (PPE) and other emergency equipment

Procedures should be developed for handling an emergency response. These response procedures should contain the following key points:

1. The senior response official will become the incident commander under an **incident command system (ICS)**. The incident command system should be National Incident Management System (NIMS) compliant. See Chapter 5, Operations Level Responders, Core Competencies: Responding to Hazardous Materials Incidents, for more information.
2. They will identify, to the extent possible, all hazardous substances and conditions present.
3. They will implement appropriate emergency operations using the appropriate PPE.
4. If the possibility of an inhalation hazard exists, **self-contained breathing apparatus (SCBA)** must be worn until appropriate air monitoring has been performed and a decreased level of respiratory protection is warranted.
5. The **buddy system** must be used in the hot zone. Only personnel actively performing emergency operations are permitted in the hot zone.
6. Backup personnel must be available for rescue of hot zone personnel. Advanced first aid support personnel must be on standby with medical equipment and transport capability.
7. An experienced and knowledgeable **safety officer** must be appointed.
8. If IDLH or imminent danger conditions exist, the safety officer has the authority to terminate or alter emergency activities immediately and unilaterally.
9. After emergency operations have terminated, appropriate decontamination procedures must be implemented.
10. When deemed necessary for meeting the tasks at hand, approved SCBA bottles may be interchanged with different manufacturers provided they are of the same capacity and pressure rating.

Skilled support personnel may be used that do not possess HAZWOPER training. During an emergency, these personnel can only perform the task(s) that cannot be performed by hazmat team members, and they must be briefed prior to entry. The briefing must encompass:

1. Instruction on use of appropriate PPE
2. Chemical hazards involved
3. Duties to be performed

Specialist employees must receive training and demonstrate competency annually in their field of specialization.

HAZWOPER defines five levels of training: an awareness level, an operations level, a technician level, a specialist level, and an incident commander level of training. *Awareness level personnel* should understand and recognize hazardous materials incidents and the associated risks, with the ultimate goal of self-evacuation and the notification of appropriate authorities that can take further action. They should be able to function within the agency's emergency response plan, understand the use of the DOT's *Emergency Response Guidebook*, and be able to implement site security and control.

Operations level personnel should protect nearby persons, property, and the environment in a defensive fashion, without coming into direct contact with the hazardous substance. In addition to the awareness competencies, they must know basic hazard and risk assessment techniques; select and use PPE; perform basic control, **containment**, and/or **confinement** operations; and implement decontamination procedures. The operations level training must be a minimum of 8 hours in length.

Technician level personnel respond to hazmat emergencies for the purpose of stopping the release. They must be able to:

1. Implement the employer's emergency response plan.
2. Use field survey instruments to verify and/or determine the nature of the release.

incident command system (ICS) ▪ A system designed to optimally use and direct resources at emergency incidents.

self-contained breathing apparatus (SCBA) ▪ A respirator consisting of a face mask and source of breathable air carried on the wearer's back; required to be used in an IDLH atmosphere.

buddy system ▪ The procedure of always entering the hot zone in groups of at least two. The buddy system is mandated by OSHA's respiratory protection standard for safety.

safety officer ▪ In an ICS, the person assigned to ensure that the strategy and tactics implemented on scene are safe. The safety officer has the authority to immediately stop unsafe actions.

containment ▪ The procedure of stopping a hazardous material from leaking out of its container. Examples of containment include patching and plugging.

confinement ▪ The procedure of limiting a hazardous materials release to a defined area. Examples of confinement include diking, damming, and booming.

3. Function within the ICS.
4. Select and use PPE. Chemical protective clothing must meet the requirements of paragraph G (see above).
5. Understand hazard and risk assessment techniques.
6. Perform advance control, containment, and/or confinement techniques.
7. Understand and implement decontamination procedures.
8. Understand termination procedures.
9. Understand basic chemical and toxicological terminology and behavior.
10. The technician level training must be a minimum of 24 hours in length.

Specialist level personnel respond with and support hazmat technician level personnel. They must know how to implement the local emergency response plan; be able to use advanced survey instruments and equipment; select and use PPE; understand in-depth hazard and risk assessment techniques; perform specialized control, containment, and/ or confinement techniques; determine and implement decontamination procedures; be able to develop a site safety and control plan; and understand chemical, radiological, and toxicological terminology and behavior. The specialist level training must be a minimum of 24 hours in length.

Incident commander level personnel will assume control of hazmat incidents beyond the awareness level response. They must be able to implement the employer's ICS and both the local and employer's emergency response plans, know of the state emergency response plan, know federal response capabilities, understand the risks associated with working in PPE, and understand the importance of decontamination procedures. The incident commander level training must be a minimum of 24 hours in length.

Trainers for hazardous materials response classes must have taken an appropriate train-the-trainer class or possess training and/or academic credentials and instructional experience to be competent. Refresher training shall consistent of receiving annual training of sufficient content and duration to maintain competency, or competency shall be demonstrated at least annually. The employer must make a statement of competency and keep a record of the methodology used to demonstrate competency.

medical surveillance ■ The process of ensuring respirator users and hazardous materials responders are able to safely wear PPE and respirators under stressful working conditions.

Medical surveillance and consultation is required for hazardous materials response team members. Members of a hazmat team and hazmat specialists shall receive a baseline physical exam and be provided with medical surveillance according to paragraph F.

HAZMAT HANDLE

Personnel trained to the awareness, operations, technician, specialist, and incident commander levels must receive at least as much training so they remain competent. A mere 8 hours of annual training is usually *not* sufficient for operations level personnel and certainly *not* sufficient for technician level personnel!

SOLVED EXERCISE 1-1

What level of training should all of the first responders in the chapter scenario have at a minimum?
Solution: According to HAZWOPER, all personnel should have a minimum of awareness level training because they could be expected to encounter hazardous materials on the job. The EMS personnel and fire department personnel that are expected to render medical care should have operations level training and the appropriate equipment to perform their expected functions.

Anyone that exhibits signs or symptoms that may have been caused by a job-related exposure to hazardous materials shall be provided with medical consultation consistent with paragraph F (see above).

Post-emergency response operations, otherwise known as cleanup, are addressed in HAZWOPER as well. If hazardous materials or contaminated materials need to be removed from the site, the employer conducting the cleanup must comply with one of the following: (1) The employer must meet all of the requirements of paragraphs B through O (the HAZWOPER hazardous waste–site worker regulations); or (2) if cleanup is done on plant property by plant employees, they must complete the following training:

1. 29 CFR 1910.38(a) (Emergency Action Plans)
2. 29 CFR 1910.134 (Respiratory Protection)
3. 29 CFR 1910.1200 (Hazard Communication)
4. Other appropriate task-specific safety and health training

As you can tell, HAZWOPER provides a comprehensive framework for hazardous materials response. The NFPA 472 (2008 edition) standard is fully consistent with paragraph Q of HAZWOPER.

OSHA RESPIRATORY PROTECTION 29 CFR 1910.134

The OSHA respiratory protection regulation, which is referenced by HAZWOPER, is designed to protect employees from hazardous atmospheres. It requires the employer to provide workers with the appropriate respiratory protection, training in how to use the respirators, and a medical evaluation to ensure the employees are fit to wear the respirator. Depending on the type of respirator and whether the use is voluntary or mandated, a written respiratory protection plan must be in place that covers respirator selection, use, maintenance, and care; **fit testing**; training; respiratory protection program evaluation; and record keeping. Anyone using a respirator should understand his or her employer's respiratory protection plan and be trained in the selection and use of the respiratory protection responders will be wearing.

fit testing ■ The process of verifying that a respirator seals properly against the wearer's face. A fit test can be qualitative or quantitative based upon the testing procedures that are used.

HAZMAT HANDLE

Your respiratory protection is your last line of defense, keeping hazardous materials from entering your lungs. Make sure you understand how to use your respirator well!

RESOURCE CONSERVATION AND RECOVERY ACT OF 1976 (RCRA)

The Resource Conservation and Recovery Act of 1976, commonly referred to as RCRA, was passed by Congress in response to several significant environmental disasters including Love Canal in New York (Box 1-1) and the Valley of the Drums in Kentucky (Box 1-2). This act imposed "cradle to grave" regulation of hazardous waste and made the generator of hazardous waste responsible for its safe and legal disposal. Any entity that generates hazardous waste must properly store, transport, and dispose of the hazardous waste. Proper disposal may include recycling, such as redistillation of waste solvent; incineration, such as burning waste hydrocarbons in specially licensed kiln ovens; or disposal in a licensed and specially engineered hazardous waste landfill. RCRA has significantly reduced the amount of illegal dumping due to stiff civil monetary penalties and criminal penalties including jail time for willful acts.

RCRA does the following: requires companies that generate hazardous waste to properly characterize the waste, limits the time hazardous waste may be stored on their premises before disposal (90–270 days, depending on the size of the generator), dictates

See OSHA Respiratory Standard (29 CFR 1910.134) from the Code of Federal Regulations for more information.

BOX 1.1 HAZMAT HISTORY: LOVE CANAL

Love Canal, one of the most appalling environmental tragedies in American history, was instrumental in bringing about significant environmental law reform in the 1970s and 1980s. At the beginning of the 20th century, William T. Love started to dig a short canal between the upper and lower Niagara rivers in order to generate electricity for a model city he envisioned. The canal was never completed, and the resulting ditch was turned into a municipal waste and industrial chemical dump site.

In 1953, the Hooker Chemical Company covered the dump site with earth and sold it to the city of Niagara Falls for one dollar. In the late 1950s, about 100 homes and a school were built at the site. Since then, over 80 different compounds, almost a dozen of them suspected carcinogens, have been leaching out of their corroded containers and migrating into the basements of the buildings built on the former dump site. After a record amount of rainfall in 1978, the toxic soup finally broke through the surface. The tips of corroding drums could be seen in backyards. Trees and gardens were turning black and dying, and the air had a faint, choking smell. Children returned from playing outside with burns on their hands and faces.

One of the most prevalent chemicals seeping through the ground and into homes at Love Canal was benzene, a known human carcinogen, and it was detected in high concentrations. A disturbingly high rate of miscarriages, along with birth defects, was reported in the area. Eventually, all of the families were evacuated from the most contaminated areas; a total of 221 families have moved or agreed to be moved. State figures show more than 200 purchase offers for homes have been made, totaling nearly $7 million. The EPA put a remediation plan in place that included removal of the hazardous waste drums, a system of trenches to contain runoff, and monitoring wells.

Adapted from the Environmental Protection Agency

BOX 1.2 HAZMAT HISTORY: VALLEY OF THE DRUMS

The A. L. Taylor site, also known as Valley of the Drums, covers 13 acres in Brooks, Kentucky, 12 miles south of Louisville, Kentucky. It was used as a refuse dump, drum recycling center, and chemical waste dump for the decade of 1967 to 1977. The chemical waste was largely from the paint and coatings industries and contaminated the air, surface water, groundwater, and soil with organic and inorganic chemicals.

The EPA conducted emergency response activities in March 1979, under Section 311 of the Clean Water Act, and in September 1981, under CERCLA, at a total cost of $650,000. Through these response activities and voluntary removal of wastes by known generators, a majority of the surface wastes (about 17,000 drums) were removed. A system was installed to control and treat contaminated runoff from the site. At one time this was the top-priority site in Kentucky.

Adapted from the Environmental Protection Agency

---

hazardous waste manifest ■ The shipping paper that accompanies hazardous waste shipments. A copy of the manifest must be returned to the generator by the disposal facility to prove that it was disposed of legally.

safe storage conditions during temporary storage, requires **hazardous waste manifests** during transport, requires the licensing of hazardous waste haulers, regulates disposal facilities, and requires that the hazardous waste manifest be returned to the original generator indicating how the hazardous waste was disposed of legally. State and federal environmental agencies regularly inspect RCRA-licensed facilities, and employees of such facilities must receive initial RCRA training and annual refresher training.

It is a good idea to periodically tour RCRA facilities within your response area. Dealing with hazardous waste releases can often be more difficult than dealing with virgin chemicals because basic information about the waste often does not exist. Sometimes hazardous waste is merely a used solvent that is just a dirtier version of the virgin material; however, other times hazardous waste can be a complicated mixture of several different chemicals that have been used in a complex industrial process. There are no practical ways for first responders to accurately and completely characterize these mixtures in the field.

COMPREHENSIVE ENVIRONMENTAL RESPONSE, COMPENSATION, AND LIABILITY ACT OF 1980 (CERCLA)

The **Comprehensive Environmental Response, Compensation, and Liability Act (CERCLA)**, commonly known as Superfund, provides federal authority to respond to releases or threatened releases of hazardous materials that may endanger public health or the environment. Additionally, CERCLA provides for liability of persons responsible for releases of hazardous waste at closed and abandoned hazardous waste sites and establishes a trust fund to provide for cleanup when a responsible party cannot be identified. The law authorizes short-term and long-term removal actions based on the Environmental Protection Agency's (EPA) National Priorities List (NPL). The trust fund is funded by a tax on the chemical and petroleum industries.

SUPERFUND AMENDMENTS AND REAUTHORIZATION ACT OF 1986 (SARA) AND THE EMERGENCY PLANNING AND COMMUNITY RIGHT-TO-KNOW ACT (EPCRA)

In 1986, CERCLA was amended by the Superfund Amendments and Reauthorization Act (SARA). Title III of SARA, also called the Emergency Planning and Community Right-to-Know Act (EPCRA), mandates that communities plan for emergencies involving hazardous materials and requires local industrial facilities to report the hazardous materials they store on-site to local first responders. Facilities that use and store large amounts of hazardous materials are therefore often called SARA Title III sites. This law is designed to help local communities protect public health, safety, and the environment from chemical hazards. The chemical inventory report, along with the site maps and the emergency procedures that are contained in the preplans are a great source of information for awareness level personnel and operations level responders.

This legislation also required the formation of state emergency response committees (SERCs) and **local emergency planning committees (LEPCs)** to plan for, and react to, hazardous materials incidents. LEPCs, usually formed at the municipal or county level, are responsible for the development of hazardous materials emergency response plans. LEPCs are comprised of local stakeholders, and they review the emergency response plan on an annual basis. The local emergency management agency and fire department usually work very closely with the LEPC to ensure that the hazardous materials response in their community will be adequate. You should become familiar with your role in the local emergency response plan.

THE HAZARDOUS MATERIALS REGULATIONS (HMR)

The United States Department of Transportation (U.S. DOT) regulates hazardous materials during transport. These regulations, known as the Hazardous Materials Regulations (HMR), can be found in 49 CFR Parts 100 to 185.

The HMR requires that hazardous materials shipments are properly characterized, safely packaged, appropriately marked and labeled, and accompanied by **shipping papers** during transportation. **Labels**, **placards**, and shipping papers are an important source of information to first responders to hazardous materials releases during transportation.

Comprehensive Environmental Response, Compensation, and Liability Act (CERCLA) ▪ The federal statute that empowers EPA to identify and clean up sites at which hazardous substances were released into the environment; commonly known as Superfund.

local emergency planning committee (LEPC) ▪ A group of community members, including members of industry, emergency responders, and the public who are responsible for local hazardous materials emergency response planning mandated by EPCRA.

shipping paper ▪ The document required by DOT that describes the contents, quantity, and hazards of the materials contained in a shipment.

label ▪ The written warnings and pictograms affixed to containers of hazardous materials designed to rapidly communicate hazard information.

placard ▪ The written warnings and pictograms affixed to vehicles transporting hazardous materials designed to rapidly communicate hazard information.

SOLVED EXERCISE 1-2

What law covers the drums containing hazardous materials at the abandoned industrial facility in the chapter scenario?

Solution: Resource Conservation and Recovery Act, RCRA.

Any employees that offer hazardous materials for transport must receive initial training that at a minimum includes:

General awareness/familiarization training (such as recognition of hazardous materials)
Function-specific training (such as proper packaging)
Safety training (such as use of personal protective equipment)
Security awareness training (such as understanding how hazardous material shipments may be stolen)
In-depth security training (when applicable, on the written transportation security plan)

Hazmat employees must receive refresher training at least every 3 years. These training requirements are meant to reduce the number of disastrous transportation accidents such as the Valujet Flight 592 plane crash in Florida that killed 110 people on May 11, 1996. This passenger aircraft was brought down by the improper packaging of oxygen generators that resulted in a fire in the cargo compartment. Had the employees of the company that shipped the oxygen generators received proper training, this accident and loss of life could have been avoided.

NATIONAL FIRE PROTECTION ASSOCIATION STANDARDS

The National Fire Protection Association (NFPA), which is a nongovernmental organization, has promulgated a set of hazardous materials response standards to which most emergency response agencies try to adhere. Standards typically do not have the force of law, but they may be enforced through civil lawsuits. Thus, NFPA standards provide a "standard of care" that hazmat responders may be held to in a court of law. Furthermore, standards may be adopted by reference, effectively making the standard a regulation in and of itself.

NFPA 472 (2008 edition)
NFPA 472, *Standard for Competence of Responders to Hazardous Materials/Weapons of Mass Destruction Incidents*, gives explicit guidance for fire service and law enforcement personnel responding to hazardous materials and weapons of mass destruction incidents. NFPA 472 follows the five levels of training stipulated by OSHA in HAZWOPER: awareness, operations, technician, specialist, and incident commander.

NFPA 473 (2008 edition)
NFPA 473, *Standard for Competencies for EMS Personnel Responding to Hazardous Materials/Weapons of Mass Destruction Incidents*, gives specific guidance to emergency medical service personnel responding to hazardous materials and weapons of mass destruction incidents. The standard defines where EMS personnel may perform patient care: Level 1 EMS personnel may perform patient care in the **cold zone**, whereas Level 2 EMS personnel may perform patient decontamination and patient care in the **warm zone** (or decontamination line).

cold zone ▪ The clean, uncontaminated area of the hazardous materials incident. Also known as the support zone. It is where the incident command post and hazardous materials team support functions are located.

warm zone ▪ The control zone where decontamination is performed.

Organizations That Need to Provide Hazardous Materials Awareness and Operations Training to Employees

Many organizations need to provide hazardous materials training to their employees. Theoretically, almost every employee should receive awareness level training because they could be expected to encounter hazardous materials during their normal job duties. On the other hand, operations level training is normally reserved for employees that work for public safety agencies, emergency response agencies, or other agencies that may respond to hazardous materials releases. Examples of employees that may require operations level training include firefighters, law enforcement officers, EMS workers, those in the military, public health workers, security guards, dispatchers, and even public works employees. Let's discuss the awareness and operations levels of hazardous materials training in more detail.

In the chapter scenario situation, could the EMS personnel with awareness level training decontaminate the patients?

Solution: No; they must be trained to a minimum of operations level according to HAZWOPER, NFPA 472 (2008), and NFPA 473 (2008).

AWARENESS LEVEL PERSONNEL

Awareness level personnel are not typically emergency responders. They are any employees that may be first on the scene of a hazardous materials or weapons of mass destruction release. They may encounter hazardous materials or weapons of mass destruction during the course of their normal job duties. Awareness level personnel should be able to recognize emergency incidents involving hazardous materials, isolate the area, warn other people that may be nearby, notify the appropriate emergency responders, and secure the area. Examples of employees that may require awareness level training are chemical industry workers, transportation industry workers, school system employees, public works employees, private security officers, and many other workers. Chapter 2 covers awareness level training.

OPERATIONS LEVEL RESPONDERS

Operations level responders are members of public safety agencies that respond to hazardous materials or weapons of mass destruction incidents in order to protect nearby persons, the environment, and property from the effects of the hazardous materials release. The goal of operations-level hazardous materials responders should be to have a safe and effective incident AFIRMED (which stands for assess, formulate, initiate, reassess, modify, extended operations, and demobilize). Initially Assess the situation using dispatch information (approach upwind, uphill, and upstream); using witness accounts and scene observation from a safe distance; and using research of the facility, container type, and released product. Gather reliable information as quickly as possible on which to develop an **incident action plan (IAP)**. Formulate an action plan consistent with your level of training, available resources (equipment, personnel, and time), and the current situation. Initiate actions that will support your incident action plan (IAP) using the appropriate personal protective equipment. Reassess the situation frequently. Modify your incident action plan accordingly, and then continue with Extended operations. Finally, Demobilize resources and turn the scene over to the appropriate agency or **authority having jurisdiction (AHJ)**.

Operations level responders will be trained to a set of core competencies that allows them to respond in a primarily defensive manner to hazardous materials or weapons of mass destruction incidents. These core competencies are covered in Chapters 3, 4, and 5. Operations level responders may be trained additionally in mission-specific competencies that specifically define limited offensive hazardous materials or weapons of mass destruction response roles:

1. Use of personal protective equipment (Chapter 6)
2. Performance of mass decontamination (Chapter 7)
3. Performance of technical decontamination (Chapter 8)
4. Evidence preservation and sampling (Chapter 9)
5. Product control (Chapter 10)
6. Air monitoring and sampling (Chapter 11)
7. Victim rescue and recovery (Chapter 12)
8. Response to illicit laboratory incidents (Chapter 13)

Your employer may choose to train you to one or more mission-specific competencies based upon your agency's assigned role in the local emergency response plan.

incident action plan (IAP) ■ A document that summarizes the goals for the current operational period and ensures all agencies and jurisdictions are operating toward the same goal.

authority having jurisdiction (AHJ) ■ The agency that sets policy and procedures concerning the subject matter in question for a jurisdiction. The AHJ will usually assume command of incidents in its jurisdiction that cover the subject matter in question. Examples of AHJs may include the FBI for WMD incidents, the local fire department for hazardous materials incidents, and the state police for clandestine methamphetamine labs.

In the chapter scenario situation, the mayor has asked the fire department to investigate the area the children were playing in to identify the hazardous material(s). What type of training should department members have already received?

Solution: According to NFPA 472 (2008), at the minimum they must receive operations level training with mission-specific competency training and operate under a set of standard operating procedures or the direct supervision of a hazmat technician.

Finally, let's look at a hypothetical situation: What would happen if your agency responded to a hazardous material incident and in the aftermath you were called to the witness stand? You should have the knowledge and ability to accurately answer the following questions:

1. Have you been trained in the use of self-contained breathing apparatus (SCBA)? Can your agency document that training?
2. Does your agency have an emergency response plan for the SARA Title III site? Were all responding employees trained in its use?
3. Did you have material safety data sheets (MSDS) available, and did you use them?
4. Were all of the responding employees trained in the use of their emergency response equipment? Can you document that?
5. Does your agency have a written policy on training, showing the amount, frequency, and type?
6. Do you know how to contact the LEPC?
7. What agency is responsible for mitigating hazardous materials incidents in your jurisdiction?
8. Does your agency use a National Incident Management System (NIMS)–compliant incident management system? Does the incident commander have formal incident command system (ICS) training?
9. Since you responded to this hazardous materials incident, do you have at a minimum operations level training?

Could you answer all of these questions honestly and accurately? Has your agency implemented all of the necessary programs? We will cover all of these issues and then some in the following chapters of this book.

Summary

HAZWOPER is the federal regulation that covers hazardous materials response to the United States. HAZWOPER covers both private sector and public sector response to both nonemergency, or incidental spills, and emergency releases of hazardous materials and hazardous waste. This regulation also establishes five levels of training for hazardous materials responders: awareness, operations, technician, specialist, and incident commander. In this book we will cover the basic information that awareness level personnel and operations level responders in the fire service, law enforcement, emergency medical services (EMS), and industry need to know.

Review Questions

1. What are the five levels of hazardous materials emergency response training according to HAZWOPER?
2. Which level of training teaches people to recognize, avoid, isolate, and notify in the case of a hazardous materials release?
3. According to NFPA 472 (2008), which level of training covers mission-specific competencies of emergency responders to a hazardous materials incident?
4. Who must receive the community right-to-know paperwork provided by industry?
5. What is the function of the LEPC?

Problem-Solving Activities

1. Find the list of SARA Title III sites in your jurisdiction, and determine what the appropriate response procedures are for the following incidents at one or more of those sites:
   a. A medical emergency in the process area
   b. A chemical exposure to three workers
   c. A small chemical release that has entered the sewer system
   d. A large chemical release that has migrated off site
   e. A large explosion with a sizable chemical plume migrating off site in the prevailing wind direction
2. Find the local emergency response plan, and determine your and your agency's role at a small-, medium-, or large-scale hazardous materials incident.
3. How would you fight a fire at a farm supply store?
4. What information do you need to know to plan a response to a mercury release at a local school?
5. What information do you need to know to plan a response to a clandestine methamphetamine laboratory?
6. What information do you need to know to plan a response to a trail derailment involving hazardous materials?

References and Further Reading

National Fire Protection Association. (2008). *NFPA 472, Standard for Competence of Responders to Hazardous Materials/Weapons of Mass Destruction Incidents.* Quincy, MA: Author.

National Fire Protection Association. (2008). *NFPA 473, Standard for Competencies for EMS Personnel Responding to Hazardous Materials/Weapons of Mass Destruction Incidents.* Quincy, MA: Author.

Occupational Safety and Health Administration. (1990). 29 CFR 1910.120, *Hazardous Waste Operations and Emergency Response (HAZWOPER).* Washington, DC: U.S. Department of Labor.

U.S. Department of Transportation. (2008). *2008 Emergency Response Guidebook.* Washington, DC: Pipeline & Hazardous Materials Safety Administration.

Weber, Chris. (2007). *Pocket Reference for Hazardous Materials Response.* Upper Saddle River, NJ: Pearson/Brady.

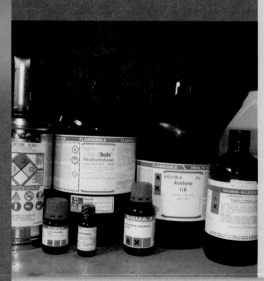

CHAPTER 2

Awareness Level Personnel

KEY TERMS

absorption, *p. 24*

acute effect, *p. 21*

air bill, *p. 37*

alkaline, *p. 21*

bill of lading, *p. 33*

biological agent, *p. 46*

blood agent, *p. 46*

carcinogen, *p. 21*

CBRNE, *p. 19*

chemical protective clothing (CPC), *p. 24*

Chemical Transportation Emergency Center (CHEMTREC), *p. 32*

chronic effect, *p. 21*

combustible liquid, *p. 26*

consist, *p. 37*

cryogenic liquid, *p. 26*

dangerous goods, *p. 20*

deflagration, *p. 26*

detonation, *p. 26*

dewar, *p. 32*

dust explosion, *p. 20*

explosive, *p. 20*

fission, *p. 52*

flammable, *p. 20*

flammable liquid, *p. 20*

flammable solid, *p. 26*

flashpoint, *p. 27*

gas, *p. 20*

hazard class, *p. 26*

ingestion, *p. 24*

inhalation, *p. 23*

inhibitor, *p. 28*

initial isolation distance, *p. 37*

injection, *p. 25*

ionizing radiation, *p. 29*

manifest, *p. 37*

marking system, *p. 25*

material safety data sheet (MSDS), *p. 32*

microorganism, *p. 29*

nerve agent, *p. 46*

odor threshold, *p. 23*

olfactory fatigue, *p. 23*

oxidizer, *p. 21*

pathogen, *p. 29*

pesticide, *p. 22*

prion, *p. 29*

pyrophoric, *p. 28*

radiological dispersal device (RDD), *p. 47*

radionuclide, *p. 30*

RAIN, *p. 20*

shelter in place, *p. 40*

site safety plan, *p. 41*

terrorism, *p. 25*

toxic industrial material (TIM), *p. 20*

toxic inhalation hazard (TIH), *p. 40*

toxic substance, *p. 21*

toxin, *p. 46*

vapor cloud, *p. 23*

vesicant, *p. 46*

virus, *p. 29*

water-reactive material, *p. 26*

waybill, *p. 37*

- Name the actions that personnel at the hazmat awareness level should perform when confronted by a hazardous materials spill or weapons of mass destruction (WMD) incident.
- Name the four components of the acronym RAIN.
- Describe five signs of a hazardous materials release.
- List the four routes of entry.
- Describe five sources of information about a hazardous materials spill or WMD incident.
- Identify an unknown material using the *Emergency Response Guidebook* (ERG).
- List four types of weapons of mass destruction (WMD).

Resource **C**entral

For additional review and practice tests, visit **www.bradybooks.com** and click on Resource Central to access book-specific resources for this text! To access Resource Central, follow directions on the Student Access Card provided with this text. If there is no card, go to **www.bradybooks.com** and follow the Resource Central link to Buy Access from there.

You Are at Work! Forklift Punctures a Fiberboard Drum

Morehead City, NC

Early in the morning of January 12, 2010, a forklift operator punctures nine cardboard drums while unloading them from an intermodal container at the port. The forklift operator knows the contents of the drums are a hazardous material, specifically an explosive in slurry form.

- What should the forklift operator do? What should first responders do?
- How can they find more information?

Let's try to answer these questions in this chapter.

Awareness level personnel are persons who may encounter hazardous materials or **CBRNE** (chemical, biological, radiological, nuclear, and explosives) agents during the course of their job duties. Awareness level personnel are not necessarily first responders; however, they should be able to recognize emergency incidents, isolate the area, warn other people that may be present, notify the appropriate emergency responders, and secure the area. In this chapter we discuss ways to recognize hazardous materials incidents and then the appropriate actions to take when a hazardous materials incident is discovered.

The information in this chapter is applicable not only to traditional first responders such as firefighters, law enforcement personnel, and emergency medical services providers but also to civilian workers in industry, transportation, school systems, public health, public works, private security, and other areas of employment. Any workers that may reasonably come into contact with hazardous materials should have an understanding of appropriate awareness level hazardous materials actions.

CBRNE ■ The acronym used to describe chemical, biological, radiological, nuclear, and explosive devices used by terrorists.

A useful acronym, developed by the Department of Homeland Security for hazardous materials and CBRNE response, is **RAIN**, which stands for recognize, avoid, isolate, and notify. This mantra closely parallels the competencies set in the National Fire Protection Association (NFPA) 472 standard, 2008 edition. The goal for personnel trained to the awareness level is to quickly recognize the release of hazardous materials, avoid the released materials, isolate the area, notify the appropriately trained response personnel, and analyze the severity of the incident from a safe distance. These steps are designed to prevent injury to yourself and others in the immediate area and to minimize further damage by quickly calling for trained personnel to mitigate the incident.

First it is important to define what we are dealing with. Hazardous materials, or **dangerous goods** as they are known in Canada, are defined as any materials that may pose an unreasonable risk to people or property. A weapon of mass destruction (WMD) is defined by the Federal Bureau of Investigation (FBI) as:

- Any **explosive** or incendiary device, as defined in Title 18 USC, Section 921: bomb, grenade, rocket, missile, mine, or other device with a charge of more than 4 ounces;
- Any weapon designed or intended to cause death or serious bodily injury through the release, dissemination, or impact of toxic or poisonous chemicals or their precursors;
- Any weapon involving a disease organism; or
- Any weapon designed to release radiation or radioactivity at a level dangerous to human life.

The primary difference between a hazardous materials incident and a weapons of mass destruction incident is intent. Hazardous materials incidents are typically accidental in nature, whereas WMD incidents are deliberate. Weapons of mass destruction incidents may involve specialized chemical, biological, or radiological weapons, or explosives of a military nature, or the misuse of common **toxic industrial materials (TIM)**, also known as toxic industrial chemicals (TIC).

Dangers of Hazardous Materials

What makes hazardous materials dangerous? Why can they pose an "unreasonable risk to people or property"? Hazardous materials can pose a threat through flammability, chemical reactivity, corrosiveness, toxicity, or a combination of two or more of these characteristics.

Flammability refers to the ability of a material to burn. Some materials are unable to burn or do not burn easily. These are referred to as noncombustible materials and are not considered to be a combustion hazard. Other materials burn, but are difficult to ignite. These are referred to as combustible materials. Yet other materials are easily ignited and are referred to as **flammable** materials. Flammable chemicals produce flammable vapors or **gases** at low temperatures, generally below 100°F (37.8°C).

Flammable liquids are not the only fire danger that may be encountered. Fine solids, such as flour, sawdust, powdered metals, and pharmaceuticals, are often highly flammable materials as well. **Dust explosions** cause a significant number of deaths, injuries, and property damage every year. Often accidents occur because employees and first responders are not aware of the hazard. Any equipment that produces a dust or powder has the potential to cause a dust explosion, often months or years after the equipment and process are originally started. Over time, the flammable dust generated during production accumulates on surfaces. If proper housekeeping is not performed, the dust can accumulate to significant thicknesses, especially in hard-to-reach places. When a sudden air current or building vibration dislodges the accumulated dust and it begins falling through the air, an explosive atmosphere may result. The only thing standing between you and tragedy is now an ignition source. It is important that employees and

RAIN ■ An acronym that summarizes the actions awareness level personnel should take at a hazardous materials or WMD incident. RAIN stands for recognize, avoid, isolate, and notify.

dangerous goods ■ Internationally accepted term for hazardous materials.

explosive ■ A substance or article capable of detonation.

toxic industrial material (TIM) ■ Poisonous substances used widely in industry that pose a significant public safety threat upon accidental or intentional release. Also known as toxic industrial chemicals (TIC).

flammable ■ Capable of readily burning.

gas ■ Matter that does not have a definite volume or shape and disperses in three dimensions when released from its container.

flammable liquid ■ A liquid having a flashpoint below 100°F (37.8°C) [per OSHA], or a liquid having a flashpoint of 141°F (60.5°C) or below [per DOT].

dust explosion ■ The sudden and energetic combustion of dust particles suspended within a confined area.

first responders be aware of this hazard on a daily basis at their workplace and during inspections.

Chemical reactivity refers to how readily a material will combine, often very energetically, with another material. For example, some materials will explode on contact with water because they interact with water (such as sodium metal). Other materials, such as **oxidizers**, react with fuels to cause them to burn rapidly (such as chlorine). When chemicals react with each other, they can produce tremendous amounts of heat, flame, or toxic by-products. Incompatible chemicals must always be kept apart! Chemical reactivity is a common cause of hazardous materials accidents.

Corrosive materials have the ability to destroy tissue, metals, and other materials. There are two types of corrosives: acids and bases. To make matters worse, acids and bases are incompatible with each other and will react violently if they come into contact. Examples of acids include muriatic acid, battery acid, phosphoric acid, and nitric acid. Vinegar is an example of a weak acid. Examples of bases, also known as **alkaline** or caustic materials, include lye, ammonia, and bleach.

Toxic substances have the ability to harm our bodies. The toxic effects of exposure can appear rapidly, also known as **acute effects**, or they can appear after prolonged exposure, also known as **chronic effects**. Toxic materials can harm the whole body, such as cyanide, or they can harm specific areas of the body, such as asbestos damaging the lungs. Some toxic materials can have latent effects, for example, **carcinogens**, which cause cancer, such as the solvent benzene.

Location of Hazardous Materials

Hazardous materials are located throughout our communities and are even found in the home (Figure 2-1). They can be found at large industrial facilities, water treatment plants, research universities, warehouses, and even local schools. They are also found

oxidizer ■ A substance that enhances or supports the combustion of other substances.

alkaline ■ A corrosive liquid whose pH is greater than 7. Also known as caustic or base.

toxic substance ■ A poisonous material that can cause death, temporary incapacitation, or permanent harm to humans or animals. A poison.

acute effect ■ Signs and symptoms that appear rapidly upon exposure to a hazardous substance.

chronic effect ■ Illness resulting from exposure to a hazardous substance that appears slowly over an extended period (weeks to decades).

carcinogen ■ A substance that causes cancer (tumor formation).

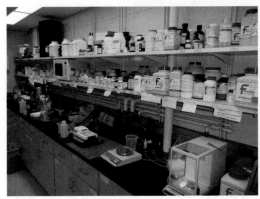

FIGURE 2-1 Four locations that often have hazardous materials on site. From the top left are a university research laboratory, a hardware store, a water treatment plant, and an industrial gas supplier. *Photos by author. Top right used with permission of WM Barr & Co., Inc., makers of Klean-Strip*

FIGURE 2-2 Hazardous materials are found at fixed site facilities and in transportation, such as on trucks, pipelines, and railways.

FIGURE 2-3 Hazardous materials are readily dispersed after they are removed from their container. On the left is a hazmat spill that has been unwittingly tracked through the facility, and on the right is an example of the intentional release of a hazardous material. In this case the truck was legitimately spraying for mosquitoes, but could a terrorist also disperse more deadly chemicals in this way?

pesticide ■ A product designed to repel or kill any human, plant, or animal pest.

at grocery stores, home improvement stores, beauty salons, auto parts stores, and most retail establishments. In the home, there are likely to be such hazardous materials as gasoline, ammonia-based cleaners, corrosive-based drain cleaners and oven cleaners, flammable solvents such as paint thinners, toxic **pesticides**, and acetone found in nail polish remover, to name a few. Although on occasion the small quantities in the home may be dangerous, hazardous materials become much more dangerous the larger the amount involved. Thus, the quantities of a hazardous material found in large fixed-site storage tanks, railroad tank cars, or highway tankers pose a much greater risk than the contents of smaller containers such as pint containers, 5-gallon pails, or even 55-gallon drums (Figure 2-2).

Recognition of Hazardous Materials Incidents

It is imperative that hazardous materials spills, terrorist incidents, or other events involving the release of substances that may pose a risk to people be recognized early (Figure 2-3). This is most important to the person that initially discovers the release. Typically, the first people

TABLE 2.1	Representative Odor Thresholds

CHEMICAL	ODOR THRESHOLD
Acetic acid	0.1–24 ppm
Acetone	0.1–699 ppm
Ammonia	0.04–55 ppm
Ammonium hydroxide	50 ppm
Benzene	1.4–120 ppm
Carbon monoxide	Odorless
Chlorine	0.01–5 ppm
Diesel fuel #2	0.08 ppm
Ethanol	1–5100 ppm
Ethyl ether	0.1–9 ppm
Ethyl mercaptan	0.00051–0.075 ppm
Gasoline	0.005–10 ppm
Hydrogen chloride	1–10 ppm
Hydrogen cyanide	0.00027–5 ppm (not everyone capable of detecting odor)
Hydrogen sulfide	0.00001–1.4 ppm (caution! olfactory fatigue)
Methyl ethyl ketone (MEK)	0.25–85 ppm
Toluene	0.02–70 ppm

Source: Data courtesy of the Environmental Protection Agency.

that discover a hazardous materials incident become the first victims if they are not aware of the situation. How then do we quickly realize that a dangerous situation exists before we are injured or killed?

Many cues to a release of a dangerous substance exist. These range from our senses—such as sight, sound, and smell—to labels and placards, to situational awareness. Naturally, the farther you are from the spill when you recognize it, the safer you will be.

Sight and sound may be very effective senses for early detection and warning. Hazardous materials spills may be visible from a distance due to smoke, a **vapor cloud**, or visible liquid or solid spilled material. If something, especially a hazardous material, looks out of place, stop and retreat until you can determine whether the situation is indeed safe. Hazardous materials spills may be heard from a distance due to the high-pressure squeal given off by a leaking gas cylinder, the sound of materials burning, or the sound of chemicals reacting. The sense of smell is not an ideal tool for early detection because most materials already have a degree of toxicity by the time they can be detected by smell. Many chemicals do not have any odor at all; others result in olfactory fatigue. **Olfactory fatigue** refers to chemicals such as hydrogen sulfide that have an odor that subsides rapidly after deadening the sense of smell. However, some chemicals have an **odor threshold** that is below the toxicity level. Witness accounts of unusual odors may also be a good indication of a hazardous materials release.

Routes of Entry

Hazardous materials may enter the body through four major routes of entry. These are inhalation, absorption, ingestion, and injection (Figure 2-4). **Inhalation** is the most dangerous and effective route of entry. The lungs are designed to absorb oxygen from the

vapor cloud ■ A mass of airborne material that travels along prevailing terrain and is moved by wind currents.

olfactory fatigue ■ The loss of ability to identify an odor after prolonged exposure. An example of a chemical that causes olfactory fatigue is hydrogen sulfide.

odor threshold ■ The level at which a chemical can be detected by the sense of smell.

inhalation ■ The introduction of a substance in the form of a gas, vapor, fume, mist, or dust into the body through the respiratory tract.

FIGURE 2-4 The four routes of entry into the human body for hazardous materials. *Art by David Heskett*

atmosphere and exchange carbon dioxide from the blood. The lungs are an organ optimized to pull chemicals out of the air we breathe. Unfortunately, the lungs cannot easily differentiate between oxygen and toxic chemicals. Furthermore, breathing is an involuntary, vital function. You can hold your breath for only so long! Therefore, it is vital to go to great lengths to protect the respiratory tract and the inhalation route of entry. Once someone has entered a toxic atmosphere, it is usually too late. Countless people have been killed and injured by unknowingly entering atmospheres of toxic gases such as carbon monoxide and hydrogen sulfide. Hazardous materials responders should use personal protective equipment (PPE) to protect their respiratory tract (respirators).

Another common route of entry is skin **absorption**. Many solids, liquids, and gases are readily absorbed through the skin; others are not. The skin is designed to be a barrier between the outside world and the body, mainly to keep water inside the body and germs outside the body. The outermost layer of skin is a waxy type of material. Many solvents such as gasoline, acetone, toluene, and hexane readily pass through the skin into the body. Certain gases such as hydrogen cyanide readily pass through the skin as well. Some of these hazardous materials are extremely toxic; others are not. Hazardous materials responders should also use personal protective equipment (PPE) to protect their skin, which is called **chemical protective clothing**, or **CPC**.

Ingestion is the route of entry that humans usually have control over. We do not have to stick things in our mouths if we do not want to! However, sometimes hazardous materials unwittingly enter the food supply. This may happen when a hazardous materials spill goes undetected and contamination is transferred to eating areas, or intentionally due to **terrorism** or a criminal event.

Injection as a route of entry can occur intentionally or accidentally. We are probably most familiar with injection as the route of entry for illegal drug use and getting those unpleasant immunizations at the doctor's office. During a hazardous materials release, injection can occur accidentally when a hazardous material coats sharp objects that may puncture our clothing and skin as we pass by.

The easiest and safest way to avoid all four routes of entry during a hazardous materials release is to avoid the area completely and keep others out as well. Hazardous materials emergency responders will have the appropriate PPE to handle the spill. Therefore, it is critical to notify these responders as early as possible and give them the most complete information we can. The rest of this chapter focuses on gathering information about the hazardous material and relaying that to the arriving first responders.

Marking Systems

The United States Department of Transportation (U.S. DOT) has devised a system of placards and labels for the quick and early identification of hazardous materials containers in the transportation system. Likewise, the National Fire Protection Association (NFPA) has devised the NFPA 704 **marking system** for fixed site facilities. This section covers each of these marking systems in detail because they can provide useful information about a hazardous materials release.

absorption ▪ One of the four routes of entry of toxic substances into the body by passing through the skin, mucous membranes, or eyes. (2) The process of using a material or media to soak up a spilled hazardous material through incorporation.

chemical protective clothing (CPC) ▪ The component of personal protective equipment that protects the skin from hazardous materials.

ingestion ▪ The introduction of a substance into the digestive tract of the body.

HAZARD CLASSES AND DIVISIONS

Nine **hazard classes** have been defined by the U.S. Department of Transportation:

Class 1	Explosives
Class 2	Gases
Class 3	Flammable liquids (and **combustible liquids** in the United States)
Class 4	**Flammable solids** and **water-reactive materials**
Class 5	Oxidizers
Class 6	Toxic and infectious materials
Class 7	Radioactive materials
Class 8	Corrosive materials
Class 9	Miscellaneous hazardous materials

Vehicles that contain hazardous materials during transport are placarded with the hazard class and, if appropriate, a four-digit United Nations (UN) or North America (NA) identification number. The placards are diamond shaped and required to be 11 3/4″ × 11 3/4″. Packages being transported that contain hazardous materials are labeled with the hazard class marking. The labels are 4″ × 4″. Figure 2-5 shows the U.S. DOT approved placard and label designs.

terrorism ■ The unlawful use of force or violence to intimidate or coerce the government, the civilian population, or any segment thereof in furtherance of political or social objectives.

injection ■ The introduction of a substance into the body through a skin puncture or transdermal wound.

marking system ■ A method of notifying first responders and the public of dangers at a particular location. Examples of marking systems include placards, labels, and the NFPA 704 marking system.

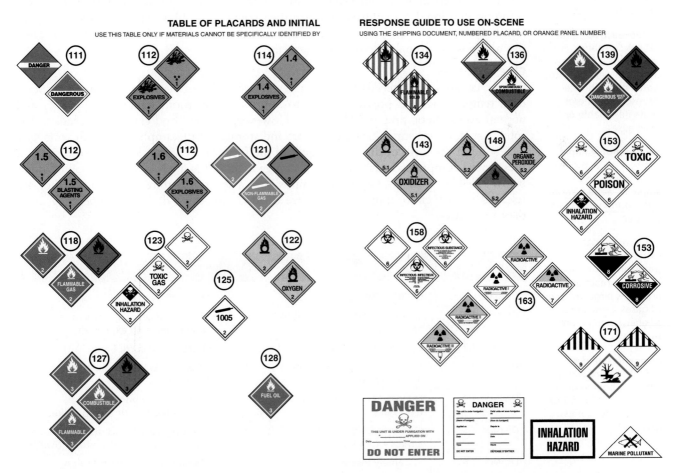

FIGURE 2-5 The DOT hazardous materials placards found in the 2008 *Emergency Response Guidebook* (ERG) on pages 16–17. *Courtesy of the U.S. Department of Transportation Pipeline and Hazardous Materials Safety Administration.*

hazard class ■ Any of the nine categories of hazardous materials as defined in the Hazardous Materials Regulations.

combustible liquid ■ A liquid having a flashpoint at or above 100°F (37.8°C) [per OSHA], or a liquid having a flashpoint above 141°F (60.5°C) and below 200°F (93°C) [per DOT].

flammable solid ■ Per DOT, any of the following types of materials: wetted explosives, thermally unstable materials that can undergo a strongly exothermic decomposition even without the participation of atmospheric oxygen, and readily combustible solids.

water-reactive material ■ A substance capable of a vigorous chemical reaction with water that produces intense heat and/or emits flammable or toxic gases.

deflagration ■ Rapid combustion without detonation. An example of deflagration is the burning of black powder.

detonation ■ An extremely rapid, supersonic oxidation-reduction reaction that produces large volumes of gases, large amounts of energy, and a pressure wave.

cryogenic liquid ■ A liquefied compressed gas having a boiling point below −130°F (−90°C) at atmospheric pressure (per DOT).

These hazard classes are often further subdivided into divisions based upon more precisely defined properties within a given class. Let's explore these nine hazard classes and their divisions in more detail and determine the dangers they pose. At the same time, some common substances and products that fall within each class will be discussed.

Class 1: Explosives

An explosion is an extremely rapid release of energy in the form of gas and heat. Explosives are substances and devices capable of yielding an explosion. Explosives can lead to devastating injuries from burns, pressure waves, and shrapnel. Class 1 explosives are divided into six divisions based upon how rapidly energy is released.

The class 1 divisions are arranged in descending order, whereby division 1.1 is most dangerous and division 1.6 is least dangerous. Division 1.1 explosives possess a mass explosion hazard that affects almost the entire load of explosives by simultaneously releasing a large amount of energy almost instantaneously, resulting in a tremendous amount of damage. Division 1.2 explosives possess a projection hazard. These materials and devices are not designed to detonate all at once. For example, detonating cord is intended to cause a **deflagration** in a controlled fashion. Division 1.3 explosives have a fire hazard and may possess a minor blast and/or projection hazard, but are significantly less hazardous than either division 1.1 or 1.2 explosives. Division 1.4 explosives present only a minor explosion hazard and contain less than 0.9 ounce (25 grams) of detonating material. An external fire would not cause a mass **detonation** of these devices. Division 1.5 explosives are very insensitive. These materials may have a mass explosion hazard, but are so insensitive that an external fire would not cause a mass detonation. The most common division 1.5 explosive is the prilled commercial blasting agent ANFO, which stands for ammonium nitrate–fuel oil mixtures. Division 1.6 explosives are very insensitive and possess no mass detonation hazard.

Class 2: Gases

Class 2 consists of materials that are gases at standard temperature and pressure. Standard temperature and pressure refers to approximately room temperature (68°F) and normal atmospheric pressure (14.7 psi). However, we may find class 2 materials in the solid, liquid, or gas state during transport and storage. For example, carbon dioxide may be a solid in the form of dry ice, a liquid in the form of a **cryogenic liquid**, or a gas in a pressurized compressed gas cylinder. A cryogenic liquid is a gas that has a boiling point below −130°F (−90°C). Cryogenic liquids pose a severe frostbite hazard when they are released from their containers. Gases are divided into three divisions, depending upon their properties.

TABLE 2.2	The Divisions of Class 1	
DIVISION	**DEFINITION**	**EXAMPLES**
Division 1.1	Explosives with a mass explosion hazard	Black powder, TNT, dynamite, and other high explosives
Division 1.2	Explosives with a projection hazard	Detonating cord, flares
Division 1.3	Explosives with predominantly a fire hazard	Propellants, liquid fueled rocket mortars
Division 1.4	Explosives with no significant blast hazard	Practice ammunition, signal cartridges, line throwing rockets
Division 1.5	Very insensitive explosives with a mass explosion hazard	Blasting agents such as ammonium nitrate and fuel oil mixtures (ANFO)
Division 1.6	Extremely insensitive articles	Certain military ammunition, bombs, and warheads

FIGURE 2-6 The drastic difference in how anhydrous ammonia is placarded in the United States versus Europe. In the United States it is placarded as a nonflammable gas (right), and in Europe it is placarded as a poisonous and corrosive gas (left).

TABLE 2.3	The Divisions of Class 2	
DIVISION	**DEFINITION**	**EXAMPLES**
Division 2.1	Flammable gases	Methane, propane, methyl chloride, inhibited butadienes
Division 2.2	Nonflammable and nontoxic gases	Anhydrous ammonia (only in the United States), nitrogen, carbon dioxide, helium, argon
Division 2.3	Toxic gases	Chlorine, methyl bromide, hydrogen fluoride, arsine

Materials placarded as class 2 may have a wide range of chemical and physical properties. They may react with each other and form polymers, such as butadiene; they may be corrosive, such as anhydrous ammonia and hydrogen fluoride; or they may be simple asphyxiating gases, such as nitrogen and argon. The property they have in common is that they are gases at room temperature and atmospheric pressure. Because gases have a wide range of chemical and physical properties, it stands to reason that they pose a wide range of hazards.

Anhydrous ammonia is placarded differently in the United States than it is in almost every other country. In the United States, it is considered a division 2.2 nonflammable and nontoxic gas. In Canada and Europe, it is considered a poisonous and corrosive gas. In Figure 2-6 are one anhydrous ammonia tank placarded in the United States and one placarded for Europe. Despite the classification in the United States, anhydrous ammonia is a corrosive, poisonous, and flammable gas. It was the cause of death of two hazardous materials responders in Louisiana, and its dangers should never be underestimated (see Chapter 6, Box 6-1).

Class 3: Flammable Liquids

Flammable liquids are defined by the U.S. DOT as liquids with a **flashpoint** below 141°F (60.5°C). Flashpoint is the lowest temperature at which a liquid gives off enough vapors to ignite, but not necessarily sustain combustion (continue to burn) when the ignition source is removed. Some examples of flammable liquids are gasoline, acetone, methanol, and toluene (Figure 2-7). In the United States, class 3 also includes combustible liquids, which are defined by the U.S. DOT as liquids with a flashpoint of 141°F (60.5°C) and higher, and below 200°F (93°C). Some examples of combustible liquids are No. 6 fuel oil, diesel fuel, kerosene, mineral oil, and peanut oil. Flammable liquids are generally more dangerous than combustible liquids due to their flash points.

flashpoint ■ The temperature of a liquid at which it gives off enough vapors to ignite, but not necessarily sustain combustion.

FIGURE 2-7 Flammable liquids in various laboratory-sized containers. Laboratory sized containers may be constructed of cardboard, plastic, glass, or metal. *Photo by author. Used with permission from Sigma-Aldrich*

pyrophoric ■ Capable of igniting on exposure to air at temperatures below 130°F.

inhibitor ■ A substance or mixture of substances used to slow or stop one or more chemical reactions.

FIGURE 2-8 Calcium carbide being transported down the highway in intermediate bulk containers.

Class 4: Flammable Solids and Dangerous When Wet Materials

Class 4 comprises a wide range of different materials within its three divisions. Division 4.1 encompasses three distinct types of materials: (1) desensitized explosives, such as wetted explosives; (2) self-reactive materials, such as thermally unstable substances; and (3) readily combustible solids, which may ignite through friction. Spontaneously combustible materials of division 4.2 consist of pyrophoric materials and self-heating materials. A **pyrophoric** material is a liquid or solid that can quickly ignite when it comes in contact with air, even in the absence of an external ignition source. A self-heating material increases in temperature upon contact with air, without input of other external energy. Division 4.3 consists of dangerous when wet materials, which tend to ignite or give off flammable vapors when in contact with water (Figure 2-8).

Class 5: Oxidizing Substances and Organic Peroxides

Oxidizing materials increase the rate of combustion of other materials (Figure 2-9). These substances can cause ordinary combustibles to burn at a much more rapid rate and spontaneously ignite without an ignition source. Some oxidizers liberate oxygen, thereby accelerating combustion. Yet other materials, such as chlorine, are more powerful oxidizers than oxygen and increase the rate of combustion by themselves. The organic peroxides are such strong oxidizers that they have explosive properties. Organic peroxides are classified as types A through G. Type A peroxides are so dangerous that they are forbidden in transport. Less dangerous organic peroxides are usually shipped under refrigeration in liquid form and contain **inhibitors**.

pathogen ■ Microorganisms (such as viruses and bacteria) or infectious substances (such as prion proteins) that have the potential to cause disease in humans or animals. Examples of pathogens include the smallpox virus, *Bacillus anthracis*, and the causative agent of mad cow disease (a prion).

virus ■ An infectious agent that requires another organism to grow and replicate. Examples of viruses include smallpox and influenza.

prion ■ An infectious protein. Examples of prion diseases include kuru, mad cow disease, and scrapie.

microorganism ■ Small, single-celled or multicellular organisms invisible to the naked eye; also called microbes. Examples of microbes include bacteria, protozoa, fungi, yeast, microscopic algae, and viruses.

TABLE 2.4	Flammable Solids	
DIVISION	**DEFINITION**	**EXAMPLES**
Division 4.1	Flammable solids	Magnesium metal, nitrocellulose
Division 4.2	Spontaneously combustible materials	Charcoal briquettes, phosphorus, aluminum alkyls, and magnesium alkyls
Division 4.3	Water-reactive substances and dangerous when wet materials	Calcium carbide, magnesium powder, sodium hydride, potassium metal alloys

TABLE 2.5	Oxidizers	
DIVISION	**DEFINITION**	**EXAMPLES**
Division 5.1	Oxidizing substances	Hydrogen peroxide, ammonium nitrate, calcium hypochlorite, bromine trifluoride
Division 5.2	Organic peroxides	Dibenzoyl peroxide, methyl ethyl ketone peroxide, peroxyacetic acid

Class 6: Toxic Substances and Infectious Substances

Poisonous materials of division 6.1 consist of materials other than gases that are toxic to humans or presumed toxic to humans, and may pose a danger during transportation. Division 6.2 consists of infectious substances that contain or are suspected of containing human or animal pathogens. **Pathogens** are bacteria, **viruses**, **prions**, or **microorganisms** that have the potential to cause human or animal disease.

Class 7: Radioactive Materials

Radioactive materials are dangerous because they emit **ionizing radiation**, which is harmful to our bodies. The ionizing radiation emitted by radioactive materials can cause cancer, skin burns, blindness, impotence, and even death in high enough doses. Ionizing radiation

ionizing radiation ■ High-energy emissions from radionuclides capable of causing harm to the human body. Examples of ionizing radiation include alpha radiation, beta radiation, gamma radiation, x-rays, and neutron radiation.

radionuclide ▪ An unstable atom capable of emitting radiation from the nucleus. Also known as a radioisotope or radioactive isotope.

TABLE 2.6	Poisonous Materials	
DIVISION	**DEFINITION**	**EXAMPLES**
Division 6.1	Toxic substances	Aniline, arsenic compounds, carbon tetrachloride, hydrocyanic acid, tear gas
Division 6.2	Infectious substances	Medical waste and laboratory waste containing pathogens such as anthrax, botulism, rabies, tetanus

FIGURE 2-10 Corrosive materials are often transported and used in intermediate bulk containers (IBC). You may find incompatible materials such as caustic soda and nitric acid being stored in close proximity to or even on top of one another at fixed site facilities.

is extremely dangerous because it cannot be seen, tasted, or smelled. Unless specialized detection equipment is used, it will not be noticed until significant damage has occurred. Radioactive materials contain radioactive elements called **radionuclides** in a quantity greater than a specified concentration and activity. Some examples of radioactive materials include cobalt-60, uranium hexafluoride, radioactive waste, and medical and laboratory isotopes such as P^{32}, S^{35}, and I^{131}.

Class 8: Corrosive Materials

Corrosive materials are liquids and solids that can cause tissue destruction and corrode steel or aluminum at a specified rate. Some examples of corrosives include sulfuric acid, hydrochloric acid (muriatic acid), sodium hydroxide (lye), phosphoric acid, and phosphorus trichloride (Figure 2-10). There are two types of corrosives: acids and bases. Bases are also known as caustic or alkaline materials. Acidic burns are usually detected quickly due to intense pain at the contact site. On the other hand, caustic burns may initially remain undetected due to a lack of pain at the contact site.

Class 9: Miscellaneous Hazardous Materials, Products, Substances, or Organisms

Class 9 materials present a hazard during transportation but do not meet the criteria of any other hazard class. Some examples of miscellaneous hazardous materials are molten sulfur, hazardous substances (such as PCBs), and some hazardous waste shipments. Class 9 materials have a wide range of hazards.

THE NFPA 704 MARKING SYSTEM

The NFPA 704 marking system (from NFPA 704, *Standard System for the Identification of the Hazards of Materials for Emergency Response*) is designed for fixed site facilities

SOLVED EXERCISE 2-1

The shipping container from the opening scenario is placarded with an orange placard that has a 1.1D at the bottom and the number 0150 in the middle. What does this mean?

Solution: The material is an explosive (hazard class 1, division 1). The "D" refers to the compatibility group, in this case a secondary detonating explosive without means of initiation. The four-digit number refers to the United Nations (UN) identification number of 0150, which is "pentaerythrite tetranitrate, wetted." (Pentaerythrite tetranitrate is commonly abbreviated PETN.) The chemical name can be determined from the hazardous materials tables in 49 CFR 172.101.

to warn occupants, employees, and responding emergency services personnel of chemical hazards located at that facility. Buildings or rooms containing hazardous materials will have a diamond with four quadrants affixed to them (Figure 2-11). Each quadrant has a different color. Blue signifies health hazards, red signifies fire hazards, yellow signifies reactivity hazards, and white signifies specific hazards. The red, blue, and yellow quadrants will have a number from 0 to 4 in them. The number 4 indicates a high degree of hazard, whereas the number 0 indicates a low degree of hazard. The specific hazard area will contain markings indicating other types of hazards such as oxidizer (OX) or water reactivity (W̶). Some facilities will also include corrosives (COR), acids (ACID), bases (BASE), radioactive materials (RAD or the trefoil propeller symbol), or other symbols in a hazard area. Even though the labeling of these specific hazards are not officially NFPA sanctioned, the information can be very useful.

PIPELINE MARKINGS

Pipeline markers are often used to mark underground pipelines, but not always. The sign will indicate the approximate location of the pipeline, the pipeline contents, an emergency contact number, and the pipeline operator. Pipelines often carry many products at different times, and the pipeline marker may indicate several materials or a class of materials, such as petroleum products. Figure 2-12 shows two examples of pipeline markers.

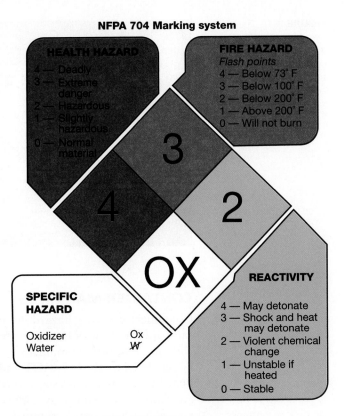

NFPA 704 Marking system

HEALTH HAZARD
4 — Deadly
3 — Extreme danger
2 — Hazardous
1 — Slightly hazardous
0 — Normal material

FIRE HAZARD
Flash points
4 — Below 73° F
3 — Below 100° F
2 — Below 200° F
1 — Above 200° F
0 — Will not burn

REACTIVITY
4 — May detonate
3 — Shock and heat may detonate
2 — Violent chemical change
1 — Unstable if heated
0 — Stable

SPECIFIC HAZARD

Oxidizer   Ox
Water   W̶

FIGURE 2-11 The NFPA 704 marking system.

FIGURE 2-12 Two examples of pipeline markers and the information they contain.

FIGURE 2-13 The label
on a 55-gallon drum of
35% hydrogen peroxide
contains much useful in-
formation in the event of
an emergency.

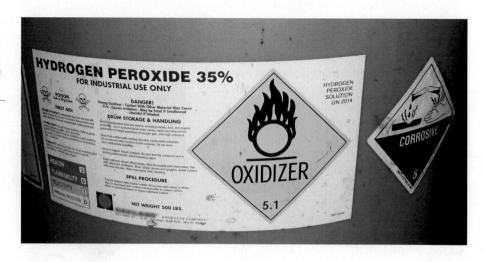

CONTAINER MARKINGS

Containers carrying hazardous materials are often marked with labels containing the same information as transportation placards (Figure 2-13). The shipping label may also contain additional manufacturer information. The hazardous materials containers may be regulated by the U.S. DOT and may be built to specific standards. Those include cardboard boxes, totes, drums, and other containers.

Container shape may indicate the presence of hazardous materials (Figure 2-14). Cylinders often contain compressed gases, **dewars** contain cryogenic liquids, metal drums may contain liquids, and plastic drums may contain corrosive liquids. Cardboard drums may contain a solid material. Heavy packaging often indicates that a pressurized material or dangerous material is being transported. But do not jump to conclusions based solely on container shape. Always check the container label and shipping papers!

MATERIAL SAFETY DATA SHEETS (MSDS)

Material safety data sheets (MSDS) are supplied by producers of chemicals to the end users per 29 CFR 1910.1200 (Hazardous Communication, or HazComm). Employers are required to have MSDS sheets for all chemicals used on site, and employees must know where these sheets are kept. Typically MSDS sheets are available in a public place at the workplace. Employees should know how to access and use MSDS sheets in case of an accidental spill or exposure. Do you? If not, you should attempt to locate the MSDS at your workplace at the earliest opportunity. MSDS sheets may also be obtained directly from the manufacturer of a product, from online databases, and from services such as **Chemical Transportation Emergency Center (CHEMTREC)**.

MSDS sheets are organized into several sections (Figure 2-15). They may have from eight to 16 sections, depending on if they are ANSI compliant, and contain the following information in this approximate order:

- General information, including a 24-hour emergency contact number, the product name, and the manufacturer.
- Hazardous ingredients.
- Hazard identification section: In this section, flammability, reactivity, and other health effects are listed.
- First aid measures for exposures by all four routes of entry.
- Firefighting procedures.
- Accidental release procedures.
- Handling and storage procedures.

dewar ▪ A container often used to carry cryogenic liquids. These vessels are designed to vent to the atmosphere because they operate on the basis of autorefrigeration.

material safety data sheet (MSDS) ▪ A technical bulletin prepared by product manufacturers to provide workers with detailed information regarding the dangerous properties and handling recommendations of commercial products containing hazardous substances.

Chemical Transportation Emergency Center (CHEMTREC) ▪ An informational hotline in the United States for emergency responders called to hazardous materials incidents.

FIGURE 2-14 ■ Examples of non-bulk hazardous materials containers. Clockwise from the top left are a metal 55-gallon drum that contains a flammable liquid, a plastic 55-gallon drum that contains a corrosive and oxidizing material (note that the corrosive label is falling off), a cardboard 55-gallon drum that contains a solid material, a cryogenic dewar that contains liquid nitrogen, and a number of compressed gas cylinders.

- Exposure control and personal protection.
- Physical and chemical properties.
- Stability and reactivity.
- Toxicological information.
- Ecological information.
- Disposal considerations.
- Transport information.
- Regulatory information.

As you can see, MSDS sheets are a great source of comprehensive information. However, just as with any other reference materials, they are only as accurate as the publisher. Always cross-reference critical information with other reference sources.

SHIPPING PAPERS

The U.S. Department of Transportation requires shippers in all modes of transportation to provide shipping papers. In over-the-road ground transportation, the shipping papers are called a **bill of lading**, which is required to be within reach of the driver while he or she is in the cab. Figure 2-16 shows an example of a bill of lading.

bill of lading ■ The shipping papers used in the road transportation industry.

MATERIAL SAFETY DATA SHEET

Hydrogen Peroxide (40 to 60%)

MSDS Ref. No.: 7722-84-1-4
Date Approved:
Revision No.:

This document has been prepared to meet the requirements of the U.S. OSHA Hazard Communication Standard, 29 CFR 1910.1200; the Canada's Workplace Hazardous Materials Information System (WHMIS), and the EC Directive, 2001/58/EC.

1. PRODUCT AND COMPANY IDENTIFICATION

PRODUCT NAME: Hydrogen Peroxide (40 to 60%)

ALTERNATE PRODUCT NAME(S): Durox® Reg. & LR 50%, Oxypure® 50%, Semiconductor Reg & Seg 50%, Standard 50%, Technical 50%, Chlorate Grade 50%, Super D® 50%

GENERAL USE: Durox® 50% Reg. and LR - meets the Food Chemical Codex requirements for aseptic packaging and other food related applications.

Oxypure® 50% - certified by NSF to meet NSF/ANSI Standard 60 requirements for drinking water treatment.

Standard 50% - most suitable for industrial bleaching, processing, pollution abatement and general oxidation reactions.

Semiconductor Reg. & Seg. 50% - conforms to ACS and Semi Specs., for wafer etching and cleaning, and applications requiring low residues.

Super D® 50% - meets US Pharmacopoeia specifications for 3% topical solutions when diluted with proper quality water. While manufactured to the USP standards or purity and to FMC's demanding ISO 9002 quality standards, FMC does not claim that its Hydrogen Peroxide is manufactured in accordance with all pharmaceutical cGMP conditions.

Technical 50% - essentially free of inorganic metals, suitable for chemical synthesis.

Chlorate Grade 50% - specially formulated for use in chlorate manufacture or processing.

SynergOx™ - combination of a proprietary catalyst and 50% hydrogen peroxide, at the point of use, for environmental applications.

MANUFACTURER

FMC CORPORATION
Hydrogen Peroxide Division
1735 Market Street
Philadelphia, PA 19103
(215) 299-6000 (General Information)

FMC of Canada Ltd.
Hydrogen Peroxide Division
PG Pulp Mill Road
Prince George, BC V2N2S6
(250) 561-4200 (General Information)

EMERGENCY TELEPHONE NUMBERS

(800) 424-9300 (CHEMTREC - U.S.)
(613) 996-6666 (CANUTEC)
(303) 595-9048 (Medical - U.S. - Call Collect)

(281) 474-8750 (Plant: Pasadena, TX, US - Call Collect)
(250) 561-4221 (Plant: Prince George, BC, Canada - Call Collect)

2. HAZARDS IDENTIFICATION

EMERGENCY OVERVIEW:

- Clear, colorless, odorless liquid
- Oxidizer.
- Contact with combustibles may cause fire.
- Decomposes yielding oxygen that supports combustion of organic matters and can cause overpressure if confined.
- Corrosive to eyes, nose, throat, lungs and gastrointestinal tract.

POTENTIAL HEALTH EFFECTS: Corrosive to eyes, skin, nose, throat and lungs. May cause irreversible tissue damage to the eyes including blindness.

3. COMPOSITION / INFORMATION ON INGREDIENTS

Chemical Name	CAS#	Wt.%	EC No.	EC Class
Hydrogen Peroxide	7722-84-1	40 - 60	231-765-0	C, R34
Water	7732-18-5	40 - 60	231-791-2	Not classified as hazardous

4. FIRST AID MEASURES

EYES: Immediately flush with water for at least 15 minutes, lifting the upper and lower eyelids intermittently. See a medical doctor or ophthalmologist immediately.

SKIN: Immediately flush with plenty of water while removing contaminated clothing and/or shoes, and thoroughly wash with soap and water. See a medical doctor immediately.

INGESTION: Rinse mouth with water. Dilute by giving 1 or 2 glasses of water. Do not induce vomiting. Never give anything by mouth to an unconscious person. See a medical doctor immediately.

INHALATION: Remove to fresh air. If breathing difficulty or discomfort occurs and persists, contact a medical doctor.

NOTES TO MEDICAL DOCTOR: Hydrogen peroxide at these concentrations is a strong oxidant. Direct contact with the eye is likely to cause corneal damage especially if not washed immediately. Careful ophthalmologic evaluation is recommended and the possibility of local corticosteroid therapy should be considered. Because of the likelihood of corrosive effects on the gastrointestinal tract after ingestion, and the unlikelihood of systemic effects, attempts at evacuating the stomach via emesis induction or gastric lavage should be avoided. There is a remote possibility, however, that a nasogastric or orogastric tube may be required for the reduction of severe distension due to gas formation.

5. FIREFIGHTING MEASURES

EXTINGUISHING MEDIA: Flood with water.

FIRE / EXPLOSION HAZARDS: Product is non-combustible. On decomposition releases oxygen, which may intensify fire.

FIREFIGHTING PROCEDURES: Any tank or container surrounded by fire should be flooded with water for cooling. Wear full protective clothing and self-contained breathing apparatus.

FLAMMABLE LIMITS: Non-combustible

SENSITIVITY TO IMPACT: No data available

SENSITIVITY TO STATIC DISCHARGE: No data available

6. ACCIDENTAL RELEASE MEASURES

RELEASE NOTES: Dilute with a large volume of water and hold in a pond or diked area until hydrogen peroxide decomposes. Hydrogen peroxide may be decomposed by adding sodium metabisulfite or sodium sulfite after diluting to about 5%. Dispose according to methods outlined for waste disposal.

Combustible materials exposed to hydrogen peroxide should be immediately submerged in or rinsed with large amounts of water to ensure that all hydrogen peroxide is removed. Residual hydrogen peroxide that is allowed to dry (upon evaporation hydrogen peroxide can concentrate) on organic materials such as paper, fabrics, cotton, leather, wood or other combustibles can cause the material to ignite and result in a fire.

7. HANDLING AND STORAGE

HANDLING: Wear chemical splash-type monogoggles and full-face shield, impervious clothing, such as rubber, PVC, etc., and rubber or neoprene gloves and shoes. Avoid cotton, wool and leather. Avoid excessive heat and contamination. Contamination may cause decomposition and generation of oxygen gas which could result in high pressures and possible container rupture. Hydrogen peroxide should be stored only in vented containers and transferred only in a prescribed manner (see FMC Technical Bulletins). Never return unused hydrogen peroxide to original container, empty drums should be triple rinsed with water before discarding. Utensils used for handling hydrogen peroxide should only be made of glass, stainless steel, aluminum or plastic.

STORAGE: Store drums in cool areas out of direct sunlight and away from combustibles. For bulk storage refer to FMC Technical Bulletins.

COMMENTS: VENTILATION: Provide mechanical general and/or local exhaust ventilation to prevent release of vapor or mist into the work environment.

8. EXPOSURE CONTROLS / PERSONAL PROTECTION

EXPOSURE LIMITS

Chemical Name	ACGIH	OSHA	Supplier
Hydrogen Peroxide	1 ppm (TWA)	1 ppm (PEL)	

ENGINEERING CONTROLS: Ventilation should be provided to minimize the release of hydrogen peroxide vapors and mists into the work environment. Spills should be minimized or confined immediately to prevent release into the work area. Remove contaminated clothing immediately and wash before reuse.

PERSONAL PROTECTIVE EQUIPMENT

EYES AND FACE: Use chemical splash-type monogoggles and a full-face shield made of polycarbonate, acetate, polycarbonate/acetate, PETG or thermoplastic.

RESPIRATORY: If concentrations in excess of 10 ppm are expected, use NIOSH/DHHS approved self-contained breathing apparatus (SCBA), or other approved atmospheric-supplied respirator (ASR) equipment (e.g., a full-face airline respirator (ALR)). DO NOT use any form of air-purifying respirator (APR) or filtering facepiece (AKA dust mask), especially those containing oxidizable sorbants such as activated carbon.

PROTECTIVE CLOTHING: For body protection wear impervious clothing such as an approved splash protective suit made of SBR Rubber, PVC (PVC Outershell w/Polyester Substrate), Gore-Tex (Polyester trilaminate w/Gore-Tex), or a specialized HAZMAT Splash or Protective Suite (Level A, B, or C). For foot protection, wear approved boots made of NBR, PVC, Polyurethane, or neoprene. Overboots made of Latex or PVC, as well as firefighter boots or specialized HAZMAT boots are also permitted. DO NOT wear any form of boot or overboots made of nylon or nylon blends. DO NOT use cotton, wool or leather, as these materials react RAPIDLY with higher concentrations of hydrogen peroxide. Completely submerge hydrogen peroxide contaminated clothing or other materials in water prior to drying. Residual hydrogen peroxide, if allowed to dry on materials such as paper, fabrics, cotton, leather, wood or other combustibles can cause the material to ignite and result in a fire.

GLOVES: For hand protection, wear approved gloves made of nitrile, PVC, or neoprene. DO NOT use cotton, wool or leather, for these materials react RAPIDLY with higher concentrations of hydrogen peroxide. Thoroughly rinse the outside of gloves with water prior to removal. Inspect regularly for leaks.

9. PHYSICAL AND CHEMICAL PROPERTIES

ODOR:	Odorless
APPEARANCE:	Clear, colorless liquid
AUTOIGNITION TEMPERATURE:	Non-combustible
BOILING POINT:	110°C (229°F) (40%); 114°C (237°F) (50%)
COEFFICIENT OF OIL / WATER:	Not available
DENSITY / WEIGHT PER VOLUME:	Not available
EVAPORATION RATE:	Above 1 (Butyl Acetate = 1)

FIGURE 2-15 An example of the material safety data sheet (MSDS) for hydrogen peroxide showing the technical information that is provided. Useful information includes hazard identification information, first aid measures, as well as firefighting and spill response advice.
Courtesy of FMC Corporation, Hydrogen Peroxide Division, Philadelphia, Pennsylvania

FLASH POINT:	Non-combustible	
FREEZING POINT:	$-41.4°C$ ($-42.5°F$) (40%); $-52°C$ ($-62°F$) (50%)	
ODOR THRESHOLD:	Not available	
OXIDIZING PROPERTIES:	Strong oxidizer	
PERCENT VOLATILE:	100%	
pH:	(as is) 1.0 to 3.0	
SOLUBILITY IN WATER:	(in H_2O % by wt) 100%	
SPECIFIC GRAVITY:	(H_2O = 1)1.15 @ $20°C/4°C$ (40%); 1.19 @ $20°C/4°C$(50%)	
VAPOR DENSITY:	Not available(Air = 1)	
VAPOR PRESSURE:	22 mmHg @ $30°C$ (40%); 18.3 mmHg @ $30°C$ (50%)	

COMMENTS:
pH (l% solution): 5.0-6.0

10. STABILITY AND REACTIVITY

CONDITIONS TO AVOID:	Excessive heat or contamination could cause product to become unstable.
STABILITY:	Stable (heat and contamination could cause decomposition).
POLYMERIZATION:	Will not occur
INCOMPATIBLE MATERIALS:	Reducing agents, wood, paper and other combustibles, iron and other heavy metals, copper alloys and caustic.
HAZARDOUS DECOMPOSITION PRODUCTS:	Oxygen, which supports combustion.

COMMENTS: Materials to Avoid: Dirt, organics, cyanides and combustibles such as wood, paper, oils, etc.

11. TOXICOLOGICAL INFORMATION

EYE EFFECTS: 70% hydrogen peroxide: Severe irritant (corrosive) (rabbit) [FMC Study Number: ICG/T-79.027]

SKIN EFFECTS: 50% hydrogen peroxide: Severe irritant (corrosive) (rabbit) FMC Study Number: I89-1079]

DERMAL LD_{50}: 70% hydrogen peroxide: > 6.5 g/kg (rabbit) [FMC Study Number: ICG/T-79.027]

ORAL LD_{50}: 50% hydrogen peroxide: > 225 mg/kg (rat) [FMC Study Number I86-914]

INHALATION LC_{50}: 50% hydrogen peroxide: > 0.17 mg/L (rat) [FMC Study Number: I89-1080]

TARGET ORGANS: Eye, skin, nose, throat, lungs

ACUTE EFFECTS FROM OVEREXPOSURE: Severe irritant/corrosive to eyes, skin and gastrointestinal tract. May cause irreversible tissue damage to die eyes including blindness. Inhalation of mist or vapors may be severely irritating to nose, throat and lungs.

CHRONIC EFFECTS FROM OVEREXPOSURE: The International Agency for Research on Cancer (IARC) has concluded that there is inadequate evidence for carcinogenicity of hydrogen peroxide in humans, but limited evidence in experimental animals (Group 3 - not classifiable as to its carcinogenicity to humans). The American Conference of Governmental Industrial Hygienists (ACGIH) has concluded that hydrogen peroxide is a Confirmed Animal Carcinogen with Unknown Relevance to Humans'(A3).

CARCINOGENICITY:

Chemical Name	IARC	NTP	OSHA	Other
Hydrogen Peroxide	Listed	Not listed	Not listed	(ACGIH) Listed (A3, Animal Carcinogen)

12. ECOLOGICAL INFORMATION

ECOTOXICOLOGICAL INFORMATION: Channel catfish 96-hour LC_{50} = 37.4 mg/L
Fathead minnow 96-hour LC_{50} = 16.4 mg/L
Daphnia magna 24-hour EC_{50} = 7.7 mg/L
Daphnia pulex 48-hour LC_{50} = 2.4 mg/L
Freshwater snail 96-hour LC_{50} = 17.7 mg/L
For more information refer to ECETOC "Joint Assessment of Commodity Chemicals No. 22, Hydrogen Peroxide." ISSN-0773-6339, January 1993

CHEMICAL FATE INFORMATION: Hydrogen peroxide in the aquatic environment is subject to various reduction or oxidation processes and decomposes into water and oxygen. Hydrogen peroxide half-life in freshwater ranged from 8 hours to 20 days, in air from 10 to 20 hours and in soils from minutes to hours depending upon microbiological activity and metal contaminants.

13. DISPOSAL CONSIDERATIONS

DISPOSAL METHOD: An acceptable method of disposal is to dilute with a large amount of water and allow the hydrogen peroxide to decompose followed by discharge into a suitable treatment system in accordance with all regulatory agencies. The appropriate regulatory agencies should be contacted prior to disposal.

14. TRANSPORT INFORMATION

U.S. DEPARTMENT OF TRANSPORTATION (DOT)

PROPER SHIPPING NAME:	Hydrogen peroxide, aqueous solutions with more than 40% but not more than 60% hydrogen peroxide.
PRIMARY HAZARD CLASS / DIVISION:	5.1 (Oxidizer)
UN/NA NUMBER:	UN 2014

PACKING GROUP:	II
LABEL(S):	Oxidizer, Corrosive
PLACARD(S):	5.1 (Oxidizer)
ADDITIONAL INFORMATION:	DOT Marking: Hydrogen Peroxide, aqueous solution with more than 40%, but not more man 60% Hydrogen Peroxide, UN 2014
	Hazardous Substance/RQ: Not applicable
	49 STCC Number: 4918775
	DOT Spec: stainless steel/high purity aluminum cargo tanks and rail cars. UN Spec: HDPE drums. Contact FMC for specific details.

INTERNATIONAL MARITIME DANGEROUS GOODS (IMDG)

PROPER SHIPPING NAME:	Hydrogen peroxide, aqueous solutions with not less than 20%, but not more than 60% hydrogen peroxide.

INTERNATIONAL CIVIL AVIATION ORGANIZATION (ICAO)/ INTERNATIONAL AIR TRANSPORT ASSOCIATION (IATA)

PROPER SHIPPING NAME:	Hydrogen peroxide (40 - 60%) is forbidden on Passenger and Cargo Aircraft, as well as Cargo Only Aircraft.

OTHER INFORMATION:
Protect from physical damage. Keep drums in upright position. Drums should not be stacked in transit. Do not store drum on wooden pallets.

15. REGULATORY INFORMATION

UNITED STATES

SARA TITLE III (SUPERFUND AMENDMENTS AND REAUTHORIZATION ACT)
SECTION 302 EXTREMELY HAZARDOUS SUBSTANCES (40 CFR 355, APPENDIX A):
Hydrogen Peroxide > 52%, RQ: 1000 lbs. Planning Threshold: 10,000 lbs.

SECTION 311 HAZARD CATEGORIES (40 CFR 370):
Fire Hazard, Immediate (Acute) Health Hazard

SECTION 312 THRESHOLD PLANNING QUANTITY (40 CFR 370):
The Threshold Planning Quantity (TPQ) for this product, if treated as a mixture, is 10,000 lbs; however, this product contains the following ingredients with a TPQ of less than 10,000 lbs.: None, (conc. <52%) (hydrogen peroxide, 1000 lbs. when conc. is >52%)

SECTION 313 REPORTABLE INGREDIENTS (40 CFR 372):
Not listed

CERCLA (COMPREHENSIVE ENVIRONMENTAL RESPONSE, COMPENSATION, AND LIABILITY ACT)
CERCLA DESIGNATION & REPORTABLE QUANTITIES (RQ) (40 CFR 302.4):
Unlisted (Hydrogen Peroxide); RQ = 100 lbs.; Ignitability, Corrosivity

TSCA (TOXIC SUBSTANCES CONTROL ACT)
TSCA INVENTORY STATUS (40 CFR 710):
Listed

RESOURCE CONSERVATION AND RECOVERY ACT (RCRA)
RCRA IDENTIFICATION OF HAZARDOUS WASTE (40 CFR 261):
Waste Number: D001,D002

CANADA

WHMIS (WORKPLACE HAZARDOUS MATERIALS INFORMATION SYSTEM):
Product Identification Number: 2014
Hazard Classification / Division: Class C (Oxidizer), Class D, Div. 2, Subdiv. B (Toxic), Class E (Corrosive)
Ingredient Disclosure List: Listed

EU EINECS NUMBERS:
008-003-00-9 (hydrogen peroxide)

INTERNATIONAL LISTINGS
Hydrogen peroxide:
China: Listed
Japan (ENCS): (1)-419
Korea: KE-20204
Philippines (PICCS): Listed

16. OTHER INFORMATION

HAZARD, RISK AND SAFETY PHRASE DESCRIPTIONS:

Hydrogen Peroxide:

EC Symbols:	C	(Corrosive)
EC Risk Phrases:	R34	(Causes burns)
EC Safety Phrases:	Sl/2	(Keep locked up and out of reach of children.)
	S3	(Keep in a cool place.)
	S28	(After contact with skin, wash immediately with plenty of water and soap.)
	S36/39	(Wear suitable protective clothing. Wear eye / face protection.)
	S45	(In case of accident or if you feel unwell, seek medical advice immediately - show the label where possible.)

FIGURE 2-15 Continued

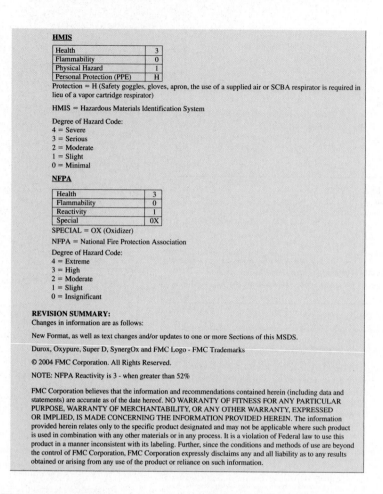

HMIS

Health	3
Flammability	0
Physical Hazard	1
Personal Protection (PPE)	H

Protection = H (Safety goggles, gloves, apron, the use of a supplied air or SCBA respirator is required in lieu of a vapor cartridge respirator)

HMIS = Hazardous Materials Identification System

Degree of Hazard Code:
4 = Severe
3 = Serious
2 = Moderate
1 = Slight
0 = Minimal

NFPA

Health	3
Flammability	0
Reactivity	1
Special	0X

SPECIAL = OX (Oxidizer)

NFPA = National Fire Protection Association

Degree of Hazard Code:
4 = Extreme
3 = High
2 = Moderate
1 = Slight
0 = Insignificant

REVISION SUMMARY:
Changes in information are as follows:

New Format, as well as text changes and/or updates to one or more Sections of this MSDS.

Durox, Oxypure, Super D, SynergOx and FMC Logo - FMC Trademarks

© 2004 FMC Corporation. All Rights Reserved.

NOTE: NFPA Reactivity is 3 - when greater than 52%

FMC Corporation believes that the information and recommendations contained herein (including data and statements) are accurate as of the date hereof. NO WARRANTY OF FITNESS FOR ANY PARTICULAR PURPOSE, WARRANTY OF MERCHANTABILITY, OR ANY OTHER WARRANTY, EXPRESSED OR IMPLIED, IS MADE CONCERNING THE INFORMATION PROVIDED HEREIN. The information provided herein relates only to the specific product designated and may not be applicable where such product is used in combination with any other materials or in any process. It is a violation of Federal law to use this product in a manner inconsistent with its labeling. Further, since the conditions and methods of use are beyond the control of FMC Corporation, FMC Corporation expressly disclaims any and all liability as to any results obtained or arising from any use of the product or reliance on such information.

FIGURE 2-15 Continued

SOLVED EXERCISE 2-2

According to the following MSDS sheet, how should the spill in the opening scenario be handled?
Solution: Section VII of the MSDS sheet from Dyno Nobel Inc states:

SECTION VII - SPILL OR LEAK PROCEDURES

Steps to be taken in Case Material is Released or Spilled: Protect from all ignition sources. In case of fire evacuate area not less than 2,500 feet in all directions. Notify authorities in accordance with emergency response procedures. Only personnel trained in emergency response should respond. If no fire danger is present, confine spill and keep wet. Pick up spilled material with non-metallic dust pan, using whisk brooms, damp sponges or damp rags as aids. If product is uncontaminated, repackage product in original packaging or other clean DOT approved container. Contamination of PETN with sand, grit or dirt will render the material more sensitive to detonation. If contaminated with sand, dirt or grit, do not repackage with uncontaminated material. ALL SPILLED PETN MUST BE RECOVERED. Follow applicable Federal, State and local spill reporting requirements.

Waste Disposal Method: Disposal must comply with Federal, State and local regulations. If product becomes a waste, it is potentially regulated as a hazardous waste as defined under the Resource Conservation and Recovery Act (RCRA) 40 CFR, part 261. Review disposal requirements with a person knowledgeable with applicable environmental law (RCRA) before disposing of any explosive material.

In rail transportation, each railcar has an individual shipping paper called a **waybill**. When the waybills of individual railcars are consolidated into one shipping paper, it is called a **consist**; this should be in the possession of the engineer or train crew at all times. The train consist lists all of the railcars on the train and their contents. Railcars containing hazardous materials are indicated in the tonnage graph. Emergency handling instructions are also included for each hazardous material on the train. Figure 2-17 shows an example of select parts of a consist. When dealing with a railroad emergency, make sure to ask for the consist, and not the manifest!

In air transportation, the shipping papers are called an **air bill**. Figure 2-18 shows an example of an air bill. In marine transport the consolidated shipping papers are called a **manifest**. Each client receives a bill of lading. The manifest is the consolidation of all of the bills of lading. An example of a bill of lading can be found in Figure 2-19.

MILITARY MARKINGS

Military organizations including the U.S. military have their own hazard identification marking systems, especially as they relate to recognition of explosives. Figure 2-20 shows some of the U.S. military identification marks.

Use of the U.S. DOT *Emergency Response Guidebook*

The general information about hazardous materials in the U.S. DOT *Emergency Response Guidebook*, or ERG, is primarily designed for use in the first 15 minutes of an incident, although it may be useful at later stages for the determination of downwind evacuation distances. Although the ERG information is very general, it points you in the right direction. The ERG contains much useful information such as potential hazards; **initial isolation distance**; evacuation information; what to do in case of a fire, spill, or leak; and first aid information in case of injuries. The ERG is quite easy to use with directions given on page 1 (Figure 2-21). The ultimate goal is to find a three-digit guide number in the orange section that will give guidance in what to do in an emergency.

If you know the name of the chemical, you can look it up alphabetically in the blue section (Figure 2-22). Or, if you know the four-digit UN/NA number, you can look it up numerically in the yellow section (Figure 2-23). If all you can see is a placard, you can

waybill ▪ The shipping papers most commonly used for individual railcars in the rail industry. The consolidated waybills are part of the train consist.

consist ▪ The consolidated shipping papers used in the rail industry. It is comprised of all of the individual railcar waybills and usually includes other administrative paperwork such as the order of railcars in the train.

air bill ▪ The shipping papers used in the airline industry.

manifest ▪ In marine transportation, the consolidated shipping papers of ship's cargo.

initial isolation distance ▪ The area around a hazardous materials release from which everyone should be quickly evacuated.

Part A: Tonnage Graph

```
                          TONNAGE GRAPH
                       ================
      TRAIN Q40629  CR TRN#  ORIG:  CFP 1  TIME:  0408290245  CONSIST#  991199

C = SHIPMENT GOVERNED BY CLEARANCE BUREAU INSTRUCTION
R = POTENTIAL RESTRICTED EQUIPMENT - SEE "RESTRICTED AND SPECIAL HANDLING LIST".
REFERENCE TIMETABLES, OPERATING RULES AND MECHANICAL INSTRUCTIONS FOR HANDLING IN
YOUR TRAIN.
L = LONG CAR
S = SHORT CAR
* = EMPTY TOFC CAR

PLATE SIZE----------------|
TYPE---------------------|
LOAD/EMPTY--------------|
```

Hazardous Material

```
            CSRR  7749  L D
            CSRR  8866  L D
    RES  TRN  ==CAR==    LCP  UN/NA  =========TONS=================          EST   TRN  A
    KEY  POS              ETS  NUMB   20   40   60   80  100  120  140  TONS  LGTH  X
R   9040  045- CNW 490978  LCC               ***** ***** ***** ***** *****  ***  3910  2920  4
R   9040  046- ATSF315464  LCC               ***** ***** ***** ***** *****  ***  4040  2975  4
          047- RBOX553070  LAE               ***** ***** ***** ***** **          4150  3025  4
R   9050  048- CLC  10002  LFC               ***** ***** ***** ***** *****       4290  3070  4
R   9050  049- TTZX 84081  LFC               ***** ***** ***** ***** *****  **   4420  3130  4
R   9250  050- CORX  4325  LTC  1824  HHHHH HHHHH HHHHH HHHHH HHHHH (HHHHH) HHH  4550  3175  4
          051- RBOX400602  LAE               ***** ***** ***** ***** *****  ***  4680  3225  4
          052- RBOX400501  LAE               ***** ***** ***** ***** *****  ***  4810  3275  4
          053- DBCX 11098  LTC               ***** ***** ***** ***** *****  ***  4940  3325  4
R   9040  054- DBEX123456  LTC  2448  HHHHH HHHHH HHHHH HHHHH HHHHH HHHHH       5060  3370  4
R   9040  055- GRAX  3333  LCC               ***** ***** ***** ***** *****  *    5185  3425  4
          056- CSXT777898  ECC               ***** ***                           5220  3470  4
R   9250  057- ABCX   002  LTC  1075  HHHHH HHHHH HHHHH HHHHH HHHHH            5340  3525  4
R   9250  058- CDUX543210  LTC  1075  HHHHH HHHHH HHHHH HHHHH HHHHH HHHHH  HHHH 5470  3585  4
R   9250  059- ABCX   001  LTC  1830  HHHHH HHHHH HHHHH HHHHH HHHHH HHHHH  HHH  5605  3645  4
          060- CSXT887700  ECC               ***** ***                           5636  3725  4
          061- DBCX109876  LTC               ***** ***** ***** ***** *****  *    5761  3780  4
R   9040  062- CNW 490835  LCC               ***** ***** ***** ***** *****       5891  3840  4

GRAND
TOTAL  MERCHAN   PIGFLAT  COAL    PERISH   AUTORAK   TOTAL  TOTAL  TOTAL
L    E  L    E   L   E    L   E   L    E   L    E    TONS  LENGTH  AXLE
51  24  51  24   0   0    0   0   0    0   0    0    7985   4250   300

RESTRICTION KEYS********************************************************************
      9040 = RE RULE 7 - HIGH CUBE CAR
      9250 = HAZARDOUS -SEE HAZ MATERIAL DESCRIPTION
      9050 = LONG CAR
      9160 = RESTRICTED TO 40 MPH - COAL WEIGHT RESTRICTION.
```

ENGINEER IS TO BE PROVIDED WITH TONNAGE GRAPH OR A LEGIBLE COPY OF THE TONNAGE
GRAPH. THE UN/NA NUMBER COLUMN CAN BE USED TO REFERENCE THE PRODUCT NAME OF THE
HAZARDOUS MATERIAL IN THE EMERGENCY RESPONSE GUIDEBOOK.

Part B: Position-in-Train Document

```
      CT-168 REPORT - NOTICE OF RAIL CARS/UNITS CONTAINING HAZARDOUS MATERIALS

      TRAIN Q40629  CR TRN#  ORIG: CFP 1  TIME:  0408290245  CONSIST#  991199
```

THE FOLLOWING RAIL CARS/UNITS CONTAINING HAZARDOUS MATERIALS ARE LOCATED IN YOUR
TRAIN. THEY MUST BE POSITIONED IN YOUR TRAIN IN ACCORDANCE WITH FEDERAL
REGULATIONS. WHENEVER THERE IS A CHANGE IN THE POSITION OF ANY HAZARDOUS MATERIAL
CAR IN THE TRAIN, THE CONDUCTOR (OR DESIGNEE) MUST IMMEDIATELY UPDATE THIS DOCUMENT
TO SHOW THE NEW POSITION OF ALL HAZARDOUS MATERIAL CARS.

```
KEY FIELD CODES:
P=POISON INHALATION HAZARD ZONE A OR ZONE B       F=FLAMMABLE GAS 2.1
E=ENVIRONMENTALLY SENSITIVE CHEMICALS             X=EXPLOSIVES 1.1 OR 1.2
```

Location of hazmat railcars from engine

```
           L  CONTAINER      L                      ID        TRN    REVISION
INIT  NUMBER  E  INIT  NUMBER  E  COMMENT          NUMB  KEY   POS   1ST  2ND  3RD
CORX    4325  L  ____  _____  _  _____      1824  __    050   ___  ___  ___
DBEX  123456  L  ____  _____  _  _____      2448  __    054   ___  ___  ___
ABCX     002  L  ____  _____  _  _____      1075  _F_   057   ___  ___  ___
CDUX  543210  L  ____  _____  _  _____      1075  _F_   058   ___  ___  ___
ABCX     001  L  ____  _____  _  _____      1830  __    059   ___  ___  ___
____  _____  _  ____  _____  _  _____      ____  __    ___   ___  ___  ___
____  _____  _  ____  _____  _  _____      ____  __    ___   ___  ___  ___
____  _____  _  ____  _____  _  _____      ____  __    ___   ___  ___  ___
____  _____  _  ____  _____  _  _____      ____  __    ___   ___  ___  ___
____  _____  _  ____  _____  _  _____      ____  __    ___   ___  ___  ___
____  _____  _  ____  _____  _  _____      ____  __    ___   ___  ___  ___
____  _____  _  ____  _____  _  _____      ____  __    ___   ___  ___  ___
____  _____  _  ____  _____  _  _____      ____  __    ___   ___  ___  ___
```

Part C: Train Listing

```
            TRAIN LISTING AND HAZARDOUS MATERIAL DESCRIPTIONS
            =================================================

      TRAIN Q40629  CR TRN#  ORIG:  CFP 1  TIME:  0408290245  CONSIST#  991199

            CARS IN THIS CONSIST COUNT FROM FRONT TO REAR
```

	CAR NUMBER	CAR TYPE	COMMODITY STCC	CODE ALPHA	DESTINATION	CONSIGNEE	YZCN
49	TTZX 84081	L	F483	2421184	LUMBER	LEXINGTKY	84LUMBER	DD002
50	CORX 4325	L	T105	4935240		ECHICAGIN	CSXTRANSFL	BI185

```
********** 01 TC //0195900 LB //SODIUM HYDROXIDE SOLUTION//8//
*  HAZMAT *                   UN1824//RQ (SODIUM HYDROXIDE)//HAZMAT  STCC=
**********    4935240

            U.S.A. EMERGENCY CONTACT:   18004249300
            FROM SHIPPER: PURECO    AUGUSTA GA
            TO CONSIGNEE:CS TRANSFER E CHICAGO IN
```

51	RBOX 400602	L	A403	2432158	PAPER	CHICAGOIL	CHICAGOTRI	BIH007
52	RBOX 400601	L	A403	2432158	PAPER	CHICAGOIL	CHICAGOTRI	BIH007
53	DBCX 11098	L	T105	2851915	LATEX	DANVILLIL	SHERWIWIL	0ZA001
54	DBEX 123456	L	T105	4961622		OTTAWAIL	BPAMOCO	BIF084

```
********** 01 TC // 0187480 LB // SULFUR, MOLTEN// 9 //NA2448//
*  HAZMAT *                   PGIII //HAZMAT STCC=4961622
**********

            U.S.A. EMERGENCY CONTACT:   18004249300
            FROM SHIPPER:  BRADCO   GLENJUNCT CO
            TO CONSIGNEE: PETROCO OTTAWA IL
```

| 55 | GRAX 3333 | L | C313 | 2041993 | GRAIN | BATTLCRMI | KELLOGGS | CD025 |

Part D: Emergency Handling Instructions (for each hazardous material on the train)

```
            HAZARDOUS SPECIAL HANDLING INSTRUCTIONS
            =======================================
--------------------------------------------------------------------------------
            IN CASE OF ACCIDENT PROVIDE THIS LIST TO RESPONSE TEAM
--------------------------------------------------------------------------------

      TRAIN Q40629  CR TRN#  ORIG:  CFP 1  TIME:  0408290245  CONSIST#  991199

EMERGENCY HANDLING INSTRUCTIONS

                                        HAZARDOUS COMMODITY    4905752

                          ABCX  0002 CAR 057 FROM ENGINE
                          CDUX 543120 CAR 058 FROM ENGINE
```

CLASS 2.1 (FLAMMABLE GAS)
LIQUEFIED PETROLEUM GAS (BUTANE/PROPANE MIXTURE) UN1075

Butane/propane mixture is a colorless gas with a petroleum-like odor.
It is shipped as a liquefied gas under pressure. For transportation it is odorized.
Contact with the material can cause frostbite. It is easily ignited.
Its vapors are heavier than air. Any leak can be either liquid or vapor. It can
asphyxiate by the displacement of air. Under prolonged exposure to fire or intense
heat the containers may rupture violently and rocket.

If material on fire or involved in fire
      Do not extinguish fire unless flow can be stopped
      Use water in flooding quantities as fog
      Cool all affected containers with flooding quantities of water
      Apply water from as far a distance as possible

If material not on fire and not involved in fire
      Keep sparks, flames, and other sources of ignition away
      Keep material out of water sources and sewers
      Attempt to stop leak if without undue personnel hazard
      Use water spray to knock-down vapors

Personnel protection
      Avoid breathing vapors
      Keep upwind
      Wear appropriate chemical protective gloves and goggles
      Do not handle broken packages unless wearing appropriate
      personal protective equipment
      Approach fire with caution

Evacuation
      If fire becomes uncontrollable or container is exposed to direct
      flame—consider evacuation of one-half (1/2) mile radius
      If material leaking (not on fire) consider evacuation from
      downwind area based on amount of material spilled, location
      and weather conditions

First aid responses
      Move victim to fresh air; call emergency medical care.
      If not breathing, give artificial respiration.

FIGURE 2-17 An example of a truncated train consist commonly used in rail transportation. *Courtesy of CSX Transportation.*

SOLVED EXERCISE 2-3

The label on the drum in the opening scenario reads "pentaerythritol tetranitrate, wetted." What resources could provide more information?

Solution: The shipping paper and the emergency response number that it contains could provide more information. Additional resources are the DOT *Emergency Response Guidebook* and an MSDS sheet that may be attached to the shipping paper or that can be obtained online.

TABLE 2.7 Shipping Papers

MODE OF TRANSPORT	NAME OF DOCUMENT	PERSON RESPONSIBLE FOR DOCUMENT	LOCATION OF DOCUMENT
Highway	Bill of lading	Driver	In cab within reach of driver
Rail	Consist / Waybill	Engineer	On person
Air	Air bill	Pilot	In cockpit
Water	Manifest	Captain	In wheelhouse or on the deck of a barge in a box

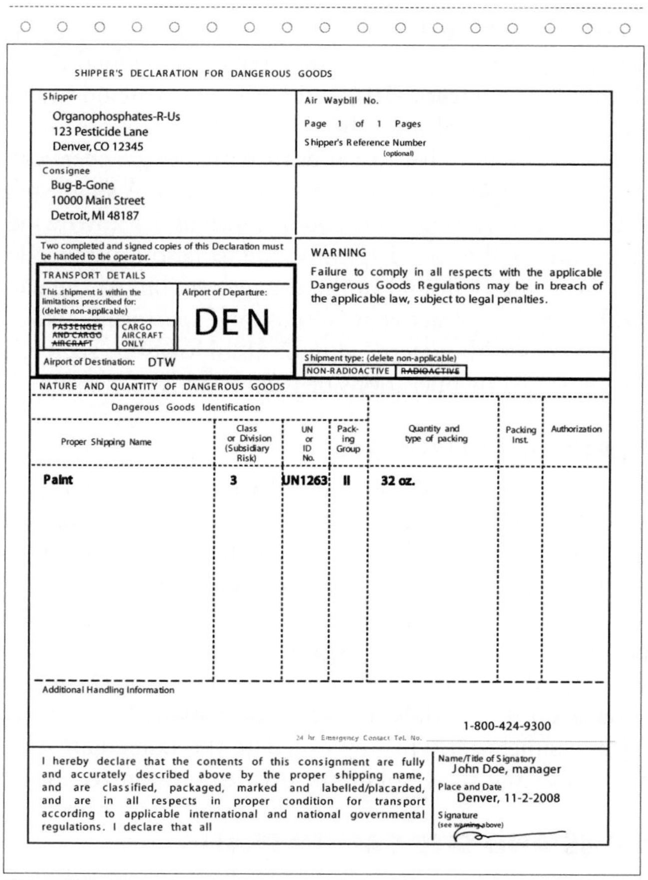

SHIPPER'S DECLARATION FOR DANGEROUS GOODS

Shipper: Organophosphates-R-Us, 123 Pesticide Lane, Denver, CO 12345

Air Waybill No.

Page 1 of 1 Pages

Shipper's Reference Number (optional)

Consignee: Bug-B-Gone, 10000 Main Street, Detroit, MI 48187

Two completed and signed copies of this Declaration must be handed to the operator.

TRANSPORT DETAILS

This shipment is within the limitations prescribed for: (delete non-applicable) ~~PASSENGER AND CARGO AIRCRAFT~~ CARGO AIRCRAFT ONLY

Airport of Departure: DEN

Airport of Destination: DTW

WARNING

Failure to comply in all respects with the applicable Dangerous Goods Regulations may be in breach of the applicable law, subject to legal penalties.

Shipment type: (delete non-applicable) NON-RADIOACTIVE ~~RADIOACTIVE~~

NATURE AND QUANTITY OF DANGEROUS GOODS

Dangerous Goods Identification						
Proper Shipping Name	Class or Division (Subsidiary Risk)	UN or ID No.	Pack-ing Group	Quantity and type of packing	Packing Inst.	Authorization
Paint	3	UN1263	II	32 oz.		

Additional Handling Information

24 hr Emergency Contact Tel. No. 1-800-424-9300

I hereby declare that the contents of this consignment are fully and accurately described above by the proper shipping name, and are classified, packaged, marked and labelled/placarded, and are in all respects in proper condition for transport according to applicable international and national governmental regulations. I declare that all

Name/Title of Signatory: John Doe, manager

Place and Date: Denver, 11-2-2008

Signature (see warning above)

FIGURE 2-18 An example of an air bill used in air transportation.

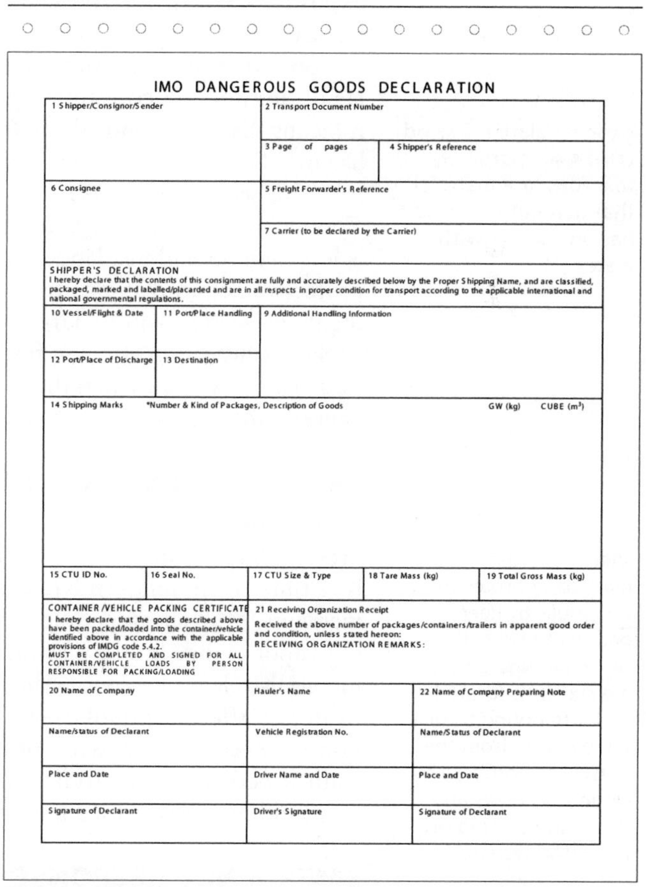

IMO DANGEROUS GOODS DECLARATION

1 Shipper/Consignor/Sender

2 Transport Document Number

3 Page of pages

4 Shipper's Reference

6 Consignee

5 Freight Forwarder's Reference

7 Carrier (to be declared by the Carrier)

SHIPPER'S DECLARATION

I hereby declare that the contents of this consignment are fully and accurately described below by the Proper Shipping Name, and are classified, packaged, marked and labelled/placarded and are in all respects in proper condition for transport according to the applicable international and national governmental regulations.

10 Vessel/Flight & Date | 11 Port/Place Handling | 9 Additional Handling Information

12 Port/Place of Discharge | 13 Destination

14 Shipping Marks | *Number & Kind of Packages, Description of Goods | GW (kg) | CUBE (m^3)

15 CTU ID No. | 16 Seal No. | 17 CTU Size & Type | 18 Tare Mass (kg) | 19 Total Gross Mass (kg)

CONTAINER/VEHICLE PACKING CERTIFICATE

I hereby declare that the goods described above have been packed/loaded into the container/vehicle identified above in accordance with the applicable provisions of IMDG code 5.4.2.

MUST BE COMPLETED AND SIGNED FOR ALL CONTAINER/VEHICLE LOADS BY PERSON RESPONSIBLE FOR PACKING/LOADING

21 Receiving Organization Receipt

Received the above number of packages/containers/trailers in apparent good order and condition, unless stated hereon:

RECEIVING ORGANIZATION REMARKS:

20 Name of Company	Hauler's Name	22 Name of Company Preparing Note
Name/status of Declarant	Vehicle Registration No.	Name/Status of Declarant
Place and Date	Driver Name and Date	Place and Date
Signature of Declarant	Driver's Signature	Signature of Declarant

FIGURE 2-19 An example of a manifest used in marine transportation.

FIGURE 2-20 An example of selected military markings that warn of explosive (on the left) and chemical agent hazards (on the right).

find the three-digit guide number for various placards on pages 16 and 17 of the guidebook (Figure 2-5). If a railcar is involved, and you can only recognize the railcar shape, you can refer to silhouettes on page 18 (Figure 2-24). Likewise, if an over-the-road trailer or tanker is involved and all you can see is a shape of the vehicle, you can refer to silhouettes on page 19 (Figure 2-25). In all cases, the ultimate goal is to find a three-digit guide number, which can help determine the action to take in an emergency. The guide numbers are found in the orange section (Figure 2-26).

The title of the guide immediately gives some information. For example, Guide 139 is titled "Substances—Water-Reactive (Emitting Flammable and Toxic Gases)." This immediately informs the reader about some of the hazards being dealt with. Under the Potential Hazards heading, two categories are listed: Fire or Explosion and Health. The one that is listed first is the primary hazard, so, in the case of Guide 139, fire or explosion is the primary hazard. Under the Public Safety heading is such useful information as an immediate isolation distance, what protective clothing is appropriate, and when to evacuate for large spills or fire. On the facing page, there is guidance on how to handle fires, spills, or leaks, and basic first aid advice for exposure victims. As you can see, you can quickly gather a lot of critical information from the ERG.

Speaking of evacuation, if an entry is highlighted in the yellow or blue section, and it is not on fire, first consult the green section (Figure 2-27). If an entry is highlighted there, it means that the chemical is a **toxic inhalation hazard (TIH)** and poses a downwind health hazard. The green section is organized in numerical order by the four-digit UN/NA number. This section contains more specific information because it deals with a single chemical. You will find that it is organized into a Small Spills and a Large Spills column. A small spill is less than 55 gallons. Under each heading there is an isolation distance and a downwind protective action distance for daytime and nighttime. Typically, the isolation distance in the green section will be larger than what will be found in the orange section. At the end of the green section you can find a table of water-reactive materials, and specifically which toxic gas is evolved when the material reacts with water (Figure 2-28). However, such advanced information will typically be consulted by operations or technician level personnel.

toxic inhalation hazard (TIH) ■ An extremely toxic gas or a material that generates a toxic gas on contact with water.

EVACUATION VERSUS IN-PLACE SHELTERING

Evacuation refers to the removal of people from a defined area based upon the hazards present. **Shelter in place** refers to protecting people at their current location. Evacuation is a time-consuming and resource-intensive action. In the case of hazardous materials incidents, evacuating people from a building may make them more vulnerable to a passing vapor cloud. In-place sheltering is often a very attractive option when the hazardous materials release is outside the occupied building or in a large building at a safe distance from the affected population. Typically, the decision whether to evacuate or shelter in place, especially downwind of the hazardous materials release, will be made by specially trained personnel. Awareness level personnel will typically evacuate the immediate area surrounding a hazardous materials release and then isolate the area to prevent reentry.

shelter in place ■ A method of protecting the public by keeping occupants inside the buildings where they are currently located and attempting to seal the building from the outside atmosphere by closing windows and doors and turning off the HVAC system.

EMERGENCY RESPONSE PLANS AND SITE SAFETY PLANS

Awareness level personnel should receive additional training regarding their agency's emergency response plans and standard operating procedures. In addition, awareness

BEFORE AN EMERGENCY—BECOME FAMILIAR WITH THIS GUIDEBOOK! In the U.S., according to the requirements of the U.S. Department of Labor's Occupational Safety and Health Administration (OSHA, 29 CFR 1910.120), and regulations issued by the U.S. Environmental Protection Agency (EPA, 40 CFR Part 311), first responders must be trained regarding the use of this guidebook.

RESIST RUSHING IN!
APPROACH INCIDENT FROM UPWIND
STAY CLEAR OF ALL SPILLS, VAPORS, FUMES, SMOKE AND SUSPICIOUS SOURCES

HOW TO USE THIS GUIDEBOOK DURING AN INCIDENT INVOLVING DANGEROUS GOODS

STEP ONE: **IDENTIFY THE MATERIAL.** USE **ANY** OF THE FOLLOWING:
- **IDENTIFICATION NUMBER (4-DIGIT ID) FROM** A PLACARD, ORANGE PANEL, SHIPPING PAPER OR PACKAGE (after UN/NA)
- **NAME OF THE MATERIAL FROM** A SHIPPING DOCUMENT OR PACKAGE

STEP TWO: **IDENTIFY 3-DIGIT GUIDE NUMBER** USE:

- ID NUMBER INDEX in yellow-bordered pages or
- NAME OF MATERIAL INDEX in blue-bordered pages

Guide number supplemented with the letter **"P"** indicates that the material may undergo violent polymerization if subjected to heat or contamination.

INDEX ENTRIES HIGHLIGHTED IN GREEN are TIH (Toxic Inhalation Hazard) material, a chemical warfare agent or a Dangerous Water Reactive Material (produces toxic gas upon contact with water). **IDENTIFY ID NUMBER AND NAME OF MATERIAL** IN TABLE 1–INITIAL ISOLATION AND PROTECTIVE ACTION DISTANCES (the green-bordered pages). **IF NECESSARY, BEGIN PROTECTIVE ACTIONS IMMEDIATELY** (see Protective Actions page 296). If no protective action required, use the information jointly with the 3-digit guide.

STEP THREE: **TURN TO THE NUMBERED GUIDE** (the orange-bordered pages) **READ CAREFULLY.**

USE GUIDE 112 FOR ALL EXPLOSIVES EXCEPT FOR EXPLOSIVES 1.4 (EXPLOSIVES C) WHERE GUIDE 114 IS TO BE CONSULTED.

NOTE: **IF ABOVE STEPS CANNOT BE COMPLETED AND PLACARD IS VISIBLE: Turn to pages 16-17;** use 3-digit guide next to placard; PROCEED TO NUMBERED GUIDE (orange-bordered pages). If shipping document is available, call emergency response telephone number listed. If document or emergency response telephone is not available, IMMEDIATELY CALL the appropriate **emergency response agency listed in the back of this guidebook.** Provide as much information as possible, such as the name of the carrier (trucking company or railroad) and vehicle number. **IF A REFERENCE TO A GUIDE CANNOT BE FOUND AND THIS INCIDENT IS BELIEVED TO INVOLVE DANGEROUS GOODS,** TURN TO **GUIDE 111** NOW, AND USE IT UNTIL ADDITIONAL INFORMATION BECOMES AVAILABLE.
AS A LAST RESORT: IF ONLY THE CONTAINER CAN BE IDENTIFIED, CONSULT THE TABLE OF RAIL CAR AND ROAD TRAILER IDENTIFICATION CHART (pages 18-19). REMEMBER THAT THE INFORMATION ASSOCIATED WITH THESE CONTAINERS IS FOR WORST CASE SCENARIOS.

FIGURE 2-21 Use of the *Emergency Response Guidebook* (ERG) is described on page 1. *Courtesy of the U.S. Department of Transportation Pipeline and Hazardous Materials Safety Administration*

personnel working at a fixed location should be trained in that location's **site safety plan.** Emergency response plans and site safety plans will vary greatly based upon the type of agency and the type of location in question. Thus, response plan and safety plan training is beyond the scope of this chapter and must be supplemented by your employer. It is especially important to understand how one's role at the awareness level is defined in the standard operating procedures and to know the proper notifications that need to be made.

site safety plan ■ A comprehensive document that enumerates the hazards and risks associated with responding to a hazmat incident, describes the methods that will be used to minimize the risk to on-site personnel, and outlines safe work practices for mitigation of the incident.

Name of Material	Guide No.	ID No.	Name of Material	Guide No.	ID No.
Chemical kit	154	1760	Chloroacetic acid, solid	153	1751
Chemical kit	171	3316	Chloroacetic acid, solution	153	1750
Chemical sample, poisonous	151	3315	Chloroacetone, stabilized	131	1695
Chemical sample, poisonous liquid	151	3315	Chloroacetonitrile	131	2668
Chemical sample, poisonous solid	151	3315	Chloroacetophenone	153	1697
Chemical sample, toxic	151	3315	Chloroacetophenone, liquid	153	1697
Chemical sample, toxic liquid	151	3315	Chloroacetophenone, liquid	153	3416
Chemical sample, toxic solid	151	3315	Chloroacetophenone, solid	153	1697
Chloral, anhydrous, stabilized	153	2075	Chloroacetyl chloride	156	1752
Chlorate and Borate mixtures	140	1458	Chloroanilines, liquid	152	2019
Chlorate and Magnesium chloride mixture	140	1459	Chloroanilines, solid	152	2018
Chlorate and Magnesium chloride mixture, solid	140	1459	Chloroanisidines	152	2233
			Chlorobenzene	130	1134
Chlorate and Magnesium chloride mixture, solution	140	3407	Chlorobenzotrifluorides	130	2234
			Chlorobenzyl chlorides	153	2235
Chlorates, inorganic, aqueous solution, n.o.s.	140	3210	Chlorobenzyl chlorides, liquid	153	2235
			Chlorobenzyl chlorides, solid	153	3427
Chlorates, inorganic, n.o.s.	140	1461	1-Chloro-3-bromopropane	159	2688
Chloric acid, aqueous solution, with not more than 10% Chloric acid	140	2626	Chlorobutanes	130	1127
			Chlorocresols	152	2669
			Chlorocresols, liquid	152	2669
			Chlorocresols, solid	152	2669
Chlorine	124	1017	Chlorocresols, solid	152	3437
Chlorine dioxide, hydrate, frozen	143	9191	Chlorocresols, solution	152	2669
Chlorine pentafluoride	124	2548	Chlorodifluorobromomethane	126	1974
Chlorine trifluoride	124	1749	1-Chloro-1, 1-difluoroethane	115	2517
Chlorite solution	154	1908	Chlorodifluoroethanes	115	2517
Chlorite solution, with more than 5% available Chlorine	154	1908	Chlorodifluoromethane	126	1018
			Chlorodifluoromethane and Chloropentafluoroethane mixture	126	1973
Chlorites, inorganic, n.o.s.	143	1462			
Chloroacetaldehyde	153	2232	Chlorodinitrobenzenes	153	1577
Chloroacetic acid, liquid	153	1750	Chlorodinitrobenzenes, liquid	153	1577
Chloroacetic acid, molten	153	3250	Chlorodinitrobenzenes, solid	153	1577

FIGURE 2-22 The blue section of the ERG lists hazardous materials in alphabetical order. *Courtesy of the U.S. Department of Transportation Pipeline and Hazardous Materials Safety Administration*

SOLVED EXERCISE 2-4

What are the hazards of the material from the opening scenario?

Solution: The UN numbers are listed numerically in the yellow section of the ERG. In this case there is no listing for any UN number that starts with a "0." This is because materials starting with a "0" are explosives. The directions under STEP THREE on page 1 of the 2008 ERG (see Figure 2-21) state to "USE GUIDE 112 FOR ALL EXPLOSIVES EXCEPT FOR EXPLOSIVES 1.4 (EXPLOSIVES C) WHERE GUIDE 114 IS TO BE CONSULTED." We should therefore use Guide 112 for a 1.1D explosive. Guide 112 lists the information.

The information under Potential Hazards lists the Fire or Explosion category before the Health category. The category listed first is the greater of the two hazards. The Fire or Explosion category states that it may explode and throw fragments 1600 meters (1 mile) or more if fire reaches cargo. The Health category states that a fire may produce irritating, corrosive, and/or toxic gases.

POTENTIAL HAZARDS

FIRE OR EXPLOSION

- MAY EXPLODE AND THROW FRAGMENTS 1600 meters (1 MILE) OR MORE IF FIRE REACHES CARGO.
- For information on "Compatibility Group" letters, refer to Glossary section.

HEALTH

- Fire may produce irritating, corrosive and/or toxic gases.

PUBLIC SAFETY

- **CALL Emergency Response Telephone Number on Shipping Paper first. If Shipping Paper not available or no answer, refer to appropriate telephone number listed on the inside back cover.**
- Isolate spill or leak area immediately for at least 500 meters (1/3 mile) in all directions.
- Move people out of line of sight of the scene and away from windows.
- Keep unauthorized personnel away.
- Stay upwind.
- Ventilate closed spaces before entering.

PROTECTIVE CLOTHING

- Wear positive pressure self-contained breathing apparatus (SCBA).
- Structural firefighters' protective clothing will only provide limited protection.

EVACUATION

Large Spill
- **Consider initial evacuation for 800 meters (1/2 mile) in all directions.**

Fire
- If rail car or trailer is involved in a fire and heavily encased explosives such as bombs or artillery projectiles are suspected, ISOLATE for 1600 meters (1 mile) in all directions; also, initiate evacuation including emergency responders for 1600 meters (1 mile) in all directions.
- When heavily encased explosives are not involved, evacuate the area for 800 meters (1/2 mile) in all directions.

* For information "Compatibility Group" letters, refer to the Glossary section.

EMERGENCY RESPONSE

FIRE

CARGO FIRE
- **DO NOT fight fire when fire reaches cargo! Cargo may EXPLODE!**
- Stop all traffic and clear the area for at least 1600 meters (1 mile) in all directions and let burn.
- **Do not move cargo or vehicle if cargo has been exposed to heat.**

TIRE OR VEHICLE FIRE
- **Use plenty of water—FLOOD it! If water is not available, use CO$_2$, dry chemical or dirt.**
- If possible, and WITHOUT RISK, use unmanned hose holders or monitor nozzles from maximum distance to prevent fire from spreading to cargo area.
- Pay special attention to tire fire as re-ignition may occur. Stand by with extinguisher ready.

SPILL OR LEAK

- ELIMINATE all ignition sources (no smoking, flares, sparks or flames in immediate area).
- All equipment used when handling the product must be grounded.
- Do not touch or walk through spilled material.
- DO NOT OPERATE RADIO TRANSMITTERS WITHIN 100 meters (330 feet) OF ELECTRIC DETONATORS.
- **DO NOT CLEAN-UP OR DISPOSE OF, EXCEPT UNDER SUPERVISION OF A SPECIALIST.**

FIRST AID

- Move victim to fresh air.
- Call 911 or emergency medical service.
- Give artifical respiration if victim is not breathing.
- Administer oxygen if breathing is difficult.
- Remove and isolate contaminated clothing and shoes.
- In case of contact with substance, immediately flush skin or eyes with running water for at least 20 minutes.
- Ensure that medical personnel are aware of the material(s) involved and take precautions to protect themselves.

* For information "Compatibility Group" letters, refer to the Glossary section.

What actions should occur? (Refer to Solved Exercise 2-4.)
   Solution: Under the Public Safety section, the ERG suggests:

PUBLIC SAFETY

- **CALL Emergency Response Telephone Number on Shipping Paper first.
  If Shipping Paper not available or no answer, refer to appropriate
  telephone number listed on the inside back cover.**
- Isolate spill or leak area immediately for at least 500 meters (1/3 mile) in all directions.
- Move people out of line of sight of the scene and away from windows.
- Keep unauthorized personnel away.
- Stay upwind.
- Ventilate closed spaces before entering.

Under the Spill or Leak section, the ERG suggests:

SPILL OR LEAK

- ELIMINATE all ignition sources (no smoking, flares, sparks or flames in immediate area).
- All equipment used when handling the product must be grounded.
- Do not touch or walk through spilled material.
- DO NOT OPERATE RADIO TRANSMITTERS WITHIN 100 meters (330 feet) OF ELECTRIC DETONATORS.
- **DO NOT CLEAN-UP OR DISPOSE OF, EXCEPT UNDER SUPERVISION OF A SPECIALIST.**

What is the evacuation distance? (Refer to Solved Exercise 2-4.)
   Solution: The area should be evacuated 800 meters in all directions for a large spill. A large spill is greater than 200 liters (approximately 55 gallons) for a liquid or 300 kilograms (approximately 600 pounds) for a solid.

Criminal or Terrorist Incidents Involving Hazardous Materials or Weapons of Mass Destruction

Criminals or terrorists may target various locations in a community. The location will vary based upon the perpetrator's intent, ideology, and ultimate goal. The common criminal may target high-value locations such as banks, jewelry stores, or other locations as a diversionary tactic. The terrorist may target locations based upon their high profile, function, high occupancy load, or significance to the community. Thus, monuments, water treatment plants, abortion clinics, schools, stadiums, malls, courthouses, and many other locations may be targeted.

ID No.	Guide No.	Name of Material	ID No.	Guide No.	Name of Material
2253	153	N, N-Dimethylaniline	2281	156	Hexamethylene diisocyanate
2254	133	Matches, fusee	2282	129	Hexanols
2256	130	Cyclohexene	2283	130P	Isobutyl methacrylate, stabilized
2257	138	Potassium	2284	131	Isobutyronitrile
2257	138	Potassium, metal	2285	156	Isocyanatobenzotrifluorides
2258	132	1, 2-Propylenediamine	2286	128	Pentamethylheptane
2258	132	1, 3-Propylenediamine	2287	128	Isoheptenes
2259	153	Triethylenetetramine	2288	128	Isohexenes
2260	132	Tripropylamine	2289	153	Isophoronediamine
2261	153	Xylenols	2290	156	IPDI
2261	153	Xylenols, solid	2290	156	Isophorone diisocyanate
2262	156	Dimethylcarbamoyl chloride	2291	151	Lead compound, soluble, n.o.s.
2263	128	Dimethylcyclohexanes	2293	128	4-Methoxy-4-methylpentan-2-one
2264	132	N, N-Dimethylcyclohexylamine	2294	153	N-Methylaniline
2264	132	Dimethylcyclohexylamine	2295	155	Methyl chloroacetate
2265	129	N, N-Dimethylformamide	2296	128	Methylcyclohexane
2266	132	Dimethyl-N-propylamine	2297	128	Methylcyclohexanone
2267	156	Dimethyl thiophosphoryl chloride	2298	128	Methylcyclopentane
2269	153	3, 3'-Iminodipropylamine	2299	155	Methyl dichloroacetate
2270	132	Ethylamine, aqueous solution, with not less than 50% but not more than 70% Ethylamine	2300	153	2-Methyl-5-ethylpyridine
			2301	128	2-Methylfuran
			2302	127	5-Methylhexan-2-one
2271	128	Ethyl amyl ketone	2303	128	Isopropenylbenzene
2272	153	N-Ethylaniline	2304	133	Naphthalene, molten
2273	153	2-Ethylaniline	2305	153	Nitrobenzenesulfonic acid
2274	153	N-Ethyl-N-benzylaniline	2305	153	Nitrobenzenesulphonic acid
2275	129	2-Ethylbutanol	2306	152	Nitrobenzotrifluorides
2276	132	2-Ethylhexylamine	2306	152	Nitrobenzotrifluorides, liquid
2277	130P	Ethyl methacrylate,	2307	152	3-Nitro-4-chlorobenzotrifluoride
2277	130P	Ethyl methacrylate, stabilized	2308	157	Nitrosylsulfuric acid
2278	128	n-Heptene	2308	157	Nitrosylsulfuric acid, liquid
2279	151	Hexachlorobutadiene	2308	157	Nitrosylsulfuric acid, solid
2280	153	Hexamethylenediamine, solid	2308	157	Nitrosylsulphuric acid

FIGURE 2-23 The yellow section of the ERG lists hazardous materials in numerical order based on their four-digit UN/NA number. *Courtesy of the U.S. Department of Transportation Pipeline and Hazardous Materials Safety Administration*

Hopper Car Dry Bulk (140)

Box Car Mixed Cargo (111)

Pressure Tank Car Compressed Liquefied Gases (117)

Low Pressure Tank Car Liquids (131)

REPORTING MARKS & CAR NUMBER
LOAD LIMIT (POUNDS OR KG)
EMPTY WEIGHT OF CAR
PLACARD HOLDER
TANK TEST & SAFETY VALVE TEST INFORMATION
CAR SPECIFICATION
COMMODITY NAME *
TC PERMIT NUMBER

REPORTING MARKS & CAR NUMBER
CAPACITY IN GALLONS OR LITERS
PLACARD HOLDER *

biological agent ▪ A disease-causing organism or toxic substance of living origin, such as a bacterium, virus, or toxin. Examples include anthrax, smallpox, and ricin.

nerve agent ▪ A substance that damages or inactivates the nervous system when inhaled or absorbed through the skin. Examples of nerve agents include sarin, soman, tabun, and VX.

vesicant ▪ A substance capable of blistering the skin and other body tissues (such as the lungs). A blister agent. Examples of vesicant include sulfur mustard and lewisite.

blood agent ▪ A chemical that causes illness by interfering with the blood's ability to absorb and carry atmospheric oxygen or the ability of the cells to use oxygen. An example of a blood agent is cyanide.

toxin ▪ A poisonous substance produced by a living organism or bacterium. Examples include ricin, botulinum toxin, and mycotoxins.

The acronym CBRNE stands for chemical, biological, radiological, nuclear, and explosives. These five categories are collectively termed weapons of mass destruction (WMD).

CHEMICAL AGENTS

Chemical agents were originally developed for military warfare. Chemical agents include **blood agents** such as cyanide, **nerve agents** such as sarin and VX, **vesicants** such as mustard, and riot control agents such as tear gas and pepper spray.

BIOLOGICAL AGENTS

Biological agents include bacteria and viruses that are living organisms, as well as **toxins** that are derived from living organisms. The biggest difference between a chemical and a biological incident is the time of onset of signs and symptoms. Chemical agents will typically produce signs and symptoms within seconds or minutes, hours at the latest (with some vesicants). Biological agents on the other hand, will typically have a delay of days or weeks, while the bacterium or virus replicates in and infects the victim. The exception to this rule is the toxin family, which typically has a quick onset of signs and symptoms.

EXPLOSIVES

Explosives, hazard class 1 materials, are the most common WMD agent used in the United States and throughout the world in terrorist attacks. This is largely due to their ease of use and wide availability. An ammonium nitrate and fuel oil mixture (ANFO) was used to bring down the Alfred P. Murrah Federal Building in Oklahoma City in 1995.

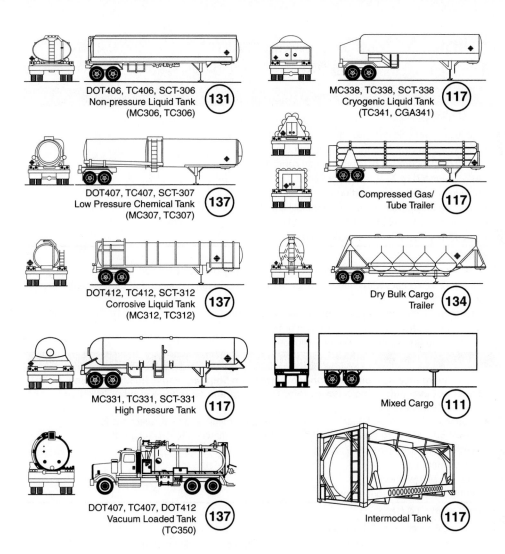

FIGURE 2-25 Highway cargo tanker silhouettes are shown on page 19 of the ERG. The encircled three digit numbers refer to the guide numbers found in the orange section of the ERG. *Courtesy of the U.S. Department of Transportation Pipeline and Hazardous Materials Safety Administration*

DOT406, TC406, SCT-306
Non-pressure Liquid Tank
(MC306, TC306) (131)

DOT407, TC407, SCT-307
Low Pressure Chemical Tank
(MC307, TC307) (137)

DOT412, TC412, SCT-312
Corrosive Liquid Tank
(MC312, TC312) (137)

MC331, TC331, SCT-331
High Pressure Tank (117)

DOT407, TC407, DOT412
Vacuum Loaded Tank
(TC350) (137)

MC338, TC338, SCT-338
Cryogenic Liquid Tank
(TC341, CGA341) (117)

Compressed Gas/
Tube Trailer (117)

Dry Bulk Cargo
Trailer (134)

Mixed Cargo (111)

Intermodal Tank (117)

If an incident is suspected to be of a criminal or terrorist nature, several extra steps and precautions should be taken. Often, secondary devices targeted at responding personnel may be deployed by the perpetrator. Extreme caution should be used in and around the release site. In addition, the release site is now a crime scene and should be disturbed as little as possible. Almost everything is potential evidence! Your local law enforcement agency and the FBI should be notified as soon as possible. Law enforcement officials will in turn notify technical experts capable of safely dealing with explosives, including an explosives ordnance disposal (EOD) team.

RADIOLOGICAL MATERIALS AND NUCLEAR INCIDENTS

Radiological materials, or hazard class 7 substances, may be combined with explosives to form **radiological dispersal devices (RDDs)**. Radiological dispersal devices have the same primary hazard as explosive devices, with the added psychological and health hazard of disseminating radioactive materials. Radioactive materials emit ionizing radiation, which at high levels can damage our bodies. Typically, the activity of the radioactive materials found in an RDD will be very low, and the primary hazard of the radioactive material will be a psychological effect. However, this psychological effect may be very powerful. The American public is extremely wary of radiation. Therefore, any release of radioactive material, or even a perceived release of radioactive material, will likely cause extensive economic damage to the region.

radiological dispersal device (RDD) ■ An explosive device that contains a radioactive material that is disseminated into the environment as the bomb detonates. Also known as a dirty bomb.

FIGURE 2-26 The or-
ange section of the ERG
gives emergency response
information for hazardous
materials including health
hazards, flammability
hazards, spill response
information, and isolation
and evacuation distances.
*Courtesy of the U.S. Depart-
ment of Transportation Pipeline and Hazardous Materials
Safety Administration*

GUIDE 139 — SUBSTANCES - WATER-REACTIVE (EMITTING FLAMMABLE AND TOXIC GASES) — ERG2008

POTENTIAL HAZARDS

FIRE OR EXPLOSION

- Produce flammable and toxic gases on contact with water.
- May ignite on contact with water or moist air.
- Some react vigorously or explosively on contact with water.
- May be ignited by heat, sparks or flames.
- May re-ignite after fire is extinguished.
- Some are transported in highly flammable liquids.
- Containers may explode when heated.
- Runoff may create fire or explosion hazard.

HEALTH

- Highly toxic: contact with water produces toxic gas, may be fatal if inhaled.
- Inhalation or contact with vapors, substance or decomposition products may cause severe injury or death.
- May produce corrosive solutions on contact with water.
- Fire will produce irritating, corrosive and/or toxic gases.
- Runoff from fire control may cause pollution.

PUBLIC SAFETY

- **CALL Emergency Response Telephone Number on Shipping Paper first. If Shipping Paper not available or no answer, refer to appropriate telephone number listed on the inside back cover.**
- As an immediate precautionary measure, isolate spill or leak area in all directions for at least 50 meters (150 feet) for liquids and at least 25 meters (75 feet) for solids.
- Keep unauthorized personnel away.
- Stay upwind.
- Keep out of low areas.
- Ventilate the area before entry.

PROTECTIVE CLOTHING

- Wear positive pressure self-contained breathing apparatus (SCBA).
- Wear chemical protective clothing that is specifically recommended by the manufacturer. It may provide little or no thermal protection.
- Structural firefighters' protective clothing provides limited protection in fire situations ONLY; it is not effective in spill situations where direct contact with the substance is possible.

EVACUATION

Large Spill

- See Table 1-Initial Isolation and Protective Action Distances for highlighted materials. For non-highlighted materials, increase, in the downwind direction, as necessary, the isolation distance shown under "PUBLIC SAFETY".

Fire

- If tank, rail car or tank truck is involved in a fire, ISOLATE for 800 meters (1/2 mile) in all directions; also, consider initial evacuation for 800 meters (1/2 mile) in all directions.

EMERGENCY RESPONSE

FIRE
- **DO NOT USE WATER OR FOAM. (FOAM MAY BE USED FOR CHLOROSILANES, SEE BELOW)**
  Small Fire
- Dry chemical, soda ash, lime or sand.
  Large Fire
- DRY sand, dry chemical, soda ash or lime or withdraw from area and let fire burn.
- **FOR CHLOROSILANES, DO NOT USE WATER;** use AFFF alcohol-resistant medium expansion foam; **DO NOT USE** dry chemicals, soda ash or lime on chlorosilane fires (large or small) as they may release large quantities of hydrogen gas that may explode.
- Move containers from fire area if you can do it without risk.
  Fire involving Tanks or Car/Trailer Loads
- Fight fire from maximum distance or use unmanned hose holders or monitor nozzles.
- Cool containers with flooding quantities of water until well after fire is out.
- Do not get water inside containers.
- Withdraw immediately in case of rising sound from venting safety devices or discoloration of tank.
- ALWAYS stay away from tanks engulfed in fire.

SPILL OR LEAK
- Fully encapsulating, vapor protective clothing should be worn for spills and leaks with no fire.
- ELIMINATE all ignition sources (no smoking, flares, sparks or flames in immediate area).
- Do not touch or walk through spilled material.
- Stop leak if you can do it without risk.
- **DO NOT GET WATER on spilled substance or inside containers**.
- Use water spray to reduce vapors or divert vapor cloud drift. Avoid allowing water runoff to contact spilled material.
- **FOR CHLOROSILANES,** use AFFF alcohol-resistant medium expansion foam to reduce vapors.
  Small Spill • Cover with DRY earth, DRY sand or other non-combustible material followed with plastic sheet to minimize spreading or contact with rain.
- Dike for later disposal; do not apply water unless directed to do so.
  Powder Spill • Cover powder spill with plastic sheet or tarp to minimize spreading and keep powder dry.
- **DO NOT CLEAN-UP OR DISPOSE OF, EXCEPT UNDER SUPERVISION OF A SPECIALIST.**

FIRST AID
- Move victim to fresh air. • Call 911 or emergency medical service.
- Give artificial respiration if victim is not breathing.
- **Do not use mouth-to-mouth method if victim ingested or inhaled the substance; give artificial respiration with the aid of a pocket mask equipped with a one-way valve or other proper respiratory medical device.**
- Administer oxygen if breathing is difficult.
- Remove and isolate contaminated clothing and shoes.
- In case of contact with substance, wipe from skin immediately; flush skin or eyes with running water for at least 20 minutes.
- Keep victim warm and quiet.
- Ensure that medical personnel are aware of the material(s) involved and take precautions to protect themselves.

INITIAL ISOLATION AND PROTECTIVE ACTION DISTANCES

ID No.	NAME OF MATERIAL	SMALL SPILLS (From a small package or small leak from a large package)						LARGE SPILLS (From a large package or from many small packages)						
		First ISOLATE in all Directions		Then PROTECT persons Downwind during-				First ISOLATE in all Directions		Then PROTECT persons Downwind during-			
				DAY		NIGHT				DAY		NIGHT	
		Meters	(Feet)	Kilometers	(Miles)	Kilometers	(Miles)	Meters	(Feet)	Kilometers	(Miles)	Kilometers	(Miles)
1005 1005	Ammonia, anhydrous Anhydrous ammonia	30 m	(100 ft)	0.1 km	(0.1 mi)	0.2 km	(0.1 mi)	150 m	(500 ft)	0.8 km	(0.5 mi)	2.3 km	(1.4 mi)
1008 1008	Boron trifluoride Boron trifluoride, compressed	30 m	(100 ft)	0.1 km	(0.1 mi)	0.6 km	(0.4 mi)	300 m	(1000 ft)	1.9 km	(1.2 mi)	4.8 km	(3.0 mi)
1016 1016	Carbon monoxide Carbon monoxide, compressed	30 m	(100 ft)	0.1 km	(0.1 mi)	0.1 km	(0.1 mi)	150 m	(500 ft)	0.7 km	(0.5 mi)	2.7 km	(1.7 mi)
1017	Chlorine	60 m	(200 ft)	0.4 km	(0.3 mi)	1.6 km	(1.0 mi)	600 m	(2000 ft)	3.5 km	(2.2 mi)	8.0 km	(5.0 mi)
1023 1023	Coal gas Coal gas, compressed	30 m	(100 ft)	0.1 km	(0.1 mi)	0.1 km	(0.1 mi)	60 m	(200 ft)	0.3 km	(0.2 mi)	0.4 km	(0.3 mi)
1026 1026	Cyanogen Cyanogen gas	30 m	(100 ft)	0.2 km	(0.1 mi)	0.9 km	(0.5 mi)	150 m	(500 ft)	1.0 km	(0.7 mi)	3.5 km	(2.2 mi)
1040 1040	Ethylene oxide Ethylene oxide with Nitrogen	30 m	(100 ft)	0.1 km	(0.1 mi)	0.2 km	(0.1 mi)	150 m	(500 ft)	0.8 km	(0.5 mi)	2.5 km	(1.6 mi)
1045 1045	Fluorine Fluorine, compressed	30 m	(100 ft)	0.1 km	(0.1 mi)	0.3 km	(0.2 mi)	150 m	(500 ft)	0.8 km	(0.5 mi)	3.1 km	(1.9 mi)
1048	Hydrogen bromide, anhydrous	30 m	(100 ft)	0.1 km	(0.1 mi)	0.4 km	(0.3 mi)	300 m	(1000 ft)	1.5 km	(1.0 mi)	4.5 km	(2.8 mi)
1050	Hydrogen chloride, anhydrous	30 m	(100 ft)	0.1 km	(0.1 mi)	0.4 km	(0.2 mi)	60 m	(200 ft)	0.3 km	(0.2 mi)	1.4 km	(0.9 mi)
1051	AC (when used as a weapon)	100 m	(300 ft)	0.3 km	(0.2 mi)	1.1 km	(0.7 mi)	1000 m	(3000 ft)	3.8 km	(2.4 mi)	7.2 km	(4.5 mi)
1051	Hydrocyanic acid, aqueous solutions, with more than 20% Hydrogen cyanide	60 m	(200 ft)	0.2 km	(0.1 mi)	0.6 km	(0.4 mi)	400 m	(1250 ft)	1.6 km	(1.0 mi)	4.1 km	(2.5 mi)
1051 1051	Hydrogen cyanide, anhydrous, stabilized Hydrogen cyanide, stabilized												
1052	Hydrogen fluoride, anhydrous	30 m	(100 ft)	0.1 km	(0.1 mi)	0.5 km	(0.3 mi)	300 m	(1000 ft)	1.7 km	(1.1 mi)	3.6 km	(2.2 mi)

FIGURE 2-27 The green section of the ERG lists downwind evacuation distances for highlighted materials, which are toxic inhalation hazards (TIH). *Courtesy of the U.S. Department of Transportation Pipeline and Hazardous Materials Safety Administration*

WATER-REACTIVE MATERIALS WHICH PRODUCE TOXIC GASES

Materials Which Produce Large Amounts of Toxic-by-Inhalation (TIH) Gas(es) When Spilled in Water

ID No.	Guide No.	Name of Material	TIH Gas(es) Produced		
1767	155	Diethyldichlorosilane	HCl		
1769	156	Diphenyldichlorosilane	HCl		
1771	156	Dodecyltrichlorosilane	HCl		
1777	137	Fluorosulfonic acid	HF		
1777	137	Fluorosulphonic acid	HF		
1781	156	Hexadecyltrichlorosilane	HCl		
1784	156	Hexyltrichlorosilane	HCl		
1799	156	Nonyltrichlorosilane	HCl		
1800	156	Octadecyltrichlorosilane	HCl		
1801	156	Octyltrichlorosilane	HCl		
1804	156	Phenyltrichlorosilane	HCl		
1806	137	Phosphorus pentachloride	HCl		
1808	137	Phosphorus tribromide	HBr		
1809	137	Phosphorus trichloride	HCl		
1810	137	Phosphorus oxychloride	HCl		
1815	132	Propionyl chloride	HCl		
1816	155	Propyltrichlorosilane	HCl		
1818	157	Silicon tetrachloride	HCl		
1828	137	Sulfur chlorides	HCl	SO_2	H_2S
1828	137	Sulphur chlorides	HCl	SO_2	H_2S
1834	137	Sulfuryl chloride	HCl		
1834	137	Sulphuryl chloride	HCl		
1836	137	Thionyl chloride	HCl	SO_2	
1838	137	Titanium tetrachloride	HCl		

Chemical Symbols for TIH Gases:

Br_2	Bromine	HF	Hydrogen fluoride	PH_3	Phosphine
Cl_2	Chlorine	HI	Hydrogen iodide	NO_2	Nitrogen dioxide
HBr	Hydrogen bromide	H_2S	Hydrogen sulfide	SO_2	Sulfur dioxide
HCl	Hydrogen chloride	H_2S	Hydrogen sulphide	SO_2	Sulphur dioxide
HCN	Hydrogen cyanide	NH_3	Ammonia		

Use this list only when material is spilled in water.

FIGURE 2-28 Table of water-reactive materials. *Courtesy of the U.S. Department of Transportation Pipeline and Hazardous Materials Safety Administration*

In the opening scenario, what other emergency responders should be notified?

Solution: Because this material involves an explosive, technical experts that have experience handling explosives, such as a bomb squad, should be notified. The PETN is shipped wetted to reduce its sensitivity. As the spilled slurry of PETN dries out, it may become extremely dangerous. Other hazardous materials are shipped with desensitizers that degrade or become less effective over time. It is very important that qualified technical experts are consulted when dealing with hazardous materials.

fission ■ The process during which an atomic nucleus is split into two smaller nuclei with the emission of neutrons and an extremely large amount of energy.

Nuclear incidents, in contrast, would cause extensive damage from the energy released and subsequently cause significant injuries from the intense ionizing radiation and radionuclides that are created during the blast. A nuclear incident, or thermonuclear event, is an extremely powerful explosion caused by a **fission** or fusion reaction that releases an immense amount of energy almost instantaneously. Examples of nuclear detonations are the nuclear weapons that were dropped on Hiroshima and Nagasaki at the end of World War II. Although it is very unlikely that terrorists or rogue nations will be able to acquire and use nuclear weapons, it is possible, especially after the downfall of the Soviet Union has made weapons experts and radioactive material available on the black market, and that North Korea, which proliferates weapons, has acquired nuclear weapons. It is also rumored that the Soviet Union lost track of several so-called "suitcase nukes." These relatively small and portable nuclear devices are reportedly almost as powerful as the Hiroshima and Nagasaki devices and could level a medium-sized city, causing a large number of fatalities as well as significant health and economic impact throughout the region for decades to come.

A nuclear incident may also occur if a nuclear power plant has an accident or is attacked by terrorists. An example of an accident involving a nuclear power plant is the core meltdown at Chernobyl in the Soviet Union in the 1980s caused by operator error. This incident caused extensive damage to the reactor itself, made a wide area around the nuclear power plant uninhabitable, and literally spread radioactive materials around the globe.

In March of 2011 a nuclear power plant accident triggered by an earthquake and subsequent tsunami disabled the cooling system at the Fukushima Daiichi nuclear power plant in Japan. This incident caused extensive damage to multiple reactors, spread contamination throughout the area, caused highly radioactive runoff to flow into the Pacific Ocean, and spread radiation around the globe. At the time of publication this incident was still not completely under control and the ultimate outcome remains up in the air.

Summary

The primary role of awareness level personnel at an emergency is to protect themselves and others in the immediate area without coming into contact with the released hazardous material or WMD agent. Awareness level personnel should always be able to recognize a release, avoid that release, isolate the immediate area, and notify the appropriate response personnel. Notification usually involves calling 911. During the response time of the trained first responders, awareness level personnel should gather as much pertinent information as possible without putting themselves in danger.

Review Questions

1. What does the acronym RAIN stand for?
2. What are five clues that would indicate the presence of hazardous materials?
3. What are the four routes of entry by which hazardous materials may enter the body? Which of these routes of entry is your greatest concern, and why?
4. What type of information can you find in an MSDS sheet?
5. What type of information can you find in the orange section of the ERG?

Problem-Solving Activities

1. List five locations in your community that you would expect to have hazardous materials on site. Rank these five locations based upon your perception of the hazards and risks.
2. List the SARA Title III sites in your jurisdiction and list the chemicals they have on site.
3. List the major transportation routes in your jurisdiction that carry hazardous materials.
4. Determine whether you have any pipelines in your jurisdiction. If so, list the products they may carry.
5. Locate the MSDS sheets at your place of employment.
6. Out of the products listed in these MSDS sheets, which one concerns you most? Why?

References and Further Reading

Klem, Thomas J. (1984, September 17). *Fire Investigations: Cold Storage Warehouse, Shreveport, Louisiana*. Quincy, MA: National Fire Protection Association.

National Fire Protection Association. (2008). NFPA 472, *Standard for Competence of Responders to Hazardous Materials/Weapons of Mass Destruction Incidents*. Quincy, MA: Author.

Occupational Safety and Health Administration. (1990). 29 CFR 1910.120, *Hazardous Waste Operations and Emergency Response (HAZWOPER)*. Washington, DC: U.S. Department of Labor.

U.S. Department of Transportation. (2008). *2008 Emergency Response Guidebook*. Washington, DC: Pipeline & Hazardous Materials Safety Administration.

Weber, Chris. (2007). *Pocket Reference for Hazardous Materials Response*. Upper Saddle River, NJ: Pearson/Brady.

3

Operations Level Responders Core Competencies: Recognizing Hazardous Materials

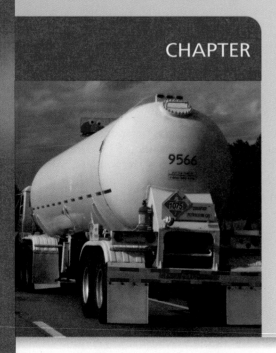

OBJECTIVES

- Read and interpret a pesticide label.
- Recognize the different types of containers used to store hazardous materials.
- Recognize the different types of intermodal containers from a distance.
- Recognize the different types of highway cargo tankers from a distance.
- Recognize the different types of railcars from a distance.
- Read and interpret shipping papers.
- List the five different types of radioactive packaging and their features.
- List the ways containers may fail and the dispersion patterns that could arise.
- List the five different categories of weapons of mass destruction, and name at least three agents in each class.

You Are on Duty! Crashed Highway Cargo Tanker (Part 1)

Paw Paw, Michigan

On the evening of January 28, 2004, a 911 call is received stating that a tanker truck and a pickup truck just collided on I-94 between Paw Paw and Mattawan in Van Buren County, Michigan. The weather conditions are a temperature of 16°F (−9°C), 10 to 15 mph winds out of the west, and light snow with 11 inches (27.5 cm) already on the ground. The 911 caller reports that the pickup truck driver appears to be severely injured and the tanker is displaying a red placard with the four-digit number 1170 in the middle. It also appears that the tanker is leaking a small amount of clear liquid. You are dispatched to respond to this incident.

- How would you approach this incident?
- What are important factors to consider?
- What other information would you like to know?

Let's try to answer some of these questions in this chapter.

Operations level personnel respond to hazardous materials and weapons of mass destruction incidents to protect the public, to stabilize the incident, and to protect property and the environment. The first step is to recognize the presence of hazardous materials and to assess the risk they pose (Figure 3-1).

A useful mnemonic for hazardous materials response at the hazardous materials operations level is AFIRMED, as mentioned in Chapter 1. This stands for **A**ssess the situation, **F**ormulate an action plan, **I**nitiate actions, **R**eassess the situation, **M**odify the plan as needed, continue with **E**xtended operations, and **D**emobilize. The next three chapters will cover the core competencies as specified in NFPA 472 (2008), which allow responders to safely operate in a defensive manner at hazardous materials incidents and weapons of mass destruction incidents. This chapter focuses on the assessment process.

Operations level responders will typically be called to the scene of a hazardous materials release by a local dispatch center. Responder safety should always come first! Carefully assess the dispatch information and cautiously approach from upwind, uphill, and upstream. It is important to approach the situation from upwind to avoid accidentally entering a vapor cloud. It is vital to approach it from uphill to avoid having liquid releases flow toward you, especially during a catastrophic container rupture. It helps to approach from upstream, when applicable, to avoid bringing hazardous materials along with the current of the falling water or the contour of the water channel. Besides that, vapor clouds have a tendency to travel along bodies of water.

Once responders have ensured their safety, they need to ensure the safety of the public. This can be accomplished by isolating the area and setting up three **control zones**: the hot zone, also known as the exclusion zone; the cold zone, also known as the support zone; and the warm zone, also known as the contamination reduction zone. Upon initial arrival at a hazardous materials incident, responders divide it into a hot zone and a cold zone. The hot zone is any area that is already contaminated or can be reasonably expected to become

control zone ■ Either the cold zone, warm zone, or hot zone. A division of the incident into uncontaminated and contaminated areas based upon response considerations.

FIGURE 3-1 Hazardous materials incidents come in a wide variety of forms, but they all are an unexpected release of a substance from its container. This hazmat incident was caused by inadequate bracing and blocking of cargo in a trailer.

contaminated. When setting up the hot zone, always remember that gases and vapors travel. Gas and vapor clouds are obviously considered contamination. The *Emergency Response Guide* (ERG) is an excellent initial source for determining the size of the hot zone. In either its orange section or its green section, it will indicate how far to isolate in all directions, depending on whether the product is highlighted (see Chapter 2 for a detailed explanation of ERG use). The cold zone is the uncontaminated area. If there are contaminated victims, or more highly trained entry teams start entering the hot zone, a warm zone or decontamination area is carved out of the cold zone on the upwind and/or uphill side. The warm zone is the location of most decontamination activities.

One of the most important functions the operations level responder will perform is to assess the severity of the incident. This means determining whether anyone has been injured or killed, the number of victims, the type and quantity of released material, and the properties of the released material. In this chapter and the next, how to gather the requisite information and determine the hazard the release poses is explored. In this chapter the focus is on recognition clues and how hazardous materials are stored and transported.

Pesticide Labels

Pesticides are chemicals used to control or destroy organisms and animals that are considered to be pests, such as some insects, rodents, and plant pathogens. Pesticides are found all around us. They are used in the home and in the workplace, and quite powerful and strictly regulated pesticides are used by professional exterminators and for many farming and industrial applications. The pesticide industry in the United States is regulated by the Environmental Protection Agency. Part of this regulatory practice is to require a standardized label on pesticide packaging (Figure 3-2). The label must include:

- Name of the pesticide
- Signal word
- EPA registration number
- Precautionary statement
- Hazard statement
- Active ingredients
- Directions for use

The name of the pesticide is the manufacturer's trade name, not the actual chemical name of the active ingredient. Many different trade names refer to the same pesticide, so do not forget to include the manufacturer's name along with the trade name when referring to the product. The signal word gives an idea of how dangerous the pesticide is. Three signal words are used:

DANGER is used for the most highly toxic pesticides.
WARNING is used for moderately toxic pesticides.
CAUTION is used for less toxic pesticides.

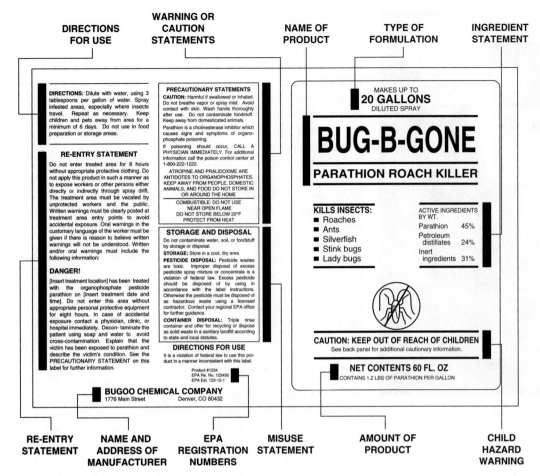

DIRECTIONS FOR USE

WARNING OR CAUTION STATEMENTS

NAME OF PRODUCT

TYPE OF FORMULATION

INGREDIENT STATEMENT

DIRECTIONS: Dilute with water, using 3 tablespoons per gallon of water. Spray infested areas, especially where insects travel. Repeat as necessary. Keep children and pets away from area for a minimum of 6 days. Do not use in food preparation or storage areas.

RE-ENTRY STATEMENT

Do not enter treated area for 8 hours without appropriate protective clothing. Do not apply this product in such a manner as to expose workers or other persons either directly or indirectly through spray drift. The treatment area must be vacated by unprotected workers and the public. Written warnings must be clearly posted at treatment area entry points to avoid accidental exposure. Oral warnings in the customary language of the worker must be given if there is reason to believe written warnings will not be understood. Written and/or oral warnings must include the following information:

DANGER!

[Insert treatment location] has been treated with the organophosphate pesticide parathion on [insert treatment date and time]. Do not enter this area without appropriate personal protective equipment for eight hours. In case of accidental exposure contact a physician, clinic, or hospital immediately. Decon- taminate the patient using soap and water to avoid cross-contamination. Explain that the victim has been exposed to parathion and describe the victim's condition. See the PRECAUTIONARY STATEMENT on this label for further information.

PRECAUTIONARY STATEMENTS
CAUTION: Harmful if swallowed or inhaled. Do not breathe vapor or spray mist. Avoid contact with skin. Wash hands thoroughly after use. Do not contaminate foodstuff. Keep away from domesticated animals.
Parathion is a cholinesterase inhibitor which causes signs and symptoms of organophosphate poisoning.
If poisoning should occur, CALL A PHYSICIAN IMMEDIATELY. For additional information call the poison control center at 1-800-222-1222.
ATROPINE AND PRALIDOXIME ARE ANTIDOTES TO ORGANOPHOSPHATES. KEEP AWAY FROM PEOPLE, DOMESTIC ANIMALS, AND FOOD DO NOT STORE IN OR AROUND THE HOME
COMBUSTIBLE: DO NOT USE NEAR OPEN FLAME
DO NOT STORE BELOW 20°F
PROTECT FROM HEAT

STORAGE AND DISPOSAL
Do not contaminate water, soil, or foodstuff by storage or disposal.
STORAGE: Store in a cool, dry area.
PESTICIDE DISPOSAL: Pesticide wastes are toxic. Improper disposal of excess pesticide spray mixture or concentrate is a violation of federal law. Excess pesticide should be disposed of by using in accordance with the label instructions. Otherwise the pesticide must be disposed of as hazardous waste using a licensed contractor. Contact your regional EPA office for further guidance.
CONTAINER DISPOSAL: Triple rinse container and offer for recycling or dispose as solid waste in a sanitary landfill according to state and local statutes.

DIRECTIONS FOR USE
It is a violation of federal law to use this product in a manner inconsistent with this label.

Product #1234
EPA Re. No. 123456
EPA Est. 123-12-1

BUGOO CHEMICAL COMPANY
1776 Main Street   Denver, CO 80432

MAKES UP TO
20 GALLONS
DILUTED SPRAY

BUG-B-GONE
PARATHION ROACH KILLER

KILLS INSECTS:
■ Roaches
■ Ants
■ Silverfish
■ Stink bugs
■ Lady bugs

ACTIVE INGREDIENTS BY WT.
Parathion   45%
Petroleum distillates   24%
Inert ingredients   31%

CAUTION: KEEP OUT OF REACH OF CHILDREN
See back panel for additional cautionary information.

NET CONTENTS 60 FL. OZ
CONTAINS 1.2 LBS OF PARATHION PER GALLON

RE-ENTRY STATEMENT

NAME AND ADDRESS OF MANUFACTURER

EPA REGISTRATION NUMBERS

MISUSE STATEMENT

AMOUNT OF PRODUCT

CHILD HAZARD WARNING

FIGURE 3-2 An example of the key features of pesticide labels. The Environmental Protection Agency requires the inclusion of specific information that can be very helpful to hazmat responders.

The signal word must be highlighted in a different color, be in bold print, be outlined, or draw attention to itself in some other way on the label.

Every legitimate pesticide has an EPA registration number that is unique to that product or trade name and manufacturer. The precautionary statement lets the end user know how to safely handle the pesticide. Examples of this statement might be "Keep out of reach of children" and "Hazardous to humans and domestic animals." The hazard statement often indicates the environmental hazards that the pesticide may pose. Some examples of environmental hazard statements are "Toxic to fish" and "Do not apply directly to water." The active ingredients are listed by their chemical name and concentration. If an exposure occurs, it is important to consult the list of active ingredients and report these to poison control and/or the hospital treating the patient. The directions for use indicate the manufacturer's recommendations for proper use of the pesticide. Improper use may endanger the user, the public, or occupants of the building days or weeks after pesticide application. An example of this is the application of outdoor pesticides indoors where residents are sickened or even killed. When responding to a building with several sick occupants with no obvious explanation, it would be a good idea to ask if pesticides have been applied recently or an exterminator has recently visited the premises.

Containers

Containers are used to store, transport, and use a wide variety of harmless and potentially harmful materials in a wide variety of quantities (Figure 3-3). The quantity of the material, the state of matter the material is in, and the type of container it is in can

FIGURE 3-3 A variety of containers in which chemicals are transported and stored.

have a profound impact on the outcome of a hazardous materials incident. Generally, the larger the quantity of the hazardous material, the bigger the problem. In other words, larger containers tend to lead to larger problems. Pressurized containers, such as those storing compressed gases or compressed liquefied gases, pose a much greater hazard than unpressurized containers. The type of container the hazardous material is in offers a lot of information as far as the dangers faced at a hazmat incident.

Containers that carry hazardous materials are usually required to be constructed to specific specifications (such as by the U.S. Department of Transportation [DOT] or the Association of American Railroads [AAR]), especially when they contain large quantities of dangerous goods or extremely hazardous materials, such as poison inhalation hazards or high specific activity radioactive materials. It is important to understand how these containers are constructed and what safety features they are equipped with in order to more effectively respond to hazardous materials releases from them. Let's explore some of these containers and how they can affect emergency operations when they fail.

NON-BULK CONTAINERS

Non-bulk containers are defined by a maximum capacity of 119 gallons (450 L) for liquids, maximum net mass of 882 pounds (400 kg) for solids, and maximum water capacity of 1000 pounds (454 kg) for gas **cylinders**. Non-bulk containers are ubiquitous in our lives. These types of containers are in the home in the form of bottles, bags, aerosol cans, and many other small containers including a 5-gallon can of gasoline. Of course, these containers can be purchased from a variety of stores, including grocery stores, home improvement stores, auto parts stores, pool supply companies, and many others. Larger sized non-bulk containers are common in industry in the form of 55-gallon drums, bags, and reagent bottles. These types of containers may be found in almost any mode of transport and in almost any type of vehicle, placarded or not placarded. Table 3-1 and Figure 3-4 list and show some of the properties of non-bulk containers.

The container's shape and size often indicate the state of matter of the material it contains. For example, fiberboard drums typically carry solids, plastic 55-gallon drums often carry corrosive liquids, and steel 55-gallon drums typically carry flammable liquids. Cylinders usually carry compressed gases, and dewars generally carry cryogenic liquids. However, there is no hard and fast rule; there are many exceptions! For example, a cylinder carrying acetylene gas actually contains a solid, porous material (such as diatomaceous earth) and acetone in which the acetylene gas is dissolved. The most reliable way to identify the contents of non-bulk containers is to look at the label and/or the shipping papers if the container is involved in transport.

Compressed gas cylinders will have the following information permanently stamped into the cylinder, typically at the neck or shoulder:

- DOT specification
- Service pressure
- Manufacturer's serial number
- Tare weight
- Test date
- Inspector's mark

non-bulk container ■ Per DOT, packaging that has (a) a maximum capacity of 119 gal (450 L) for a liquid; (b) a maximum net mass of 882 lb (400 kg) and a maximum capacity of 119 gal (450 L) for a solid; or (c) a maximum water capacity of 1000 lb (454 kg) for a gas. Also known as non-bulk packaging.

cylinder ■ A cylindrical pressure vessel designed to store and transport nonliquefied and liquefied compressed gases at pressures greater than 40 psi (2.75 bar).

TABLE 3-1	Non-bulk Packaging	
CONTAINER	**CHARACTERISTICS**	**TYPICAL CONTENTS**
Carboy	Glass or plastic 1- to 10-gallon capacity	Liquids of all hazard classes
Bottle	Glass, plastic, or metal May be coated on the interior and/or exterior	Liquids or solids of all hazard classes
Bag	Paper, fiber, or plastic Possibly lined Supersacks or tote sized available	Solids of all hazard classes
Cardboard box	Should be DOT specification for most hazardous materials Often contains the primary hazmat container (such as glass, plastic, or metal bottles)	Solids, liquids, and gases of all hazard classes
Drum	Steel, plastic, fiberboard, stainless steel, or aluminum Up to 95 gallon capacity	Liquids and solids of all hazard classes
Cylinder	200–6000 psi (13.8-415 bar) Have pressure-relief devices such as fusible links or frangible disks Steel or aluminum	Compressed gases, or liquefied compressed gases of any hazard type
Dewar	Stainless steel Low pressure Insulated	Cryogenic gases such as liquid nitrogen

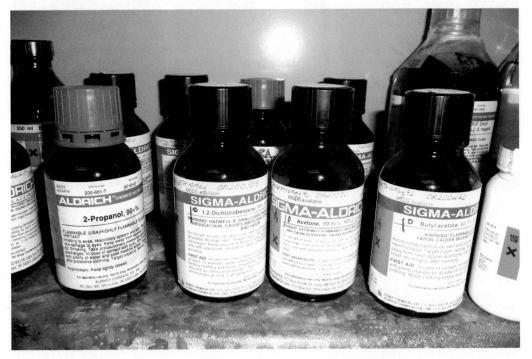

FIGURE 3-4 Examples of non-bulk chemical containers often found in research and industrial laboratories. *Photo by author. Used by permission from Sigma-Aldrich*

INTERMEDIATE BULK CONTAINERS

Intermediate bulk containers contain larger quantities than does non-bulk packaging; however, they are still mobile and used in transport (Figure 3-5). Some of the most common **intermediate bulk containers**, or **IBCs** as they are known, are 1-ton containers carrying materials such as chlorine or sulfur dioxide and totes that typically carry liquids. This type of packaging is very popular in industries using intermediate amounts of materials. IBCs are cheaper to acquire, transport, and store than smaller 55-gallon drums, for example; and the facilities do not need to install expensive fixed storage tanks. The product can generally be used directly from the intermediate bulk container. Once again, there are clues about the contents from the container shape and size, but the label or shipping papers are more reliable. Table 3-2 shows examples of intermediate bulk containers.

intermediate bulk container (IBC) ■ Packaging that typically weighs between 90 and 1200 pounds (40 and 550 kg) and can be moved using a standard forklift; also known as tote.

FIXED SITE STORAGE TANKS

Fixed site storage tanks come in many sizes and shapes, and they are not generally regulated by the U.S. DOT because they are not used in transportation (Figure 3-6). The storage tanks may be aboveground or belowground, as are found commonly at gasoline stations. At industrial facilities, fixed site storage tanks are usually found aboveground with secondary containment walls around them. The **secondary containment system** should be able to hold the larger of 100% of the largest container or at least 10% of the aggregate container volume when there are multiple storage tanks in the area. Aboveground storage tanks may be vertical or horizontal, depending on the facility. Storage tanks used for chemical processes may be funnel shaped in order to facilitate product movement. Spherical storage tanks are used for high-pressure gas or liquefied gas storage (such as methane).

In the petroleum industry, extremely large storage tanks able to hold tens of millions of gallons may be used for crude oil, gasoline, diesel fuel, and other petroleum products. As you can imagine, the catastrophic failure of one of these tanks could spell disaster. It

secondary containment system ■ A capturing device surrounding a container that is used to prevent the release of a hazardous material into the environment.

FIGURE 3-5 Examples of intermediate bulk containers (IBC). On the left is a one-ton container filled with liquid chlorine. In the middle is a bag designed to hold solids. On the right is a plastic tote designed to hold liquids. Metal totes often contain flammable solvents whereas plastic totes often contain corrosive materials.

TABLE 3-2	Intermediate Bulk Containers	
CONTAINER	**CHARACTERISTICS**	**TYPICAL CONTENTS**
Totes	Steel, plastic, stainless steel, or aluminum May be lined May be collapsible Up to 500 gallon (1900 L) capacity	Liquids of all hazard classes
One-ton containers	Steel Pressurized	Liquefied compressed gases such as chlorine and sulfur dioxide

FIGURE 3-6 Examples of fixed site storage tanks. Pictured at the top left is a spherical tank designed to hold high vapor pressure liquids and liquefied compressed gases. This particular tank holds liquefied propane and has been marked with a placard normally used during transportation even though it is a fixed site storage tank. The top right pictures an insulated tank designed to hold cryogenic liquids. The product name and the NFPA 704 marking are visible on the tank. The bottom shows several different types of large petroleum storage tanks.

is extremely important to preplan and routinely visit industrial facilities in your response area in order to be adequately prepared in the event of an emergency. Fixed site storage tank properties are summarized in Table 3-3.

INTERMODAL TANKS

As their name implies, **intermodal containers** can be used in many modes of transportation. Let's follow a shipment of one of the most heavily produced and used hazardous materials in the world: sulfuric acid. The chemical plant that produces the sulfuric acid places it directly in an intermodal tank. This intermodal tank is loaded on a semitruck, which transports it to a railroad terminal; and without unloading the acid, the intermodal tank is moved onto a railcar. The railcar then transports the intermodal tank to a ship-yard where, again, without unloading the product, the intermodal tank is loaded onto a container ship. The process continues through various modes of transportation until the tank gets to its destination, at which point the product is used directly from the intermodal tank by the end user (Figure 3-7).

Intermodal containers are gaining wide popularity in industry due to their safety and low transportation costs. Often accidents occur when products are transferred from one container to another or incompatible materials are inadvertently mixed. For example, in 1999 in Whitehall, Michigan, a truck driver was killed by the hydrogen sulfide gas

intermodal container ■ Bulk packaging designed to be used in multiple modes of transportation.

TABLE 3-3	Fixed Site Storage Tanks	

TANK TYPE	TYPICAL HAZARDOUS SPECIFICATIONS	CONTENTS
Cone roof tank	Atmospheric pressure Pressure-vacuum valve for breathing Welded steel tank (*Caution*: older tanks may still be riveted and present a greater risk of failing under fire.) Diameter: 20–300 feet Capacity: 40,000–4,000,000 gallons Vapor space between liquid level and roof Weak roof-to-shell seam for emergency venting Possibly lined	Flammable liquids Combustible liquids Solvents Oxidizers Corrosives
Open floating roof tank Note the wind girder around the top.	Atmospheric pressure Steel tank Roof floats on the surface of product using pontoon or honeycomb system Seal area between tank shell and roof No vapor space (unless nearly empty) Limited drainage system to carry water off roof Diameter: 40–400 feet Capacity: 40,000–20,000,000 gallons Issues: seal fires from lightning strikes, roof sinking from incorrect water application or excessive precipitation	Flammable liquids
Covered floating-roof tank Note the vent system at the top of the tank.	Atmospheric pressure Welded steel tank Floating roof on liquid Fixed flat, geodesic, or conical roof supported by cylindrical tank walls No vapor space (unless nearly empty) Diameter: 40–400 feet Capacity: 40,000–20,000,000 gallons Issues: Internal tank fires can be difficult to extinguish owing to inaccessible void spaces.	Flammable liquids Combustible liquids
Vertical Storage tank	Low pressure (2.5–15 psi) Pressure-vacuum valve Welded steel tank (*Caution:* Older tanks may still be riveted and present a greater risk of failing under fire.) Some tanks may be fiberglass Weak roof-to-shell seam for emergency venting Diameter: up to 25 feet Capacity: 4000–400,000 gallons Possibly lined	Flammable liquids Combustible liquids Solvents Oxidizers Poisonous liquids Corrosives

TABLE 3-3	Fixed Site Storage Tanks (*Continued*)	
TANK TYPE	**TYPICAL HAZARDOUS SPECIFICATIONS**	**CONTENTS**
Horizontal storage tank	Low pressure (up to 0.5 psi)	Flammable liquids
	Pressure-vacuum valve	Combustible liquids
	Wide range of construction: welded/riveted steel; often, tank within-a-tank design (double walled)	Solvents
		Oxidizers
	Diameter: up to 20 feet	Poisonous liquids
	Capacity: 300–100,000 gallons	Corrosives
	Issues: Tank supports must be protected during fires.	
Pressure vessels	High pressure	Flammable gases
Pressure relief valve		

generated when he accidentally put sodium hydrosulfide solution into the pickle acid (ferrous sulfate) tank. One way this accident could have been prevented is if the sodium hydrosulfide had been delivered and used in an intermodal tank.

The width, height, and length of intermodal containers are standardized so they can be easily moved and stacked during transportation and storage. They are usually 8 feet wide and 8 to 9½ feet tall. Domestic intermodal containers, which are used only in rail and road transportation, are often 8½ feet wide. The 8½-foot wide intermodal containers do not conform to international standards; their lengths range from 20 feet to 53 feet. In addition, the **corner castings**, by which the intermodal containers are lifted and moved, are also standardized internationally (Figure 3-8). Intermodal containers must always be lifted and moved by their corner castings using the appropriate heavy machinery. If loaded intermodal containers are lifted by a conventional industrial forklift along their midsection, they will almost certainly fail catastrophically.

corner casting ■ The lifting point located on all eight corners of intermodal containers. Intermodal containers should only be lifted using the corner castings with the appropriate heavy equipment to prevent severe damage.

FIGURE 3-8 Corner casting of an intermodal container. Intermodal containers should only be lifted and moved using the corner castings; otherwise, catastrophic container failure may occur.

Intermodal tanks are a tank mounted inside a metal frame of standardized dimensions. These tanks were developed to facilitate transport of goods, including hazardous materials. The maximum capacity of intermodal tanks is typically less than 24,000 L (6340 gal). Four types of intermodal tanks are commonly encountered in the United States: the IMO type 1 (also known as IM-101), the IMO type 2 (also known as IM-102), the IMO type 7, and the IMO type 5 (also known as DOT Spec 51). Table 3-4 lists the characteristics of these four types of intermodal containers. The main difference between these four types of intermodal containers is the **maximum allowable working pressure (MAWP)** of each, and consequently the products that each is capable of carrying. The tank type and capacity, as well as other construction features of the tank, may be found on the **specification plate** on the front of intermodal tanks near the valves (Figure 3-9).

In addition to the specification plate, the tank will also have container markings that typically include the following information:

- **Reporting marks**
- Country code, tank size, type code
- Approvals, compliance, and conformity information
- DOT or AAR specification markings
- DOT exemption marking (if applicable)
- Hazardous materials markings and placards

maximum allowable working pressure (MAWP) ▪ The maximum amount of pressure a container is able to withstand based upon the weakest component.

specification plate ▪ A marking located on tanks that shows the construction features, capacity, and number of compartments of intermodal tanks and highway cargo tanks.

reporting mark ▪ A sequence of two to six identification letters assigned by the Association of American Railroads (AAR) to rail carriers that operate in North America.

TABLE 3-4	Intermodal Containers
INTERMODAL CONTAINER TYPE	**SPECIFICATIONS**
Nonpressure or pressure tank Corner casting Beam type Box type	Non-pressure: 90% of tank containers in use
	Capacity: 5000–6300 gallons IM 101 or IMO Type 1 (25.4–100 psi)
	Pressure-relief valves
	Vacuum-relief valves
	Flammables with flashpoint less than 32°F
	IM 102 or IMO Type 2 (14.5–25.4 psi)
	Flammables with flashpoint between 32 and 140°F.
	Pressure: (LPG, anhydrous ammonia)
	Capacity: 4500–5500 gallons
	DOT Specification 51 or IMO Type 5 (100–600 psi)
	Enclosed fittings
Cryogenic tank	IMO Type 7
	Liquefied compressed gases (cryogenic liquids) such as liquid nitrogen, liquid oxygen, liquid helium, liquid hydrogen, liquid carbon dioxide, and liquid argon
	These containers will vent periodically under normal operating conditions, especially warm weather
Tube module	3000–5000 psi
	T3 cylinders permanently mounted in an IMO frame
	Compressed gases such as nitrogen, oxygen, helium, and hydrogen

FIGURE 3-9 The information contained on the specification plate of an intermodal tank (DOT 51) can be very useful.

The meaning of the codes for containers manufactured after 1996 can be found in Table 3-5A, and the meaning of the codes for containers manufactured before 1996 can be found in Table 3-5B. Intermodal containers have become more prevalent in North America as international commerce increases in the global economy.

IM-101

IM-101 intermodal tanks, or IMO type 1 tanks when used internationally, are designed to accommodate pressures between 25.4 psig (1.75 bar) and 100 psig (6.9 bar) (Figure 3-10). Their maximum capacity is typically 6300 gallons (24,000 L). These tanks have a housing

FIGURE 3-10 An example of an IM-101 intermodal tank. The GHS intermodal container code on the container is 22T6 which indicates the container is 20 feet long, 8 feet wide, 8.5 feet tall, and contains a dangerous liquid with a minimum pressure of 6 bar.

TABLE 3-5A — New Intermodal Codes (Constructed After 1996)

1st Char	1	2	3	4	A	B	C	D	E
LENGTH	2.991 m (10 feet)	6.068 m (20 feet)	9.125 m (30 feet)	12.192 m (40 feet)	7.150 m	7.315 m (24 feet)	7.430 m (24.5 feet)	7.450 m	7.820 m

2nd Character	0	2	4	5	6	8	9
WIDTH	2.438 m (8 feet)	2.438 m (8 feet)	2.438 m (8 feet)	2.438 m (8 feet)	2.438 m (8 feet)	2.438 m (8 feet)	2.438 m (8 feet)
HEIGHT	2.438 m (8 feet)	2.591 m (8.5 feet)	2.743 m (9 feet)	2.895 m (9.5 ft)	>2.895 m (>9.5 ft)	1.295 m (4.25 ft)	≤1.219 m (≤ 4 ft)

3rd Character	A (AS-Air/surface)	B (BU/BK-Dry bulk)	G (GP-General purpose w/o ventilation)	H (HI/HR-Thermal)	P (PC/PF/PL/PS-Platform)
4th Char (Digit) 0		Closed, non-pressure	Openings at one or both ends	Refrigeration and/or heating with removable, external equip. (k=0.4 W/m²K)	Plain platform
1		Airtight, non-pressure	Passive vents near top	Refrigeration and/or heating with removable, internal equip.	Two complete and fixed ends
2		-	Openings at one or both ends and full openings one or both sides	Refrigeration and/or heating with removable, external equip (k=0.7 W/m²K)	Fixed posts, either freestanding or with removable top member
3		Horizontal discharge (1.5 bar test pressure)	Openings at one or both ends and partial openings one or both sides	-	Folding complete structure
4		Horizontal discharge (2.65 bar test pressure)		-	Folding posts, either freestanding or with removable top member
5		Tipping discharge (1.5 bar test press)		Insulated (k=0.4 W/m²K)	Open top, open ends
6		Tipping discharge (2.65 bar test press)		Insulated (k=0.7 W/m²K)	
7		-			
8		-			
9		-			

(m = meters; ft = feet; k = heat transfer coefficient)

F	G	H	K	L	M	N	P
8.100 m	12.500 m (41 feet)	13.106 m (43 feet)	13.600 m	13.716 m (45 feet)	14.630 m (48 feet)	14.935 m (49 feet)	16.154 m

C	D	E	F	L	M	N	P
>2.438 m & <2.5 m	>2.438 m & <2.5 m	>2.438 m & <2.5 m	>2.438 m & <2.5 m	>2.5 m	>2.5 m	>2.5 m	>2.5 m
2.591 m (8.5 feet)	2.743 m (9 feet)	2.895 m (9.5 ft)	>2.895 m (>9.5 ft)	2.591 m (8.5 feet)	2.743 m (9 feet)	2.895 (9.5 ft)	>2.895 m (>9.5 ft)

R (RE/RS/RT-Thermal)	S (SN-Named)	T (TD/TG/TN-Tanks)	U (UT-Open top)	V (VH-General purpose with ventilation)
Mechanical refrigeration	Livestock carrier	Non-dangerous liquids (minimum pressure 0.45 bar)	Openings at one or both ends	Non-mechanical vents in lower and upper space
Mechanical refrigeration and heating	Automobile carrier	Non-dangerous liquids (minimum pressure 1.50 bar)	Openings at one or both ends with removable top members in end frames	-
Self-powered mechanical refrigeration	Live fish carrier	Non-dangerous liquids (minimum pressure 2.65 bar)	Openings at one or both ends and openings one or both sides	Internal mechanical ventilation system
Self-powered mechanical refrigeration and heating		Dangerous liquids (minimum pressure 1.50 bar)	Openings at one or both ends, openings one or both sides, and removable top members in end frames	-
		Dangerous liquids (minimum pressure 2.65 bar)	Openings at one or both ends and partial openings one side and full opening other side	External mechanical ventilation system
		Dangerous liquids (min press 4.00 bar)	No doors	-
		Dangerous liquids (min press 6.00 bar)		-
		Gas (min press 9.10 bar)		-
		Gas (min press 22.0 bar)		-
		Gas (min press TBD)		-

TABLE 3-5B   Old Intermodal Codes (Constructed After 1996)

1st Char	1	2	3	4
LENGTH	2.991 m (10 feet)	6.068 m (20 feet)	9.125 m (30 feet)	12.192 m (40 feet)
2nd Character	1	2	3	4
HEIGHT	2.438 m (8 feet)	2.438 m (8 feet) TUNNEL	2.591 m (8.5 feet)	2.591 m (8.5 feet) TUNNEL

3rd Char	0 General purpose container	1 Ventilated closed container	2 Dry bulk (0–4) and named (5-9) containers	3 Thermal containers	4 Thermal containers
4th Char (Digit) 0	Openings at one or both ends	Passive vents near top ($<25 cm^2/m$ vent cross section)	Closed, non-pressure box type	Refrigeration, expendable refrigerant	Refrigeration and/or heating with removable, external equip.
1	Openings at one or both ends and full openings one or both sides	Passive vents near top ($\geqslant 25\ cm^2/m$ vent cross section)	Vented, non-pressure box type	Mechanical refrigeration	Refrigeration and/or heating with removable, internal equip.
2	Openings at one or both ends and partial openings one or both sides	-	Ventilated, non-pressure box type	Refrigeration and heating	Refrigeration and/or heating with removable, external equip.
3	Openings at one or both ends and opening roof	Non-mechanical vents in lower and upper space	Airtight, non-pressure box type	Heated	Refrigeration and/or heating with removable equip., spare
4	Openings at one or both ends, partial openings one or both sides and openings roof	-	-	Heated	Refrigeration and/or heating with removable equip., spare
5	-	Internal mechanical ventilation system	Livestock carrier	Heated	Insulated
6	-	-	Automobile carrier	Self-powered mechanical refrigeration	Insulated
7	-	External mechanical ventilation system	-	Self-powered refrigeration and heating	Insulated, spare
8	-	-	-	Self-powered heating	Insulated, spare
9	-	-	-	-	Insulated, spare

5	6	7	8	9
-	-	-	10.668 m (35 feet)	13.716 m (45 feet)
5	**6**	**7**	**8**	**9**
>2.591 m(>8.5 ft)	>2.591 m (>8.5 ft) TUNNEL	>4′ and <4.25′	>4′ and <4.25′ TUNNEL	>4.24′ and <8′
5 Open top containers	**6** Platform containers	**7** Tank containers	**8** Dry bulk containers	**9** Air/surface containers
Openings at one or both ends	Plain platform	Non-dangerous liquids (test pressure 0.45 bar)	Closed, non-pressure hopper type	
Openings at one or both ends and removable top members in end frames	Incomplete superstructure with complete and fixed ends	Non-dangerous liquids (test pressure 1.50 bar)	Vented, non-pressure hopper type	
Openings at one or both ends and openings one or both sides	Incomplete superstructure with fixed free-standing posts	Non-dangerous liquids (test pressure 2.65 bar)	Ventilated, non-pressure hopper type	
Openings at one or both ends, openings one or both sides, and removable top members in end frames	Incomplete superstructure with folding complete and structure	Dangerous liquids (test pressure 1.50 bar)	Airtight, non-pressure hopper type	
Openings at one or both ends and partial openings one side and full opening other side	Incomplete superstructure with folding free-standing posts	Dangerous liquids (test pressure 2.65 bar)	Non-pressure hopper type, spare	
-	Complete superstructure with roof	Dangerous liquids (test pressure 4.00 bar)	Horizontal discharge (1.5 bar test pressure)	
-	Complete superstructure with open top	Dangerous liquids (test pressure 6.00 bar)	Horizontal discharge (2.65 bar test pressure)	
-	Complete superstructure with open top and open ends	Dangerous gases (test pressure 10.5 bar)	Tipping discharge (1.5 bar test press)	
-	-	Dangerous gases (test pressure 22.0 bar)	Tipping discharge (2.65 bar test press)	
HT, spare	-	Dangerous gases (test pressure unassigned)	Pressurized, spare	CO, spare

FIGURE 3-11 A drawing of an IM-102 intermodal tank.

on top that contains the product piping. Typical contents of IM-101 tanks include food grade products, nonflammable liquids, and mild corrosives.

IM-102

IM-102 intermodal tanks, or IMO type 2 tanks when used internationally, are designed to accommodate lower pressures than the IM-101 tanks (Figure 3-11). Tank pressures will be between 14.5 psig (1 bar) and 25.4 psig (1.75 bar) with a maximum capacity of 6300 gallons (24,000 L). An indication that it is an IMO type 2 tank is that product piping will be visible on the outside of the tank. All valves must be contained within the structural framework of the frame. Typical contents of IM-102 tanks include food grade commodities, alcohols, pesticides and insecticides, industrial solvents and flammable liquids, and several corrosives.

IMO Type 7

IMO type 7 intermodal tanks carry cryogenic liquids and are constructed in a similar manner to the MC 338 highway cargo tankers (Figure 3-12). These tanks carry refrigerated liquid gases such as nitrogen, oxygen, helium, ethylene, and argon.

DOT Spec 51

DOT Spec 51 intermodal tanks, or IMO type 5 tanks when used internationally, are designed to accommodate high pressures from 100 psig (6.9 bar) to 500 psig (34.5 bar) (Figure 3-13). These tanks carry similar materials to the MC 331 highway cargo tanker, such as the liquefied compressed gases propane, ammonia, and chlorine. Their maximum capacity is typically 6300 gallons (24,000 L).

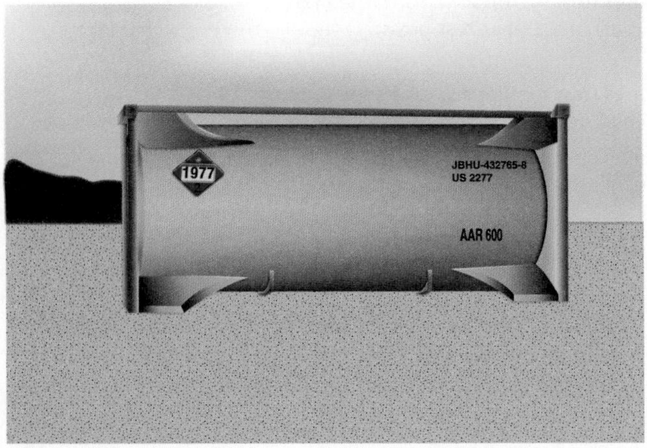

Intermodal Tube Containers

Intermodal tube containers consist of high-pressure cylinders mounted within an 8′ × 8′ ISO frame (Figure 3-14). The high-pressure cylinders are typically tested to 3000 to 5000 psi (207 to 345 bar) and are permanently mounted within the framework. The cylinders are typically 12 inches to 48 inches in diameter and constructed of steel. The cylinder valves are contained in a compartment at one end of the frame. Typical contents include liquefied compressed gases such as nitrogen, oxygen, helium, and argon.

FIGURE 3-12 A drawing of an IMO Type 7 intermodal tank.

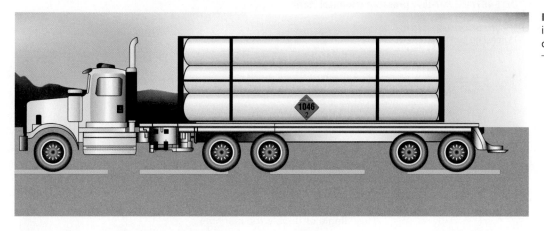

FIGURE 3-13 An example of a DOT specification 51 intermodal tank.

FIGURE 3-14 A drawing of an intermodal tube container.

Highway Cargo Tankers

Several different highway cargo tank configurations that U.S. DOT has approved for over-the-road transportation are currently in widespread use. The Motor Carrier (MC) 300 series—which includes the MC 306, MC 307, MC 312, MC 331, and MC 338 cargo tankers—was the standard highway cargo tank design for several decades. In August 1995 the highway cargo tanker specifications were modernized into the DOT 400 series, which created the DOT 406, DOT 407, and DOT 412 tankers. The primary differences between the 300 series and the 400 series are thicker tank shells, improved rollover protection, and improved manhole assemblies that can withstand higher static pressures. Although MC 306, MC 307, and MC 312 cargo tankers may no longer be manufactured, existing cargo tankers will continue to be used in transport on a regular basis for many years to come.

Highway cargo tanker properties are summarized in Table 3-6. Although similar to intermodal tanks, the main differences between these tanks are the material of construction

TABLE 3-6	Highway Cargo Tankers

DOT 406/MC 306-non-pressure liquid tank

Tank pressure:	Less than 5 psi
Maximum capacity:	9000 gallons
Tank construction:	Aluminum (new); steel (old)
	Multiple compartments
	Recessed manholes/rollover protection
	Bottom valves
	Vapor recovery system (newer)
Typical contents:	Flammable liquids
	Combustible liquids
	Gasoline
	Diesel fuel
	Fuel oils
	Alcohols
	Solvents

DOT 407/MC 307-low pressure chemical tank

Tank pressure:	25–40 psi
Maximum capacity:	6000 gallons
Tank construction:	Steel with stiffening rings (typically double shell)
	May be rubber lined
	Single or double top manhole
	Single outlet discharge for each compartment at bottom (midship or rear)
	Discharges are sometimes customized by owner
	Often pressure unloaded (pneumatic)
	Rollover protection
	May have multiple compartments
	Horseshoe or round profile as viewed from rear
Typical contents:	Flammable liquids
	Combustible liquids
	Corrosives
	Poisons
	Hazardous waste
	Food-grade materials/products

| TABLE 3-6 | Highway Cargo Tankers (*Continued*) |

DOT 412/MC 312-corrosive liquid tank

Round profile — Rear or midship top valving — Ribbed reinforcements

Tank pressure:	Less than 75 psi
Maximum capacity:	6000 gallons
Tank construction:	May be lined (butyl rubber or polyethylene)
	Recessed manhole
	Rollover protection around valving
	Corrosive-resistant paint band below valving
	Steel/stainless steel/aluminum with stiffening rings
	Top loading at center rear
	Usually single compartment
	Baffled compartment to slow product surges during transport
Typical contents:	Corrosive liquids
	Strong acids
	Strong bases

MC 331: high-pressure tanker

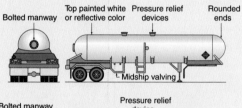

Bolted manway — Top painted white or reflective color — Pressure relief devices — Rounded ends — Midship valving

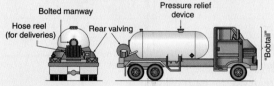

Hose reel (for deliveries) — Bolted manway — Rear valving — Pressure relief device — "Bobtail"

Tank pressure:	300 psi
Maximum capacity:	11,500 gallons
Tank construction:	Steel single compartment/noninsulated
	Bolted manhole at front or rear
	Internal and rear-outlet valves
	Excess flow shutoff valve
	Typically painted white or other reflective color
	May be marked "Flammable Gas" and "Compressed Gas"
	Round, dome-shaped ends
Typical contents:	Propane
	Anhydrous ammonia
	Butane
	Pressurized gases and liquids

(continued)

TABLE 3-6	Highway Cargo Tankers (*Continued*)

MC 338—cryogenic liquid tanker

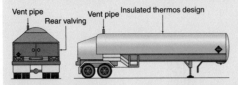

Tank pressure:	Less than 22 psi
Maximum capacity:	9000 gallons
Tank construction:	Well-insulated steel tank
	May have vapor discharging from vent pipe under normal operation
	Loading/unloading valves enclosed at rear
	Pressure relief valve
	May be marked "Refrigerated Liquid"
	Round tank with cabinet at rear
Typical contents:	Liquid nitrogen
	Liquid carbon dioxide
	Liquid hydrogen
	Liquid oxygen

Compressed gas/tube trailer

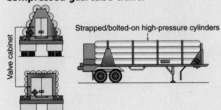

Tank pressure:	3000–6000 psi
Maximum capacity:	1500 cubic feet per cylinder
Tank construction:	Steel cylinders stacked and banded together
	Overpressure device for each cylinder
	Protected valving at rear
	Flat truck with multiple cylinders stacked in modular or nested shape
Typical contents:	Helium
	Hydrogen
	Methane
	Oxygen
	Argon
	Nitrogen

Dry bulk cargo tanker

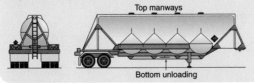

TABLE 3-6	Highway Cargo Tankers (*Continued*)
Tank pressure:	Not under pressure
Maximum capacity:	2500 cubic feet
Tank construction:	Top-side manholes
	Bottom valves
	Air-assisted loading/unloading
	Shapes vary but will have hoppers
	Aluminum shell
	Multiple compartments
Typical contents:	Calcium carbide
	Oxidizers
	Corrosive solids
	Cement
	Plastic pellets
	Fertilizers
	Cyanides

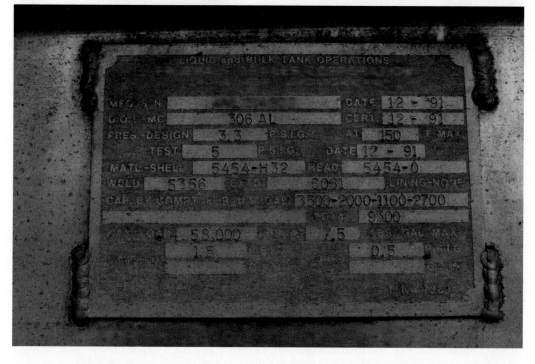

FIGURE 3-15 The specification plate from a highway cargo tanker. This "spec plate" from an MC 306 shows the tank construction, number of tank compartments, and their capacities, among other useful information. This cargo tanker has four compartments, with tank capacities of 3500 gal, 2000 gal, 1100 gal, and 2700 gal from front to rear.

and the maximum allowable working pressure (MAWP), which consequently determine the types of products each tank can carry. Once again, highway cargo tankers have a specification plate, typically located on the frame rail on either side of the vehicle (Figure 3-15). The specification plate lists the important features of the tank, including tank type, construction material, number of compartments, and tank compartment capacity.

DOT 406/MC 306 CARGO TANKERS

Some tankers that you may be very familiar with, at least from seeing them go down the road, are gasoline tankers (UN 1203), which are DOT 406 or MC 306 specification tanks (Figure 3-16). Visually, their most striking feature is an elliptical (oval) cross section when

FIGURE 3-16 DOT406 highway cargo tanker commonly used to transport petroleum products. The black dots on either side of the tank indicate the internal baffles have holes at the 9 o'clock and 3 o'clock positions (as well as the top and bottom) which facilitates product unloading when the tank is on its side.

BOX 3.1 GASOLINE TANKER FIRES

When a DOT 406 or an MC 306 tanker carrying hydrocarbons such as gasoline, gasohol, or E85 catches fire, it presents a special problem. Trying to extinguish these tankers using water will exacerbate the problem tremendously. Hydrocarbons such as gasoline are lighter than water. The water will sink to the bottom of the tanker and cause the burning gasoline to overflow. Now you have a river of burning gasoline seeking the lowest point in the immediate area. The lowest points are usually sanitary and storm sewers as well as basements. Now, instead of just having a gasoline tanker fire, you may very well have multiple structure fires and grass fires due to flowing fuel. Foam must be used to fight hydrocarbon fires, but is imperative that you have enough foam on hand to extinguish the fire and maintain a foam blanket to suppress vapors before you begin operations.

Mixtures of gasoline and ethanol, such as gasohol and E85, can pose problems as well. Gasohol usually contains 10% to 20% ethanol and behaves more like a hydrocarbon. E85 (which contains 85% ethanol, 15% gasoline) behaves more like an alcohol and dissolves completely in water. The higher the ethanol content, the more soluble the fuel will be in water. This requires alcohol-resistant foam to fight the fire (see Chapter 10, Operations Level Responders Mission-Specific Competencies: Product Control). As with hydrocarbon fires, it is imperative to have enough foam on hand to extinguish the fire and maintain a foam blanket to suppress vapors before beginning operations.

Sometimes a better way to solve the problem is to let the tanker burn. The risk of a BLEVE is minimal due to the construction safety features of a DOT 406 and an MC 306 tanker. The aluminum shell of the tanker will burn down as the gasoline is consumed. Environmental agencies usually prefer this option rather than risk having the gasoline overflow, enter the soil, and migrate into the groundwater table. As with all decisions at hazardous materials incidents, it is up to the incident commander to make an informed decision based upon his or her training and experience.

viewed from either end. Often these tankers will have multiple compartments. When they are involved in accidents, one or more of these compartments may be damaged or leaking. They are typically constructed of aluminum or carbon steel, and transport petroleum products such as gasoline, diesel fuel, fuel oil, motor oil, jet fuel (JP 8), kerosene,

E 85, and a variety of hydrocarbon-based solvents. Multiple compartments are separated by bulkheads, and baffles are located periodically inside the tank to increase structural stability. The baffles have four holes to allow movement of product within a given tank compartment. Tank capacities range from 1500 gallons to 13,000 gallons (5675 L to 50,000 L). The largest capacities are only allowed in certain states.

DOT 407/MC 307 CARGO TANKERS

Another common tank is the DOT 407 or MC 307 chemical tank, which is the workhorse of the chemical industry (Figure 3-17). These tankers are often insulated, which gives them a horseshoe-shaped appearance from behind. The tanks themselves are circular cylinders and have **stiffening rings**. Stiffening rings are included because these tankers often carry heavier products, such as corrosives, water-based solutions, and halogenated hydrocarbons. These tankers may transport just about any hazard class material, including flammable and combustible liquids, corrosives, poisons, and miscellaneous dangerous goods. The tanks are typically constructed of stainless steel to withstand corrosion and provide greater structural stability.

Both the 406/306 series and 407/307 series of tankers are considered low-pressure cargo tankers, although the DOT 406 tank can have up to 3 psi (0.2 bar) residual pressure and the DOT 407 tanker may have up to 40 psi (2.75 bar) residual pressure.

stiffening ring ■ A component of a cargo tank that adds structural strength to the tank. They can be observed welded onto the outside of MC 312 cargo tankers and uninsulated MC307 and DOT 407 cargo tankers.

DOT 412/MC 312

Corrosives are carried in specialized tankers due to their reactivity with most metals. These tanks have circular cross sections, a relatively small diameter, and many stiffening rings. Stiffening rings are necessary because most corrosives are quite heavy (Figure 3-18).

FIGURE 3-17 An MC307/DOT407 highway cargo tanker.

FIGURE 3-18 An MC312/DOT412 highway cargo tanker commonly used to transport corrosive materials.

Corrosives generally weigh significantly more than water. For example, water weighs 8.3 pounds per gallon (3.76 kg/gal) at room temperature, whereas pure sulfuric acid weighs 15.35 pounds per gallon (7 kg/gal). In comparison, gasoline weighs only 6.2 to 6.3 pounds per gallon (2.8 to 2.85 kg/gal). The 412/312 series tankers transport acids and bases such as sodium hydroxide, sulfuric acid, hydrochloric acid, phosphoric acid, and nitric acid. These tanks are typically constructed of mild steel and are usually lined with a polymer coating that resists the corrosive properties of the tank contents.

MC 331

Another cargo tank you may be familiar with is the high pressure MC 331 cargo tanker that often carries liquefied petroleum gas (LPG) and anhydrous ammonia (Figure 3-19). The MC 331 may have pressures of up to 300 psi (20.7 bar) and is made of hardened steel. Visually, these tanks have rounded ends that distribute the high internal pressure more evenly across the tank shell than the low-pressure vessels that have 90° bends in their tank construction. These tankers usually transport propane, liquefied petroleum gas (LPG), butane, anhydrous ammonia, and chlorine.

MC 338

Cryogenic liquid tankers, the MC 338 series, have circular cross sections with slightly rounded ends (Figure 3-20). These tankers have a thermos bottle design, meaning they contain a tank inside a tank. In addition, they are heavily insulated on the outside, which keeps the very cold liquefied compressed gases from evaporating. These tankers have pressures below 22 psi (1.5 bar), but may be seen releasing small amounts of vapor as they go down the road due to autorefrigeration. **Autorefrigeration** involves using evaporative cooling, by boiling off some of the contents, to keep the rest of the tank contents extremely cold. Sometimes the public assumes these tankers are leaking and calls 911. However, this is normal operation for the MC 338 tank, and there is usually no leak to be concerned about. This is why MC 338 tankers can carry only nontoxic gases such as oxygen, nitrogen, carbon dioxide, argon, helium, hydrogen, and the like. These tankers typically have an enclosed valve compartment at the rear.

There is also a class of non-specification cryogenic trailers that carries atmospheric gases such as nitrogen, oxygen, carbon dioxide, helium, and argon. Non-specification

autorefrigeration ■ The cooling of a liquid, usually cryogenic, through the evaporation of that liquid; also known as evaporative cooling. Containers that carry cryogenic liquids (such as liquefied oxygen and liquefied nitrogen) usually maintain relatively low operating pressures through this process.

FIGURE 3-20 An MC 338 highway cargo tanker used to transport cryogenic liquids.

FIGURE 3-22 A drawing of a tube trailer that consists of several compressed gas cylinders in series.

FIGURE 3-21 A non-specification highway cargo tanker (top). It is impossible to tell this tanker apart from other tankers without looking at its specification plate (bottom).

cryogenic trailers may transport only nonflammable atmospheric gases, and the internal tank pressures may not exceed 25.3 psig (1.75 bar) during transport. These tanks are not required to have any emergency discharge controls. They also do not have to display a placard for the product (Figure 3-21).

TUBE TRAILERS

Tube trailers fall under the category of non-specification cargo tanks and trailers because DOT does not classify them as cargo tanks (Figure 3-22). However, the components of the tube trailer are DOT specification cylinders (which are regulated) that are manifolded together and mounted to the vehicle. Typically there will be a control box at one end or the other of the tube trailer assembly. The cylinders carry compressed gases at up to 6000 psi (414 bar). Each cylinder will have a **safety relief device**. Safety relief devices for flammable compressed gases must discharge upward and not impinge on the other cylinders in the tube trailer assembly. Compressed gases such as nitrogen, helium, argon, oxygen, hydrogen, and methane are typically carried in tube trailers.

safety relief device ■ Equipment designed to safely relieve excess pressure in a closed system such as a cylinder, cargo tank, or stationary tank.

DRY BULK COMMODITY TRAILERS

Dry bulk commodity trailers, also known as pneumatic trailers or hopper trailers, also fall under the category of non-specification cargo tanks (Figure 3-23). These trailers are constructed of aluminum or steel, which typically transport dry, flowable products. These trailers may be unloaded through gravity or pneumatic means, and are designed to operate to a pressure of up to 15 psi (1 bar). These trailers may carry such diverse products as foodstuffs (for example, grains), water-reactive metals such as magnesium powder, or even explosives such as ammonium nitrate–fuel oil (ANFO) mixtures.

Railroad Tank Cars

With the exception of oil tankers and liquefied natural gas tankers, railroad tank cars are the largest tanks found in transportation. The extremely large quantities that are dealt with make these very dangerous vehicles. There is a world of difference between the pint

FIGURE 3-23 A drawing of a dry bulk commodity trailer. Usually these trailers carry nonhazardous cargo such as grains; however, they also can carry hazardous materials such as ammonium nitrate, aluminum chloride, and hazardous waste.

Based upon the picture, what type of highway cargo tanker is involved in the opening scenario accident?

Solution: Based upon the horseshoe shape, this is an MC 307 or DOT 407 insulated cargo tanker.

of acetone that may be around the house being used as nail polish remover, and 40,000 gallons of acetone in a railcar.

As seen with highway cargo tankers, railroad tankers have a number of different railcar specifications, which are listed in Table 3-7. Railroad tank cars can be divided into two general classes: low-pressure tanks and high-pressure tanks.

TABLE 3-7	Railcar Tanker Specifications and Shapes	
LOW-PRESSURE TANK CARS (NON-PRESSURE)	**PRESSURE TANK CARS**	**CRYOGENIC**
DOT **103**, 104, **111**, 113, and 115	DOT 105, 107, 109, 112, and 114	DOT 113
AAR 201, 203, 204, 206, 207, 208, and 211	DOT 106 and 110 (ton container or 1-ton cylinders)	AAR 204

Boldface indicates most commonly used railcars.

TANK FEATURES	SAFETY FEATURES	TYPICAL CONTENTS
Class 103 General Service Car 35–60 psi safety valves 10,000–40,000 gal. capacity Low pressure Bottom outlets allowed **Class 104** Low pressure		Benzene Corrosives Phosphorus trichloride Hydrogen peroxide Acrylonitrile
Class 105 High pressure	DOT 105A500W insulated carbon dioxide tank car	
75–450 PSI safety valve	Insulated Top and bottom shelf couplers Thermal protection Head protection	Ammonium nitrate solution Anhydrous ammonia Anhydrous hydrofluoric acid Carbon dioxide Chlorine Ethylene oxide Hydrogen cyanide Liquefied hydrocarbon gas LPG Metallic sodium Sulfur dioxide Vinyl chloride

| TABLE 3-7 | Railcar Tanker Specifications and Shapes (*Continued*) |

TANK FEATURES	SAFETY FEATURES	TYPICAL CONTENTS
Hydrogen cyanide $1\frac{1}{8}$-inch inner shell	DOT 105A500W or DOT 105J500W or DOT 105A600W or DOT 105J600W	
4-inch cork insulation ¼-inch outer shell Enclosed valves		
Chlorine 90-ton capacity 375 psi safety valve	DOT 105A500W insulated	
Class 106A High pressure DOT 106A500X 375-600 psi safety valve		Chlorine Anhydrous ammonia Sulfur dioxide Butadiene Refrigerant gases
Class 107A High pressure		
Class 109A High pressure		
Class 111 Low pressure		Hydrochloric acid Phosphoric acid Aluminum sulface Sulfuric acid Vegetable oil Phosphorus Gasoline Ethyl ether Ammonia solution Acetic acid Whiskey
Class 112 High pressure Bottom outlets	Thermal protection Head protection 10,000–40,000 gal. capacity 75–450 psi safety valve	LPG Sulfur dioxide Vinyl chloride

(*continued*)

TABLE 3-7 | Railcar Tanker Specifications and Shapes (*Continued*)

TANK FEATURES	SAFETY FEATURES	TYPICAL CONTENTS
Class 114 High pressure No bottom outlets		
Class 115 Low pressure		Methyl methacrylate
Tube car (high pressure)	3000–5000 psi	Gases Helium, hydrogen, methane, oxygen
Cryogenic tank car Class 113	16 psi outer shell 75 psi tank 30,000–40,000 gal. capacity	Liquefied gases Liquid hydrogen, liquid oxygen, liquid helium, liquid argon, liquid nitrogen, liquid carbon dioxide
Flat bed car with intermodal tanks		
Covered hopper car		Calcium carbide Cement Grain
Open-top hopper car		Materials are exposed to weather Coal, rock, sand
Pneumatic hopper car	15 psi 2600-5900 cubic feet	Fine powdered materials Plastic pellets, flour
Gondola car (open roof)		Materials are exposed to weather Sand, rolled steel
Box car		Varied May contain tank inside!

LOW-PRESSURE TANK CARS

General-service tank cars are commonly referred to as "non-pressure" tank cars in the industry, although this is an extremely dangerous misnomer (Figure 3-24). There are no non-pressure tank cars being used in the railroad industry today. Low-pressure tank cars can have test pressures as high as 100 psi (6.9 bar) and can transport materials with vapor pressures of up to 25 psi (1.7 bar). Tank capacities range from 4000 to 45,000 gallons (15,000 L to 170,000 L). Low-pressure tank cars make up approximately 75% of tank cars in service. In order to prevent over pressurization, these tank cars have **rupture discs** (for hazardous materials with a low vapor pressure) or spring activated, re-seating safety relief valves (for flammable liquids and poisonous materials with a high vapor pressure). Safety relief valves are typically set to discharge at 75% of the tank test pressure. Examples of common low-pressure tank cars include the DOT 103, the DOT 111, and the AAR 211 tank cars with typical tank test pressures of 60 psi (4.1 bar) or 100 psi (6.9 bar).

rupture disc ■ A safety relief device that relieves pressure by breaking a replaceable seal on the container. Rupture discs do not reset automatically and will continue to leak once activated.

HIGH-PRESSURE TANK CARS

High-pressure cars have tank test pressures from 100 to 600 psi (6.9 to 41.4 bar) and typically carry hazardous materials such as flammable liquids and poison gases (Figure 3-25). Approximately 25% of the tank cars in service today are high-pressure tank cars. Tank capacities range from 4000 to 45,000 gallons (15,000 L to 170,000 L). These tank cars have a single compartment, are constructed of steel or aluminum, and have rounded ends to help contain the pressure. Unlike low-pressure tank cars, high-pressure tank cars will have all of their fittings (such as valves, thermometer wells, sampling ports, and safety relief devices) located inside a protective housing that provides rollover protection. But there are exceptions to every rule, such as the DOT 111J low-pressure tank car that has all of its fittings located within a protective housing. Therefore, it is always best to locate the specification markings and determine exactly which type of tank car you are dealing with!

High-pressure tank cars typically have safety relief valves set to 75% of tank test pressure. One exception to this rule is the 340 psi (23.5 bar) high-pressure tank car that has a safety relief valve set at 280.5 psi (19.3 bar). Examples of common high-pressure tank cars include the DOT 105, the DOT 112, and the DOT 114 tank cars.

TANK CAR MARKINGS

Railcars will have markings that indicate their contents, tank construction, and ownership. This information is crucial to a safe and effective risk-based response to a derailment involving railcars containing hazardous materials. Figure 3-26 shows typical stenciling found on the sides and ends of railcars that indicate the company name, reporting mark, load limits, and tare weights. The reporting mark indicates the owner of the railcar as well as the five-digit numerical serial number unique to that tank car. The reporting mark is a crucial piece of information that can be used to gather other information such as tank

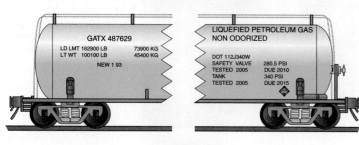

WATER CAPACITY
145144 LBS
65836 KGS
(LOCATED IN AREA OF DOME)

GATX 487629
LD LMT 162900 LB   73900 KG
LT WT 100100 LB   45400 KG
NEW 1 93

LIQUEFIED PETROLEUM GAS
NON ODORIZED

DOT 112J340W
SAFETY VALVE   280.5 PSI
TESTED 2005   DUE 2010
TANK   340 PSI
TESTED 2005   DUE 2015

GATX 487629
CAPY 16090 GAL US
60907 LITERS

FIGURE 3-26 Railcar stenciling requirements. *Source: "First Hazardous Materials Guide for First Responders," FEMA, USFA*

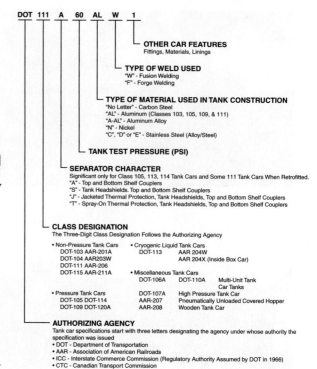

DOT 111 A 60 AL W 1

OTHER CAR FEATURES
Fittings, Materials, Linings

TYPE OF WELD USED
"W" - Fusion Welding
"F" - Forge Welding

TYPE OF MATERIAL USED IN TANK CONSTRUCTION
"No Letter" - Carbon Steel
"AL" - Aluminum (Classes 103, 105, 109, & 111)
"A-AL" - Aluminum Alloy
"N" - Nickel
"C", "D" or "E" - Stainless Steel (Alloy/Steel)

TANK TEST PRESSURE (PSI)

SEPARATOR CHARACTER
Significant only for Class 105, 113, 114 Tank Cars and Some 111 Tank Cars When Retrofitted.
"A" - Top and Bottom Shelf Couplers
"S" - Tank Headshields. Top and Bottom Shelf Couplers
"J" - Jacketed Thermal Protection, Tank Headshields, Top and Bottom Shelf Couplers
"T" - Spray-On Thermal Protection, Tank Headshields, Top and Bottom Shelf Couplers

CLASS DESIGNATION
The Three-Digit Class Designation Follows the Authorizing Agency
• Non-Pressure Tank Cars
  DOT-103 AAR-201A
  DOT-104 AAR203W
  DOT-111 AAR-206
  DOT-115 AAR-211A
• Pressure Tank Cars
  DOT-105 DOT-114
  DOT-109 DOT-120A
• Cryogenic Liquid Tank Cars
  DOT-113   AAR 204W
                    AAR 204X (Inside Box Car)
• Miscellaneous Tank Cars
  DOT-106A DOT-110A   Multi-Unit Tank
                         Car Tanks
  DOT-107A High Pressure Tank Car
  AAR-207   Pneumatically Unloaded Covered Hopper
  AAR-208   Wooden Tank Car

AUTHORIZING AGENCY
Tank car specifications start with three letters designating the agency under whose authority the specification was issued
• DOT - Department of Transportation
• AAR - Association of American Railroads
• ICC - Interstate Commerce Commission (Regulatory Authority Assumed by DOT in 1966)
• CTC - Canadian Transport Commission

FIGURE 3-27 Railcar specification markings. *Source: "First Hazardous Materials Guide for First Responders," FEMA, USFA*

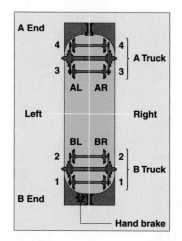

contents, load level, origination, and destination by contacting the owner. From a safe distance, every effort should be made to locate the reporting marks to relay this information to the incident commander or the railroad representative.

Specification markings indicate information about tank construction. Figure 3-27 illustrates typical specification markings for a railcar. These markings point out the authorizing agency, tank car class, safety features such as jackets, tank test pressure, tank construction material, type of welds, and even fittings and tank liners. This information will be important to hazmat technicians that need to do damage assessments and off-loading of the tank contents. Figure 3-28 shows the standard railcar layout and terminology for damage assessment and reporting. Table 3-8 lists tank shell thicknesses for common commodities.

FIGURE 3-28 Common terminology for describing railcar construction for damage assessments. The end of the railcar with the brake is always referred to as the "B end," regardless of its orientation with respect to direction of travel. *Source: "First Hazardous Materials Guide for First Responders," FEMA, USFA*

TABLE 3-8	Tank Shell Thickness Requirements	
Tank thickness:	$1\frac{1}{8}$ inch	Hydrogen sulfide
	¾ inch	Chlorine
	11/16 inch	Liquified petroleum gas (LPG)
	7/16 inch	Sulfuric acid
	¼ inch	Container material
	$\frac{1}{8}$ inch	Jacket material

What would be the most accurate way to identify the highway cargo tanker mentioned in the opening scenario?
Solution: If it is safe to do so, look at the specification plate attached to the frame rail of the cargo tanker. If it is not safe to approach the vehicle, from a distance determine the license plate of the vehicle, trace the owner, and determine the vehicle specifications from the owner.

How would you determine whether the tanker referenced in the opening scenario had multiple compartments?
Solution: Look at the specification plate. Also the number of valves and/or dome covers can be an indicator, although the specification plate is much more reliable.

Shipping Papers

Bulk containers carrying hazardous materials will typically be marked from the outside with a placard, stenciling, or a UN/NA number. This labeling offers very limited information, even if the *Emergency Response Guide* is at your disposal. How can more detailed information be obtained at the scene of the hazardous materials incident involving transportation vehicles? All modes of transportation are required to carry shipping papers. The shipping papers give much more detailed information regarding contents of the shipment, including the name of the product, the quantity of the material, origination and destination locations, an emergency contact number, the hazard class, and often packaging requirements. As you can see, getting hold of the shipping papers early can be very beneficial.

The types of information you can expect to find in the shipping papers include:

1. *Proper shipping name.* The shipping name as stated in the federal DOT regulations.
2. *Hazard class and division.* The US DOT hazard class and division.
3. *UN/NA number.* The four-digit United Nations identification number.
4. *Type of hazard and other notations.* Such as "CORROSIVE" or "Marine Pollutant."
5. *Total quantity and* **reportable quantity (RQ)**, *if applicable.* The reportable quantity is the amount of material that, if spilled, must be reported to the National Response Center.
6. **Packing group.** Classification of the material based upon its danger in transportation and the consequent packaging requirements:
   I   Great danger
   II   Medium danger
   III   Minor danger
7. Inhalation hazard or poison inhalation hazard (PIH):
   A LC_{50} = 0–200 ppm
   B LC_{50} = 200–1000 ppm
   C LC_{50} = 1000–3000 ppm
   D LC_{50} = 3000–5000 ppm
   The LC_{50}, known as the lethal concentration 50%, is the concentration at which 50% of the test population (such as laboratory animals) would be expected to die.
8. *EPA waste stream number.* The waste stream number must be included for RCRA hazardous waste shipments. This number will be required for some class 9 materials per 40 CFR 263.20.

reportable quantity (RQ) ■ The release of an amount of a hazardous material or waste that triggers mandatory notification of the National Response Center (NRC).

packing group ■ Per DOT, a classification based upon the degree of danger a hazardous material presents. Packing Group I indicates great danger; Packing Group II, medium danger; and Packing Group III, minor danger.

But remember, the information contained in the shipping papers is only as good as the person completing the form! Shipping papers have been known to contain significant errors. Each year the U.S. DOT issues fines to the transportation industry for improper shipping papers.

HIGHWAY: BILL OF LADING

The bill of lading, or freight bill, must remain in the cab of the vehicle within reach of the driver at all times. Two common locations where you may find the shipping papers are the front seat of the cab and in the driver's side door compartment. On occasion, an extra copy of the shipping papers may be found on the trailer located in a tube on the chassis or within the valve compartment. Figure 2-16 shows an example of typical shipping papers found in highway transportation.

RAILROAD: CONSIST

The train consist, or the waybills, will be with a member of the train crew, most commonly the conductor or the engineer. The conductor is responsible for the train consist. The consist typically is comprised of four parts: the tonnage graph, the position-in-train document, the train listing, and the emergency handling instructions. The tonnage graph shows the lading of each railcar in a train. Railcars containing hazardous materials will be specially indicated often by marking the lading in tons with "HHHH." The position-in-train document specifically lists railcars containing hazardous materials and their position in the train. The train listing contains information about each railcar in the train.

Railcars that contain hazardous materials contain extra information and are highlighted in the consist. This extra information includes the chemical name, the UN/NA number, the standard transportation commodity code (STCC) number, the origin and destination locations, and emergency contact information. STCC codes are unique seven-digit identification numbers. STCCs for hazardous materials start with 49, and for hazardous waste they start with 48. The emergency handling instructions will be included for all hazardous materials that are carried on the train. Different railroads may have slightly different formats but typically include all of this information. Figure 2-17 shows a typical consist used by CSX Transportation.

AIR: AIR BILL

The air bill will be found in the cockpit, and the pilot is the responsible person. Copies of individual waybills will usually also be attached to the outside of the respective packages. Figure 2-18 shows an example of an air bill.

dangerous cargo manifest ■ The shipping papers used in marine transportation. The manifest lists the hazardous materials and their locations aboard watercraft.

MARINE: DANGEROUS CARGO MANIFEST

The **dangerous cargo manifest** will be found in the wheelhouse, or a pipe-like container on a barge. The captain or master of the vessel is responsible for the dangerous cargo manifest. Due to the size of the ship, it may be almost impossible to match individual bills of lading to cargo. Also be aware that there is a difference between the bill of lading and a ship's manifest. The bill of lading is a contract between the shipper and consignee that

SOLVED EXERCISE 3-4

Where would you find the shipping papers for the cargo tanker referenced in the opening scenario?
Solution: The shipping papers should be in the cab or with the truck driver (if he had the presence of mind to take them out of the cab after the accident).

Based upon the following shipping papers, what product is the cargo tanker from the opening scenario carrying?

SHIPPER PLEASE NOTE	FREIGHT CHARGES ARE PREPAID ON THIS BILL UNLESS MARKED COLLECT		Page 1

STRAIGHT BILL OF LADING–NOT NEGOTIABLE

Carrier: ETOH HAULERS

PLACE PRO LABEL HERE

Shipper's Bill or Lading No.

Consignee's Reference/PO No.

To: CANADIAN FUEL MIXERS	TRAILER	B/L DATE
CONSIGNEE	Route	
STREET 1 E85 BLVD		
DESTINATION TORONTO ON M3R 1A1 CA	SPECIAL INSTRUCTIONS	

FROM: IOWA BIOFUELS
SHIPPER
STREET 1 GRAIN ALLEY

ORIGIN CITY/ST/ZIP DES MOINES IA 50307 US

FOR PAYMENT SEND BILL TO:
NAME

STREET
CITY/ST/ZIP

Collection Delivery _____ and mail to _____
Steet _____ City _____ State_____
Carrier must collect cash, money order, bank cashier's check, or bank certified check unless
shipper signs here to accept company check:
Signed:_____

C.O.D. Change   Shipper ☐

to be paid by   Consignee ☐

Units No. Type	Packages No. Type	kind of Package, Description of Articles, Special Marks and Exceptions (Subject to Correction)	Weight Subl. to Correction) (LBS)	Class or Rate Per. (For info. only)	Cube (Optional) Cuft
1	1 TL	UN 1170, ETHANOL, 3, BULK PACKAGE EMERGENCY CONTACT: (800) 424-9300 Ethanol Additional Services: Arrival Notification Customs Administrative Fee Single Shipment	56000		

Total Handling Pieces: 1	Individual Pieces: 1	Weight: 56000	Cube:

NOTE(1) Where the rate is dependent on value, shippers are required to state specifically in writing the agreed or declared value of the property as follows:
"The agreed or declared value of the property is specifically stated by the shipper to be not exceeding _____
per _____."
Note (2) Liability Limitation for loss and damage on this shipment may be applicable. See 49 U.S.C 14708 (c)(1)(A) and (B).
Note (3) Commodities requiring special or additional care or attention in handling or stowing must be so marked and packaged as to ensure safe transportation with ordinary care. See Sec. 2(e) of NMFC item 380.
Freight Charges are PREPAID unless marked collect

SHIPPER IOWA BIOFUELS

CARRIER

PER (SIGNATURE REQUIRED)

PER   DATE

Solution: Ethanol, also known as ethyl alcohol.

How much ethanol is the cargo tanker from the opening scenario carrying?
Solution: 56,000 pounds. This is 8500 gallons based upon the density of ethanol.

grants title. There is one bill of lading per client. The ship's manifest consolidates all these bills of lading and lists the entire ship's cargo. Thus, when you are involved in a shipping incident, make sure to obtain the ship's manifest with a complete listing of the hazardous materials on board. Figure 2-19 shows an example of a marine bill of lading as found in a dangerous cargo manifest.

Radioactive Materials Packaging

Due to the danger of radioactive materials, five types of packaging are used: excepted, industrial, Type A, Type B, and Type C (Figure 3-29). Table 3-9 lists the characteristics of each of these regulated package types. Excepted packaging is used to ship very low-level radioactive materials such as smoke detectors, which pose a minimal hazard if released. Industrial packaging is used to ship low specific activity materials such as concrete-solidified radioactive waste, which pose little hazard if released (non-life-threatening amounts).

Type A packaging is used to ship higher activity materials than industrial packaging but still non-life-threatening amounts of radioactive materials. Type A packages are constructed of steel, wood, or fiberboard with an inner containment vessel for the radioactive material made of glass, plastic, or metal. Radiopharmaceuticals and research isotopes are commonly shipped in Type A packages. Certified packaging materials are marked with "US DOT 7A—Type A." This type of packaging should withstand normal shipping and handling but not vehicle accidents or fires.

Type B packaging is used to ship high activity, potentially life-threatening amounts of radioactive materials. Type B packaging ranges from shielded 55-gallon drums all the way to heavily shielded steel casks that weigh over 100 tons (90,000 kg). This type of packaging is tested to withstand severe accidents and fires. Spent nuclear fuel and high-level radioactive waste is shipped in Type B packages. Type C packaging is used for international shipments of radioactive materials on aircraft, and its requirements are even more stringent than those of Type B packaging. Both Type B and Type C packaging

FIGURE 3-29 Examples of industrial packaging containing radioactive materials. The left and middle images show examples of a Type A package containing radioactive materials. On the right is an example of a Type B package containing highly radioactive waste. *The leftmost image is courtesy of Brian Wagner, National Institutes of Health Fire Department*

TABLE 3-9	Radioactive Packaging
RADIOACTIVE PACKAGING TYPE	**CHARACTERISTICS**
Excepted Packaging	Limited quantity of radioactive material
	Maximum external radiation 0.5 mR/hr
Industrial Packaging	Higher radioactivity than excepted packaging
	Low specific activity (LSA) material or surface contaminated object (SCO)
	May be boxes, freight containers, tanks, etc.
	There are Type 1 (IP-1), Type 2 (IP-2), and Type 3 (IP-3) industrial packages
Type A	Low-level radioactive material shipments
	Designed to retain integrity and shielding under normal transport conditions
	Must have tamper-evident security seal (or the vehicle or cargo compartment must have it)
Type B	High-level radioactive material shipments (within limits)
	Designed to retain integrity and shielding under normal transport conditions and accident conditions
Type C	Air transport of radioactive materials

must be permanently marked with the trefoil symbol to ensure that they can be identified even after a severe accident. Any type of radioactive materials packages should only be approached after proper screening with radiation detection equipment by qualified personnel.

INTERPRETING RADIOACTIVE PLACARDS AND LABELS

There are three types of radioactive package labels based upon the activity of the radioactive material contained within it (Figure 3-30). Activity refers to the strength of the radioactive material. The higher the activity, the more danger it poses to the first responder. The various types of radiation and their hazards will be discussed in Chapter 4.

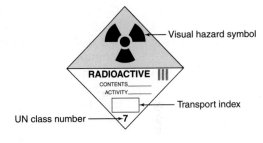

FIGURE 3-30 The U.S. DOT radioactive III label.

Radioactive I — Label has a completely white background with black and red writing.
Label indicates the radioactive isotope and its activity (for example, Cs^{137} and 2 Bq).
Radiation levels up to 0.5 mrem/hr on the outside surface of the package.

Radioactive II — Label has a yellow and white background with black and red writing. Label indicates the radioactive isotope and its activity (for example, Cs^{137} and 2 Bq), as well as the **transport index (TI)**.
Radiation levels between 0.5 mrem/hr and 50 mrem/hr on the outside surface of the package.
Note: Radiation levels may be up to 1 mrem/hr at 1 meter (3.3 feet) from the package (maximum TI = 1).

transport index (TI) ▪ A dimensionless number used to designate the degree of control to be exercised during transportation; indicates the maximum amount of radiation permitted at 1 m (3.3 feet) from the package in mrem/hr.

Radioactive III   Label has a yellow and white background with black and red writing.

Label indicates the radioactive isotope and its activity as well as the transport index (TI).

Radiation levels between 50 mrem/hr and 200 mrem/hr on the outside surface of the package.

Note: Radiation levels may be up to 10 mrem/hr at 1 meter (3.3 feet) from the package (maximum TI = 10).

It may come as a surprise that radioactive packaging is legally permitted to emit radiation, and under certain circumstances this can be dangerous. Needless to say, it would not be a wise idea to use a Radioactive III package as a makeshift seat! This is why it is very important to understand the meanings of the various radioactive labels.

Pipelines

Pipeline emergencies can be difficult to respond to because they are often discovered and reported late, the actual location of the leak may be hard to locate, or the product being transported may be difficult to identify. Most pipelines in the United States are buried belowground and are hard to locate. Buried pipelines are required to be marked using pipeline markers (Figure 2-12). Pipeline markers are required to have the owner's name, emergency telephone number, and product it transports listed. However, pipelines often carry multiple materials. For example, a pipeline that carries gasoline and diesel fuel, among other materials, may just be labeled "petroleum products." Liquid pipelines may be up to 42 inches (107 cm) in diameter and may carry everything from petroleum products to liquefied gases. Gas pipelines may be up to 48 inches (122 cm) in diameter and commonly carry natural gas. However, they may also carry many other products such as propane, ethylene, and even chlorine and ammonia.

Pipelines have aboveground valving stations positioned periodically. If one of these stations is in your jurisdiction, it is very important to remember not to open or close valves unless directed to do so by an authorized representative of the pipeline company. Given the large diameter and pressures of pipelines, unauthorized valve movement (opening or closing) may lead to over pressurization and subsequent rupture or explosion of the pipeline. The effects of unauthorized opening or closing of valves may be felt many miles away.

Container Failure

Hazardous materials incidents are often caused by container failure. Containers may fail in many different ways. For example, there may be a manufacturing defect in the container that causes it to fail unexpectedly under normal conditions, or the container may fail due to abuse or an accident during normal operations. Containers may also fail under extraordinary conditions such as fires, floods, uncontrolled chemical reactions, or other emergencies.

CONTAINER STRESS

There are three types of stress that could cause a container to catastrophically fail and release its contents: thermal stress, mechanical stress, and chemical stress.

Thermal Stress

Thermal stress, or heat, causes a container to fail by reducing the strength of the material of which it is made. For example, under increased heat, plastics tend to melt and metals tend to distend and deform. In addition, the internal pressure of the container increases under thermal stress due to expansion of the contents.

One subcategory of thermal stress is the **BLEVE**, which stands for "boiling liquid expanding vapor explosion" (Figure 3-31). This occurs when the contents of a container are heated above their boiling point and the increased pressure causes the container to fail catastrophically. When the container fails, the liquid contents almost instantaneously vaporize, typically expanding several hundredfold in volume. If the container contains a flammable liquid, such as gasoline or liquefied propane, the sizable vapor almost always spontaneously ignites in the presence of an ignition source. The energy released during this process usually leads to catastrophic damage and significant loss of life if the proper isolation distances have not been instituted.

Mechanical Stress

Mechanical stress causes the container to fail at its weakest point. Mechanical stresses include too much weight, weight imbalances, torque, deformation (denting), and intrusion of foreign objects. Most containers have weak areas such as seams, closures, and chimes. The chime is where the ends of the container are fastened to the sides. Containers tend to fail at their weakest link when subjected to mechanical stress due to uneven force distribution. When an external mechanical stress is applied, it will puncture the container if it imparts more force than the container can withstand. For example, a forklift tine can go through most containers like butter (Figure 3-32). This is a very common cause of hazardous materials releases at fixed site facilities. Usually plant personnel will handle this type of release themselves. If the hazardous material enters the sewer system or leaves plant property, local, state, and federal authorities must be notified. If this occurs in your jurisdiction, you will likely be notified of this release as a member of an agency trained to the hazardous materials operations level.

Chemical Stress

Chemical stress can be caused by placing a chemical into an incompatible container, or through the mixing of incompatible materials that leads to a chemical reaction that damages the container (Figure 3-33). In either case, the presence of a chemical causes

BLEVE ■ The abbreviation of "boiling liquid expanding vapor explosion." The process by which the buildup of internal pressure within a container is catastrophically relieved by its violent rupture, accompanied by the release of a vapor or gas and fragmentation of the container.

FIGURE 3-31
Catastrophic container rupture due to thermal stress from a boiling liquid expanding vapor explosion (BLEVE). Many emergency responders have been killed and injured due to BLEVEs that occur minutes to hours after their arrival. *Art by David Heskett*

FIGURE 3-32 A drawing of a container rupture due to the mechanical stress of a forklift puncture.

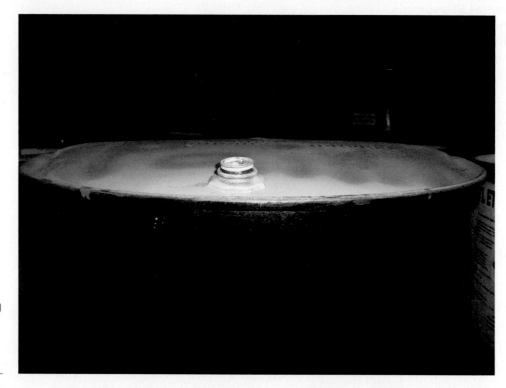

FIGURE 3-33 A bulging container due to chemical stress on this 55-gallon drum. Be cautious of these already weakened containers.

the weakening of the container. Once again, the container will likely fail at its weakest point. Weak points include thinner portions of the container, areas where the container has macro- or micro-imperfections, or seam areas. Chemical stress often creates heat and can be located using a thermal imaging camera (TIC) or an infrared temperature gun. Areas where chemical reactions are occurring will usually appear hot, but occasionally may appear cooler.

CONTAINER BREACHING

When the stresses on a container exceed its design parameters, it will breach. There are five ways a container could breach and release its contents: The closures could open up, the container could split or tear, the container could be punctured, runaway cracking could occur, or it could disintegrate. Closures and seams are some of the weakest areas in containers. As mentioned, when containers are stressed, they tend to tear or split at the openings, seams, and chimes. Runaway cracking typically occurs at a location where the container has been previously weakened, either through a manufacturing defect, through normal wear and tear, or from damage through a previous accident. When additional stressors are applied to the container, such as thermal, chemical, or mechanical stresses, a runaway crack may form and lead to catastrophic failure of the container. Disintegration of the container typically occurs under extreme stresses. One example of container disintegration is the BLEVE phenomenon discussed previously.

Container Release

When the contents of the container are released by a breach or failure, the release rate will vary (Figure 3-34). There are four different types of releases: spill or leak (slow), rapid release (moderate), violent rupture (fast), or detonation (instantaneous).

Dispersion Patterns

The pattern the escaping product will form depends on the type of stress that caused the container to release, the type of chemical, the state of matter the chemical was originally in, any phase changes the material may undergo, the rate of the release, and

FIGURE 3-34 Types of container releases. *Art by David Heskett*

environmental factors such as weather and terrain. There are seven types of dispersion patterns: pool, stream, cone, plume, cloud, hemisphere, and irregular. Most releases will have a combination of dispersion patterns depending upon the length of the release, the changing weather conditions, the chemical and physical properties of the material, and the actions taken by first responders.

A pool tends to form when a liquid or a gas is released from its container under little pressure. The material then tends to settle or pool. For a gas, it may pool in low-lying areas if the vapor density is greater than 1, or it may pool at the ceiling if the vapor density is less than 1. If the vapor pressure of a liquid is significant, it will evaporate from the pool and may lead to the formation of a vapor cloud. A stream tends to form when a solid, liquid, or liquefied gas is released from its container under moderate pressure and geographic and atmospheric conditions allow it to flow. A cone tends to form when a solid, liquid, or gas is released from its container under high pressure.

A cloud is generally formed when a gas or liquefied gas is released almost instantaneously, and it rises and disperses in all directions simultaneously. On the other hand, a plume is formed when a gas, liquefied gas, or high vapor pressure liquid is released or evaporates over time. An outdoor plume will be driven by the weather (especially the wind) and geography. In practice, a cloud and a plume may be difficult to distinguish initially. However, a plume will have to be much more aggressively managed because it is a continuous release. In all likelihood, if the initial release is an outdoor cloud, it will have run its course before first responders have had time to act. On the other hand, if a cloud forms due to catastrophic failure of the container after first responders have arrived, it may be very dangerous if proper isolation distances and staging locations were not used from the beginning.

A hemispheric release occurs when there is a rapid release of energy, generally from a single point of origin. The chemical or energy—in the case of an explosion or radiation—travels in a semicircular pattern in all directions. The exact dispersion pattern will be affected by terrain, buildings and other structures, and cloud cover in the case of explosions. Parts of the container may show an irregular dispersion pattern.

Irregular releases occur when the material is released over time and container stresses, container integrity, atmospheric conditions, and dispersion patterns do not remain constant over time. For example, after a container breach, a cloud may initially form due to over pressurization of the container; then liquid streams out and eventually pools in low-lying areas. Due to the relatively high vapor pressure of the material, a plume starts to form as the surface area of the released material increases and the sun heats the liquid. Irregular releases are very common when offensive mitigation actions cannot be promptly started. Irregular releases can also be caused when hazardous materials are dispersed by contaminated people or animals.

Recognizing Terrorist Incidents and Criminal Events

Criminals or terrorists may target various locations in your community. The location will vary based upon the intent, ideology, and ultimate goal of the perpetrator. The common criminal may target high-value locations such as banks, jewelry stores, or other locations

SOLVED EXERCISE 3-8

What type of dispersion pattern would you expect in case of container failure in the opening scenario?

Solution: A pool of liquid. MC 307 and DOT 407 tankers are low-pressure containers and would most likely not release a cloud of hazardous material. Special circumstances that could produce a vapor cloud include superheated contents or a buildup of pressure due to heating or container compression.

as a diversionary tactic. Criminal incidents may also include clandestine drug manufacturing labs, chemical releases used as diversionary tactics during crimes such as bank robberies, toxic chemical releases during a murder or attempted murder, or environmental crimes such as illegal dumping (Figure 3-35).

The terrorist may target locations based upon their high profile, function, high occupancy load, or significance to the community. Thus, monuments, water treatment plants, abortion clinics, schools, stadiums, malls, courthouses, and many other locations may be targeted. Weapons of mass destruction incidents may involve military or improvised chemical or biological agents deployed in order to hurt or kill civilians and responders alike.

FIGURE 3-35 Criminal events involving hazardous materials. These drums containing hazardous waste were illegally dumped and constitute an RCRA violation.

HISTORY OF TERRORISM USING CHEMICAL AND BIOLOGICAL AGENTS

A wide variety of hazardous materials has been used in warfare over the millennia. More recently, these hazardous materials, including **chemical agents**, explosives, radioactive materials, and biological agents, have been increasingly used by terrorist organizations against civilian targets.

Chemicals have been employed in warfare as early as 1000 B.C. in China with the use of arsenic smoke. Modern chemical agents were first employed in World War I. Initially, toxic industrial chemicals such as chlorine were used and eventually agents were specifically developed for warfare such as sulfur mustard. In World War II nerve agents were developed. More recently terrorist organizations have employed these chemical agents, such as the Aum Shinrikyo cult's use of the nerve agent sarin in the Tokyo subway attacks of 1995 in which 12 people died.

Explosives, including gunpowder, have been used in warfare ever since their invention in China in the ninth century A.D. Over the last several centuries, ammunition, bombs, mines, and other incendiary weapons have been developed using explosives. Explosives are the first choice of terrorists due to the ease of use and ready availability of explosive materials. In 1995 Timothy McVeigh used a homemade mixture of ammonium nitrate–fuel oil (ANFO) to kill 186 people at the Alfred P. Murrah Federal Building in Oklahoma City. The passenger planes fully loaded with jet fuel that were used in the September 11, 2001, attacks in New York City and Washington, DC, can be considered incendiary device attacks.

When explosives are combined with radioactive materials, it is called a dirty bomb, or radiological dispersion device (RDD). Dirty bombs were used by Chechen terrorists in Russia in the 1990s. Dirty bombs can be considered weapons of mass disruption rather than weapons of mass destruction because the psychological effects of this type of terrorism will likely far outweigh the immediate damage caused by either the explosive itself or the dissemination of the radioactive material. Of much greater concern is the potential creation of an improvised nuclear device by terrorists. If terrorists were to obtain a critical mass of a fissile radioactive material, they could potentially create a nuclear weapon that could destroy a small city. Therefore, most world governments have gone to great lengths to closely guard fissile material. Nevertheless, since the breakup of the former Soviet Union, there have been numerous seizures of weapons-grade plutonium and uranium from unauthorized individuals.

Biological agents had been used in warfare before pathogens were even discovered. As early as 400 B.C., Scythian archers used arrows dipped in the blood of decomposing bodies and manure. In the mid-14th century A.D. Tartars catapulted bodies of bubonic plague victims over the walls of besieged Kaffa. Smallpox was allegedly used by the British during the French and Indian wars in the mid-18th century when smallpox-infested blankets were given to the enemy.

More recently, terrorists have seized on the bioavailability of biological agents as well. In 2001, weapons-grade anthrax was mailed to prominent people in Washington DC,

chemical agent ■ An extremely toxic chemical substance that has the capability to cause mass casualties when disseminated. Examples include sarin, hydrogen cyanide, and mustard gas.

New York City, and Boca Raton, Florida. As a result five people lost their lives from contracting inhalational anthrax. Toxins, such as ricin and botulinum toxin, can be produced from living organisms. There have been several ricin attacks in the United States due to the ready availability of the castor bean plant and the relative ease with which ricin can be extracted from the plant.

WEAPONS OF MASS DESTRUCTION AND TOXIC INDUSTRIAL MATERIALS

The acronym CBRNE stands for chemical, biological, radiological, nuclear, and explosive weapons. These five categories are collectively termed weapons of mass destruction (WMD). Weapons of mass destruction are not the only tool criminals and terrorists may use. They can misuse a host of very deadly chemicals that exist in our homes, stores, and workplaces.

Toxic Industrial Materials

Toxic industrial materials (TIM), also known as toxic industrial chemicals (TIC), are potentially dangerous chemicals that are used routinely in industry and even in the home (Figure 3-36). These chemicals include chlorine, hydrogen fluoride, strong corrosives, and industrial explosives. Many of these chemicals can be mixed together to form other hazardous materials, such as hydrogen sulfide.

Chemical Warfare Agents

Chemical warfare agents (CWA) were originally developed for military warfare. Chemical agents include blood agents such as cyanide, nerve agents such as sarin and VX, vesicants (blister agents) such as mustard, **choking agents** such as chlorine and phosgene, and riot control agents such as tear gas and pepper spray. The chemical agents and their symptoms, as well as those of biological agents, are summarized in Table 3-10.

choking agent ■ A substance used to incapacitate or kill people by making it difficult to breathe. Also known as pulmonary agent. Examples include chlorine and phosgene.

Biological Warfare Agents

Biological warfare agents (BWA) include bacteria and viruses that are living organisms, as well as toxins derived from living organisms. The biggest difference between a chemical and a biological incident is the time of onset of signs and symptoms. Chemical agents will typically produce signs and symptoms within seconds or minutes, hours at the latest (with some vesicants). Biological agents, on the other hand, will typically have a delay of days or weeks while the bacterium or virus replicates in and infects the victim. The exception to this rule is the toxin family, which typically has a relatively quick onset of signs and symptoms.

FIGURE 3-36 Common toxic industrial chemicals (TIC) found across the country. On the left is a shipment of liquefied chlorine gas being delivered to a water treatment plant in one-ton containers. On the right is a tanker full of hydrogen fluoride, a highly toxic and corrosive volatile liquid.

TABLE 3-10	Examples of Chemical and Biological Agents		

COMMON NAME	MILITARY ABBREVIATION	DOT HAZARD CLASS	SIGNS AND SYMPTOMS
Nerve Agents			SLUDGEM—drooling, tearing, urination, defecation, gastrointestinal upset, vomiting, and pinpoint pupils
Tabun	GA	6.1	
Sarin	GB	6.1	
Soman	GD	6.1	
VX	VX	6.1	
Blister Agents			Skin lesions, skin redness, skin irritation, blistering
Mustard	H	6.1	
Distilled mustard	HD	6.1	
Nitrogen mustard	HN	6.1	
Lewisite	L	6.1	
Blood Agents			Difficulty breathing, reduced level of consciousness
Hydrogen cyanide	AC	6.1	
Cyanogen chloride	CK	2.3	
Choking Agents			Difficulty breathing, throat irritation, respiratory distress
Chlorine	CL	2.3	
Phosgene	CG	2.3	
Irritants			Difficulty breathing, throat irritation, coughing
Tear gas, Mace®	CS	6.1	
Dibenzoxazepine	CR	6.1	
Chloroacetophone	CN	6.1	
Pepper spray, Mace®	OC	2.2 (subsequent risk 6.1)	
Mace®, phenylchloromethylketone, chloropicrin	PS	6.1	
Biological Agents and Toxins			Flu-like symptoms, fever
Anthrax	n/a	6.2	Possible skin lesions (cutaneous), possible gastrointestional upset (ingestion)
Mycotoxin	n/a	6.1 or 6.2	
Plague	n/a	6.2	Skin lesions
Viral hemorrhagic fevers	n/a	6.2	Bleeding
Smallpox	n/a	6.2	Skin lesions
Ricin	n/a	6.2	Loss of consciousness, coma

n/a - not applicable
Adapted from NFPA 472 (2008) Annex A.

Explosives

Explosives, hazard class 1 materials, are the most common WMD agent used in the United States and throughout the world. This is largely due to the ease of use and wide availability of explosives.

Radiological Materials

Radiological materials, or hazard class 7 substances, may be combined with explosives to form radiological dispersal devices (RDDs). These devices have the same primary hazard as explosive devices, with the added psychological and health hazard of disseminating radioactive materials. Radioactive materials emit ionizing radiation, which at high levels can damage our bodies. Typically, the activity of the radioactive materials found in an RDD will be very low, and the primary hazard of the radioactive material will be a psychological effect. However, this psychological effect may be very powerful. The American public is extremely wary of radiation. Therefore, any release of radioactive material, or even a perceived release of radioactive material, will likely cause extensive economic damage to the region.

INDICATORS OF A TERRORIST INCIDENT

Signs of a terrorist incident involving hazardous materials or weapons of mass destruction may include all of the following:

- Reports of an unexplained explosion
- Many sick people with unusual symptoms, complaining of similar symptoms, and/or at an unusual time of year
- Many injured people with an unexplained cause
- Crop dusting or spraying activities in unusual locations or at unusual times (for example, in the evening or early morning)
- Unusual devices or equipment near building air intakes
- An unusual number of dead animals, insects, or vegetation in an area

If an incident is suspected to be of a criminal or terrorist nature, several extra steps and precautions should be taken. Often, secondary devices targeted at responding personnel may be deployed by the perpetrator (Figure 3-37). Extreme caution should be used in and around the release site. In addition, because the release site is now a crime scene, it should be disturbed as little as possible. Almost everything is potential evidence. Your local law enforcement agency and the FBI should be notified as soon as possible.

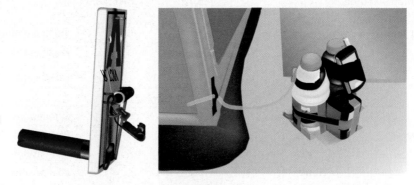

FIGURE 3-37 Examples of secondary devices used by terrorists and criminals to target law enforcement and other first responders.

Summary

It is very important for the operations level responder to quickly recognize the presence of hazardous materials and weapons of mass destruction at the incidents to which they are called. Placards, labels, and shipping papers all provide clues that potentially hazardous chemicals are present. The operations level responder should be able to quickly assess the situation, isolate the immediate area, determine the presence of the hazardous materials, and be able to determine the primary hazards these hazardous materials pose. In the next chapter there will be an examination of how hazardous materials and weapons of mass destruction are dangerous, and a discussion of how to interpret their chemical and physical properties in order to make sound decisions at the incident.

Review Questions

1. Name three common methods of operation of emergency shutoff valves.
2. What type of information can be found on a railroad tank car?
3. What type of information can typically be found in shipping papers?
4. What are the three types of stress that may lead to container failure?
5. Name five indicators of a terrorist incident.

Problem-Solving Activities

1. You have responded to an emergency medical incident. As you enter the front door of the occupancy, you find an unconscious patient covered in white powder and see a box next to him. Written on the box is an EPA registration number as well as the word DANGER. What is the most likely cause of this medical emergency, and how would you proceed?
2. How would you respond to reports of a tire fire on the following tanker? Justify your actions.

3. How would you respond to reports of a vehicle rollover involving the following tanker? Justify your actions.

4. You have been assigned to formulate the pre-incident plans for a facility in your jurisdiction containing the stock room at right. What are your concerns based upon the types of containers found in the stockroom? What type of response recommendations would you make to your department?

5. Between now and the next class, keep an eye out for highway cargo tankers. Record the type of cargo tanker based on a visual identification, and its contents based on external placarding for five tankers.

References and Further Reading

National Fire Protection Association. (2008). NFPA 472, *Standard for Competence of Responders to Hazardous Materials/Weapons of Mass Destruction Incidents*. Quincy, MA: Author.

Occupational Safety and Health Administration. (1990). 29 CFR 1910.120, *Hazardous Waste Site Operations and Emergency Response (HAZWOPER)*. Washington, DC: U.S. Department of Labor.

U.S. Department of Transportation. (2008). *2008 Emergency Response Guidebook*. Washington, DC: Pipeline & Hazardous Materials Safety Administration.

Weber, Chris. (2007). *Pocket Reference for Hazardous Materials Response*. Upper Saddle River, NJ: Pearson/Brady.

4

Operations Level Responders Core Competencies: Understanding Hazardous Materials

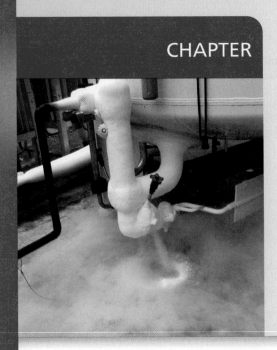

OBJECTIVES

- Use the current version of the NIOSH guide to research a hazardous material.
- Use an MSDS sheet to research a hazardous material.
- Use the program WISER to research a hazardous material.
- Interpret the chemical and physical properties of any given hazardous material.
- Determine how any given hazardous material will behave on contact with air.
- Determine how any given hazardous material will behave on contact with water.
- Determine the flammability properties of any given hazardous material.
- Determine the chemical reactivity of any given hazardous material.

You Are on Duty! Crashed Highway Cargo Tanker (Part 2)

Paw Paw, Michigan

On the evening of January 28, 2004, a 911 call is received stating that a tanker truck and a pickup truck just collided on I-94 between Paw Paw and Mattawan in Van Buren County, Michigan. The weather conditions are a very cold temperature of 16°F (−9°C), 10 to 15 mph winds out of the west, and light snow with 11 inches (27.5 cm) already on the ground. The pickup truck driver is critically injured. Based upon the placard and the shipping papers, the tanker contains 8500 gallons of ethanol. It also appears that the tanker is leaking a small amount of clear liquid.

> **physical property** ■ A change that does not result in an alteration of the chemical identity of a substance, such as vaporization, sublimation, melting, and boiling.

- What are the hazards of ethanol?
- How will the ethanol behave?
- Does the weather factor into the behavior of the ethanol?
- Will the snow affect the ethanol?

Let's try to answer these questions in this chapter.

Operations level response personnel respond to hazardous materials and weapons of mass destruction (WMD) incidents to protect the public, stabilize the incident, and protect property and the environment. In order to make sound decisions at hazardous materials incidents, it is critical to be able to understand the behavior of hazardous materials. In order to predict how a hazardous material will behave, you must understand its chemical and **physical properties**. In this chapter we take a look at what the chemical and physical properties of the hazardous materials can tell us about what they are likely to do, and explore where to find this information.

Using Reference Sources

In order to make sound decisions on a hazmat incident, it is necessary to find the values of chemical and physical properties for the hazardous material involved. This is accomplished using reference sources and technical experts. Let's explore some of these reference materials and learn how to use them.

USING THE NIOSH GUIDE

The National Institute for Occupational Safety and Health (NIOSH), part of the Centers for Disease Control and Prevention, has developed an excellent pocket guide containing detailed information about hazardous materials (Figure 4-1). This information

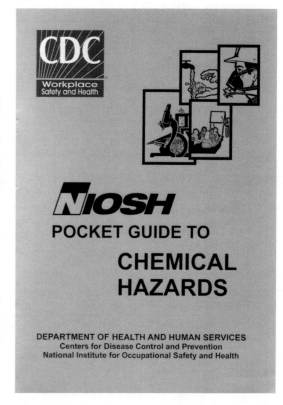

FIGURE 4-1 The NIOSH guide is an excellent source of information regarding the chemical and physical properties of hazardous materials. *Courtesy of the National Institute for Occupational Safety and Health*

includes chemical and physical properties, chemical reactivity, exposure limits, personal protective actions, health effects, and first aid measures. Almost 700 chemicals are listed in alphabetical order, with a synonym and trade name index in the back. Figure 4-2 shows an example of the NIOSH guide entries for ammonia and chlorine. The NIOSH guide is a first-rate source of information when the identity of the released hazardous material is known.

Some of the NIOSH guide contents include:

Pages iv–xxx	Explanatory material (tables of abbreviations, definitions)
Pages 1–340	Data on hazardous materials in alphabetical order
Pages 341–372	Appendices (Carcinogens, Supplementary Exposure Limits, Respirator Requirements, and 1989 Vacated PELs)
Pages 374–378	Chemical Abstracts Services (CAS) number index
Pages 379–382	DOT United Nations ID number index
Pages 383–424	Chemical, Synonym, and Trade Name index

INTERPRETING MSDS SHEETS

Material safety data sheets (MSDS) are usually associated with fixed site facilities. However, they may be also be found in transportation vehicles attached to shipping papers in lieu of carrying the current edition of the DOT *Emergency Response Guidebook*. Industrial

Ammonia	Formula: NH_3	CAS#: 7664-41-7	RTECS#: BO0875000	IDLH: 300 ppm
Conversion: 1 ppm = 0.70 mg/m³	**DOT:** 1005 125 (anhydrous); 2672 154 (10-35% solution); 2073 125 (>35-50% solution); 1005 125 (>50% solution)			

Synonyms/Trade Names: Anhydrous ammonia, Aqua ammonia, Aqueous ammonia
[**Note:** Often used in an aqueous solution.]

Exposure Limits: **NIOSH REL:** TWA 25 ppm (18 mg/m³) ST 35 ppm (27 mg/m³) **OSHA PEL †:** TWA 50 ppm (35 mg/m³)	Measurement Methods (see Table 1): **NIOSH** 3800, 6015, 6016 **OSHA** ID 188

Physical Description: Colorless gas with a pungent, suffocating odor.
[**Note:** Shipped as a liquefied compressed gas. Easily liquefied under pressure.]

Chemical & Physical Properties: **MW:** 17.0 **BP:** −28°F **Sol:** 34% **Fl.P:** NA (Gas) **IP:** 10.18 eV **RGasD:** 0.60 **VP:** 8.5 atm **FRZ:** −108°F **UEL:** 28% **LEL:** 15%	Personal Protection/Sanitation (see Table 2): **Skin:** Prevent skin contact **Eyes:** Prevent eye contact **Wash skin:** When contam (solution) **Remove:** When wet or contam (solution) **Change:** N.R. **Provide:** Eyewash (>10%) Quick drench (>10%)	Respirator Recommendations (see Tables 3 and 4): **NIOSH** **250 ppm:** CcrS*/Sa* **300 ppm:** Sa:Cf*/PaprS*/CcrFS/ GmFS/ScbaF/SaF **§:** ScbaF: Pd, Pp/SaF: Pd, Pp:AScba **Escape:** GmFS/ScbaE
[**Note:** Although NH_3 does not meet the DOT definition of a Flammable Gas (for labeling purposes), it should be treated as one.]		

Incompatibilities and Reactivities: Strong oxidizers, acids, halogens, salts of silver & zinc
[**Note:** Corrosive to copper & galvanized surfaces.]

Exposure Routes, Symptoms, Target Organs (see Table 5): **ER:** Inh, Ing (solution), Con (solution/liquid) **SY:** Irrit eyes, nose, throat; dysp, wheez, chest pain; pulm edema; pink frothy sputum; skin burns, vesic; liquid: frostbite **TO**: Eyes, skin, resp sys	First Aid (see Tabel 6): **Eye:** Irr immed (solution/liquid) **Skin:** Water flush immed (solution/liquid) **Breath**: Resp support **Swallow:** Medical attention immed (solution)

FIGURE 4-2 Anhydrous ammonia and chlorine entries in the NIOSH guide. *Courtesy of the National Institute for Occupational Safety and Health.*

Chlorine	Formula: Cl$_2$	CAS#: 7782-50-5	RTECS#: FO2100000	IDLH: 10 ppm
Conversion: 1 ppm=2.90 mg/m^3	DOT: 1017 124			

Synonyms/Trade Names: Molecular chlorine

Exposure Limits: **NIOSH REL:** C 0.5 ppm (1.45 mg/m^3) [15-minute] **OSHA PEL†:** C 1 ppm (3 mg/m^3)	**Measurement Methods** **(see Table 1):** **NIOSH** 6011 **OSHA** ID101, ID126SGX
Physical Description: Greenish-yellow gas with a pungent, irritating odor. [**Note:** Shipped as a liquefied compressed gas.]	

Chemical & Physical Properties:	**Personal Protection/Sanitation**	**Respirator Recommendations**
MW: 70.9	**(see Table 2):**	**(see Tables 3 and 4):**
BP: −29°F	**Skin:** Frostbite	**NIOSH**
Sol: 0.7%	**Eyes:** Frostbite	**5 ppm:** CcrS*/Sa*
Fl.P: NA	**Wash skin:** N.R.	**10 ppm:** Sa:Cf*/PaprS*/CcrFS/GmFS/
IP: 11.48 eV	**Remove:** N.R.	ScbaF/SaF
RGasD: 2.47	**Change:** N.R.	**§:** ScbaF:Pd, Pp/SaF:Pd,Pp:AScba
VP: 6.8 atm	**Provide:** Frostbite wash	**Escape:** GmFS/ScbaE
FRZ: −150°F		
UEL: NA		
LEL: NA		
Nonflammable Gas, but a strong oxidizer.		

Incompatibilities and Reactivities: Reacts explosively or forms explosive compounds with many common substances such as acetylene, ether, turpentine, ammonia, fuel gas, hydrogen & finely divided metals.

Exposure Routes, Symptoms, Target Organs (see Table 5):	**First Aid (see Table 6):**
ER: Inh, Con **SY:** Burning of eyes, nose, mouth; lac, rhin; cough, choking, subs pain; nau, vomit; head, dizz; syncope; pulm edema; pneu; hypox; derm; liquid: frostbite **To:** Eyes, skin, resp sys	**Eye:** Frostbite **Skin:** Frostbite **Breath:** Resp support

FIGURE 4-2 Continued

facilities that use hazardous materials are required to keep MSDS sheets for the hazardous materials that they use. This is required by both the hazard communication standard for plant workers (29 CFR 1910.1200) and the Emergency Planning and Community Right-to-Know Act for the public and emergency responders (EPCRA). MSDS sheets contain the following information, which may be of use to operations level responders:

- Manufacturer and emergency contact information
- Product identification
- Components and hazardous ingredients
- Primary hazard identification
- First aid measures
- Firefighting procedures
- Accidental release procedures
- Handling and storage procedures
- Exposure control and personal protection
- Chemical and physical properties
- Stability and reactivity
- Toxicological information
- Ecological information
- Disposal considerations
- Transport information
- Regulatory information

molecular weight ■ The sum of the atomic weights of all the atoms in a molecule.

Using the NIOSH guide entry for ethanol, answer the following questions for the spilled product in the opening scenario:

What is the **molecular weight** of ethanol?
What is the solubility of ethanol?
What is the ionization potential of ethanol?
What is the specific gravity of ethanol?
What is the vapor density of ethanol?
What is the freezing point of ethanol?
What is the IDLH of ethanol?
What is the PEL of ethanol?

Ethyl alcohol	Formula: CH_3CH_2OH	CAS#: 64-17-5	RTECS#: KQ6300000	IDLH: 3300 ppm[10%LEL]
Conversion: 1 ppm = 1.89 mg/m^3	DOT: 1170 127			

Synonyms/Trade Names: Alcohol, Cologne spirit, Ethanol, EtOH, Grain alcohol

Exposure Limits: **NIOSH REL:** TWA 1000 ppm (1900 mg/m^3) **OSHA PEL:** TWA 1000 ppm (1900 mg/m^3)	**Measurement Methods (see Table 1):** **NIOSH** 1400 **OSHA** 100

Physical Description: Clear, colorless liquid with a weak, ethereal, vinous odor.

Chemical & Physical Properties: **MW:**46.1 **BP:**173°F **Sol:**Miscible **Fl.P:**55°F **IP:**10.47 eV **Sp.Gr:**0.79 **VP:**44 mmHg **FRZ:**-173°F **UEL:**19% **LEL:**3.3% Class IB Flammable Liquid	**Personal Protection/Sanitation (see Table 2):** **Skin:**Prevent skin contact **Eyes:**Prevent eye contact **Wash skin:**When contam **Remove:**When wet (flamm) **Change:**N.R.	**Respirator Recommendations (see Tables 3 and 4):** **NIOSH/OSHA** **3300 ppm:**Sa/ScbaF **§:**ScbaF:Pd,Pp/SaF:Pd,Pp:AScba **Escape:**ScabE

Incompatibilities and Reactivities: Strong oxidizers, potassium dioxide, bromine pentafluoride, acetyl bromide, acetyl chloride, platinum, sodium

Exposure Routes, Symptoms, Target Organs (see Table 5): **ER:**Inh, Ing, Con **SY:** Irrit eyes, skin, nose; head, drow, lass, narco; cough; liver damage; anemia; repro, terato effects **TO:** Eyes, skin, resp sys, CNS, liver, blood, repro sys First Aid (see Table 6):	**First Aid (see Table 6):** **Eye:** Irr immed **Skin:** Water flush prompt **Breath:** Fresh air **Swallow:** Medical attention immed

Courtesy of the National Institute for Occupational Safety and Health

Solution: Chemical and physical properties are found in the leftmost column about halfway down the entry. The molecular weight of ethanol is 46.1 amu and can be found under the abbreviation "MW."

The solubility of ethanol is miscible and can be found under the abbreviation "Sol."

The ionization potential of ethanol is 10.47 eV and can be found under the abbreviation "IP."

liquid ■ A material with a melting point of 68°F (20°C) or lower at one atmosphere of pressure (101.3 kPa). Liquids take on the shape of their container and possess a definite volume.

The specific gravity of ethanol is 0.79 and can be found under the abbreviation "Sp.Gr."

The vapor density of ethanol is not listed because it is a **liquid** at room temperature. We will learn how to calculate this value from the molecular weight later in this chapter. For gases, the vapor density is listed under the abbreviation "RGasD."

The freezing point of ethanol is −173°F (−114°C) and can be found under the abbreviation "FRZ."

The IDLH of ethanol is 3300 ppm and can be found in the upper right corner of the entry. The [10% LEL] notation next to the IDLH value indicates that the IDLH is determined by flammability and not toxicity for ethanol.

The PEL of ethanol is 1000 ppm and can be found under the abbreviation "OSHA PEL."

Although an ANSI standard for MSDS sheets exists, a wide variety of formats are used. Figure 2-15 shows an example of an MSDS sheet for hydrogen peroxide. It is crucial to become familiar with the hazardous materials stored at the fixed site facilities in your response area. The facility preplans should include the appropriate MSDS sheets. In addition, if you are called to respond to one of these facilities, the MSDS sheets should be on location a safe distance away from where the hazardous material is stored or used. MSDS sheets can also be obtained directly from the manufacturer, from CHEMTREC, or from online databases in the event of an emergency. As global harmonization with international standards progresses, MSDS will eventually be called Safety Data Sheets (SDS) here in the United States as well, as they are already called internationally. Over the next several years the term SDS will become more prevalent. SDS contain essentially the same information as MSDS.

WISER ELECTRONIC DATABASE

Electronic databases are becoming increasingly used in this day and age. WISER, which stands for Wireless Information System for Emergency Responders, is an electronic database developed by the National Library of Medicine (NLM) that contains detailed chemical and physical property information, spill and fire procedures, personal protective equipment suggestions, and health information for more than 400 hazardous materials. Its detailed medical information cannot easily be found elsewhere and therefore EMS providers find this resource especially useful. The free electronic database is available for use online (WebWISER) and for download at www.wiser.nlm.nih.gov.

The database is intuitive and easy to use:

Quick Start
1. Open the program (double-click the WISER icon, or go into START, All Programs).
2. Select the appropriate user profile from the drop-down list on the toolbar. The values are 1st Responder, Hazmat Specialist, and EMS.
3. The application takes two paths, depending on whether the substance is known or not, as described next.
4. Help is available for each of the windows of the application via the Help menu.

Known Substance
1. Select the substance by scrolling through the list or entering the substance name or number. Use the "Search by" pull-down menu to indicate whether you are entering a name or one of the available identification numbers (UN/NA, CAS, or STCC).
2. Once the desired substance is located, double-click on it to go to the Data window. By default, the Data window displays Key Info, a very brief summary of the most critical information about the substance.
3. Select the menu item or the button drop-down on the Data window to select more information about the substance. The top section of the pop-up menu represents quick, convenient links to the most pertinent information for the current user profile.
4. The bottom section of the menu provides access to all information available for the substance. Select the desired category, and a submenu pops up presenting each of the data elements available in that category. Select one of the data elements, and the Data window updates to present the selected data.
5. Close the Data window, or select the Home icon to return to the main window.

Unknown Substance
1. Press the "Help Identify . . ." button on the main window to go to the Identify Substance (search) window. You will be brought to that window's Start tab.
2. If you can observe the substance's physical properties, select the "Properties" tab. If you know symptoms of patients affected by the substance, select the "Symptoms" tab. If you know both, select either one first.

3. On the Properties tab, press one of the property category buttons, and select the appropriate value from the list.

4. On the Symptoms tab, click the body graphic in the affected areas: eyes, ears, nose, mouth/throat, neurological (brain), respiratory, cardiovascular, gastro/urinary (stomach or kidney), skin (arm), or body temperature (thermometer). Then select the appropriate value from the list.

5. To unselect a property or symptom, click on the property or symptom from the selected list, and uncheck the box. Alternately, click on the category the property or symptom is from, and uncheck the box.

6. As properties or symptoms are added, the progress bar at the bottom indicates the decreasing number of possible matching substances.

7. To see matching substances at any point, select the "Results" tab.

8. On the Results tab, select a substance by double-clicking the name, or single-click the substance name in the list and press the Details button. You are taken to the Data window for that substance. Close the Data window, or select the back arrow on the toolbar to return to the Results tab.

9. Use the "Group by" menu to group the results by any of the symptom and property categories, or use the "Sort by" pull-down menu to order them by name or one of the identification numbers.

10. To remove a substance from the list, right-click the substance and a context menu appears. Select "remove." Alternately, you can single-click the substance in the list and press the "Remove" button.

11. To view removed substances, check the Show Removed box. Removed substances can be reinserted similar to how they were removed.

12. To start a new search, select the "New Search" button on the Start tab.

13. To recall previous searches, press the "Previous" button on the Start tab. Double-click one of the previous searches to replace the current search with the previous search, or single-click the search in the list and press the "OK" button.

CONTACTING CHEMTREC (UNITED STATES) OR CANUTEC (CANADA)

Often shipping papers and MSDS sheets will list CHEMTREC as the emergency response contact number. CHEMTREC is a free service to emergency responders and can be reached by calling 1-800-424-9300. The CHEMTREC call center is staffed 24 hours a day, 7 days a week by trained personnel. Often smaller companies and shippers will use CHEMTREC as a more economical alternative to providing their own technical experts around the clock as required by the U.S. DOT. CHEMTREC has the MSDS sheets and contact numbers for technical experts at its disposal. When you call the CHEMTREC operator, he or she will expect you to have gathered certain information:

- Your name and title
- Your company or organization
- Your location
- A callback number
- Dispatch center number if available or appropriate
- Fax number
- Location of the incident
- Current and predicted weather conditions
- Time the incident occurred (estimates are OK)
- Shipping papers
- UN number or STCC number of product
- Trade name of product
- Carrier, shipper, and point of origin

- Consignee and destination
- Container or package information
- Brief description of incident and actions taken
- Number and type of injuries and exposures
- Amount of product involved and released
- Are industry representatives on the scene?
- What information do you need ASAP?

Although having all of this information is not absolutely necessary, having as much of it as possible will expedite the process.

CONTACTING THE MANUFACTURER AND SPEAKING WITH TECHNICAL EXPERTS

If the number listed is not CHEMTREC (1-800-424-9300), you can call a listed number and speak directly with a manufacturer's representative. Often this number is staffed by a call center that will transfer you to a technical expert who may be a chemist, chemical engineer, plant manager, or other qualified personnel. Technical experts will also ask for much of the same information that CHEMTREC requires. It is always a good practice to be well prepared when calling technical experts because they will need to know details about the hazardous materials release in order to help you.

Chemical and Physical Properties

The chemical and physical properties of a material are the road map to its behavior. The many different chemical and physical properties should be studied in conjunction with each other because these individual properties do not operate in a vacuum. This section details what these properties indicate about the behavior of the hazardous material in the real world. What is going to happen if the product contacts air? if the product contacts water? if the product gets out of its container? By understanding the chemical and physical properties, and knowing where to find this information for a particular hazardous material, it is possible to make an informed decision and favorably affect the outcome of the incident.

STATES OF MATTER

Hazardous materials are commonly found in one of three states of matter: solid, liquid, or gas (Figure 4-3). **Solids** are the least mobile form of matter, tend to stay in one location, and retain their shape and size. Solids can melt and are potentially volatile (through **sublimation**). Liquids take the shape of the container they are in, or once they are released tend to flow in two dimensions and follow gravity downhill. Liquids can freeze and are often volatile (through **evaporation**). The **vapors** of liquids behave like gases. Gases are the most mobile of the three states of matter and disperse readily in three dimensions. Gases may be compressed or condensed into liquids with temperature and/or pressure changes. Depending upon the scenario being faced, the product's state of matter can be either an advantage or a disadvantage. For example, if we need to clean up a hazardous material and get it back in its container, it may be an advantage if it is a solid because it can be shoveled readily. On the other hand, if it is an extremely dangerous material, and there is a strong wind blowing, it may be an advantage if it is a gas. The wind can disperse the gas and dilute it to nontoxic levels. Knowing the state of matter a product is in is crucial to making sound decisions at the scene of a release.

Transitions between states of matter occur through temperature and pressure changes. The transition from a liquid to a solid is called the **freezing point**. The transition from a solid to a liquid is called the **melting point**. The temperature of the freezing point and the melting point are identical; the only difference is whether the temperature

solid ■ A material with a melting point above 68°F (20°C) at one atmosphere of pressure (101.3 kPa); a material that has a defined shape and volume.

sublimation ■ The phase transition of a solid directly to a vapor without first passing through the liquid state.

evaporation ■ The process by which molecules in a liquid escape into the vapor state; the process of a liquid becoming a vapor.

vapor ■ The gaseous form of a material that is primarily a solid or liquid at 68°F (20°C) and at one atmosphere of pressure (101.3 kPa).

freezing point ■ The temperature at which a liquid turns into a solid at atmospheric pressure.

melting point ■ The temperature at which a solid becomes a liquid.

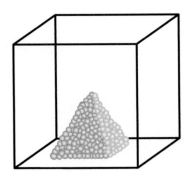 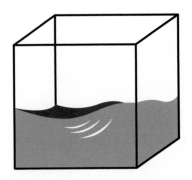 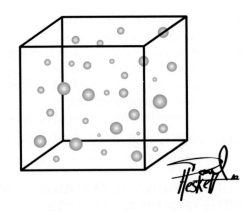

FIGURE 4-3 The three states of matter. *Art by David Heskett*

boiling point ■ The temperature at which the vapor pressure of a substance equals the atmospheric pressure. The temperature at which a liquid evaporates at its most rapid rate.

vapor pressure ■ The pressure exerted on a container by a solid or liquid in equilibrium with its vapors; a measure of a material's tendency to evaporate.

condensation point ■ The temperature at which a gas or vapor turns into a liquid at its most rapid rate.

is increasing or decreasing as the material passes through the transition point. The transition from a liquid to a gas is called the **boiling point**. The boiling point is the temperature at which a liquid transitions to a gas at its most rapid rate. Keep in mind that at almost any temperature liquids transition to vapor through evaporation. The evaporation rate is based on the **vapor pressure** of the liquid, discussed in more detail later. The transition from a gas to a liquid is called the **condensation point**. Once again, the temperature of the boiling point and condensation point are identical, and the only difference is in which direction a material passes through the transition point.

GAS AND VAPOR BEHAVIOR

How will a hazardous material behave if it is released from its container and contacts the air (Figure 4-4)? It is important to know this information because the material may react violently with the air and explode. Or, we may want to know how quickly vapors are

FIGURE 4-4 A vapor cloud generated by the release of liquid nitrogen from a fixed storage tank. The visible white cloud is a mixture of nitrogen gas and water vapor that has been condensed into a fog due to the extremely low temperature of the cryogenic liquid. Nitrogen gas is colorless. *Courtesy of Brian Wagner, National Institutes of Health Fire Department*

generated by a flammable liquid to determine the fire hazard. Or, we may want to know how much of a toxic material the victim may have inhaled.

The first piece of information we would like to know is whether the material reacts with the air, whether it is a pyrophoric material. A pyrophoric material is capable of igniting spontaneously when exposed to air. These types of materials are extremely dangerous and can release a lot of energy very quickly.

Vapor Pressure

Next, we would like to know how quickly the material evaporates. The evaporation rate of a material is defined by its vapor pressure. Vapor pressure is typically measured in millimeters of mercury (mmHg) or atmospheres (atm). The higher the vapor pressure, the more rapidly the material evaporates when it gets out of its container (Figure 4-5). For

FIGURE 4-5 The effects of vapor pressure. *Art by David Heskett*

TABLE 4-1	Vapor Pressures of Materials Listed in Order of Increasing Volatility
CHEMICAL	**VAPOR PRESSURE**
Sodium cyanide	0 mmHg (approximate)
Methylene bisphenyl isocyanate (MDI)	0.000005 mmHg (at 77°F)
VX	0.0001 mmHg
Sulfuric acid	0.001 mmHg
Sarin	2 mmHg
Kerosene	5 mmHg (at 100°F)
Water	18–25 mmHg
Ethanol	44 mmHg
Methyl ethyl ketone (MEK)	78 mmHg
Acetone	180 mmHg
Gasoline	38 to 300 mmHg (depending upon blend)
Ethyl ether	440 mmHg
Hydrogen fluoride	783 mmHg
Chlorine	6.8 atm
Ammonia	8.5 atm

example, sulfuric acid has a vapor pressure of 0.001 mmHg, whereas hydrofluoric acid has a vapor pressure of 783 mmHg at room temperature. But what do these numbers really mean? It is important to be able to put these numbers into context. The easiest way to understand the significance of vapor pressure is to remember the vapor pressure of water and acetone, because most of us have a good idea of how quickly these materials evaporate at room temperature. Water has a vapor pressure of roughly 20 mmHg, whereas acetone has a vapor pressure of 180 mmHg at room temperature. Acetone is a fairly volatile solvent found in nail polish remover and paint thinner. By examining these two materials and their relative vapor pressures, we quickly realize that lower numbers mean less evaporation and higher numbers mean greater evaporation. So, looking at our first example of sulfuric acid and hydrofluoric acid, we see that sulfuric acid evaporates very slowly. Such slowly evaporating materials are known as being **persistent**. On the other hand, hydrofluoric acid evaporates readily and is **nonpersistent**.

The boiling point of the material is closely related to the vapor pressure. In fact, they are inversely related; as the boiling point increases, the vapor pressure decreases. Thus, if a material has a high boiling point, it evaporates slowly; and vice versa, if it has a low boiling point, it evaporates quickly.

When a solid material evaporates, it is known as sublimation. Sublimation occurs when a solid becomes a gas directly without first becoming a liquid. Examples of materials that sublimate readily include water (in the form of ice and snow), carbon dioxide (dry ice), naphthalene (mothballs), and iodine.

Vapor Density

Once we know how quickly the material evaporates, the next thing we would like to know is where it is going to go. Whether a gas or vapor rises or sinks is indicated by its **vapor density** (Figure 4-6). Vapor density compares the weight of a volume of gas

persistent ▪ Not volatile; not prone to evaporate.

nonpersistent ▪ Volatile; prone to evaporate.

vapor density ▪ The mass of a gas or vapor compared with the mass of an equal volume of air.

FIGURE 4-6 Vapor density. The top drawing illustrates the movement of a gas with a vapor density less than one, while the bottom drawing illustrates a gas or vapor with a vapor density greater than one. *Art by David Heskett.*

TABLE 4-2	Vapor Density of Selected Chemicals
CHEMICAL	**VAPOR DENSITY (RGASD)**
Hydrogen	0.07
Helium	0.14
Methane	0.55
Ammonia	0.59
Hydrogen cyanide	0.93
Air	1
Ethanol	1.59
Chlorine	2.47
Gasoline	3.8 (approximate)
Sulfur hexafluoride	5.11
Tetrachloroethylene (PERC)	5.7
Mercury	6.9

or vapor to an equal volume of air. The vapor density of air is therefore set at 1.0. The NIOSH guide refers to vapor density as **RGasD,** which stands for **relative gas density.** Note that the vapor density is a unitless number because it involves comparing the density of a gas or vapor to the density of air. If the vapor density of the material is greater than 1, such as 2.47 for chlorine, the material will eventually sink in air. Conversely, if the vapor density of the material is less than 1, such as 0.6 for ammonia, it will eventually rise. Why did I say "eventually" rise? Ammonia escaping from a liquefied compressed gas cylinder will initially fall due to its extremely cold temperature. The cold ammonia gas is actually denser (heavier) than the much warmer ambient air surrounding it. This is the same principle that allows hot air balloons to float. Heated air is significantly less dense than cold air, which allows the balloon and its loaded basket to rise.

relative gas density (RGasD) ■ The term used in the NIOSH guide for the vapor density of gases.

Most gases and vapors are heavier than air. Two mnemonics can be used to remember the lighter-than-air gases. The first is HAHA MICE, which is incomplete but lists all of the significantly lighter-than-air gases:

H—Hydrogen (vapor density = 0.07, the lightest gas)
A—Ammonia (0.59)
H—Helium (0.14)
A—Acetylene (0.91)

M—Methane (0.55)
I—Illuminating gases (includes natural gas [0.6] and neon [0.7])
C—Carbon monoxide (0.97)
E—Ethylene (0.97)

The more complete, although slightly misleading, mnemonic is 4H MEDIC ANNA:

H—Hydrogen (0.07)
H—Helium (0.14)
H—Hydrogen cyanide (0.93)
H—Hydrogen fluoride (0.69 – theoretical; **1.92 – actual due to strong aggregation of HF molecules**)

M—Methane (0.55)
E—Ethylene (0.97)

D—Diborane (0.96)

I— Illuminating gases (in this case natural gas [0.6], which is roughly 90% methane and 10% ethane)

C—Carbon monoxide (0.97)

A—Acetylene (0.91)

N—Neon (0.70)

N—Nitrogen (0.97)

A—Ammonia (0.59)

There are relatively few lighter-than-air gases because the density of gases and vapors is directly proportional to their molecular weight. The molecular weight of air is 29 Daltons, which is a relatively small number. In comparison, acetone, with a molecular weight of 58.1, is twice as heavy as air (58 divided by 29 equals a vapor density of 2). The molecular weight of a chemical can be found in most reference materials that list chemical and physical properties, such as the NIOSH guide. Table 4-3 lists the lighter-than-air gases and their principal characteristics.

Be aware that air currents and the wind can have a much greater role than vapor density in determining where gases and vapors will go. This is especially true for gases that are barely lighter than air such as nitrogen, carbon monoxide, ethylene, and the like. Remember, the atmosphere is actually a mixture of roughly 80% nitrogen and 20% air. All of the nitrogen does not rise and leave us with a 100% oxygen atmosphere on the ground just because its vapor density is slightly less than oxygen! However, in an undisturbed, sealed room, heavy gases such as chlorine will be found near the ground, and lighter gases such as ammonia will be found near the ceiling. Vapor density indicates where the safest areas are during a release, how the gas or vapor will spread, and where to sample during air-monitoring activities.

TABLE 4-3	Lighter-Than-Air-Gases		
CHEMICAL NAME	HAZARDS	VAPOR DENSITY/RGASD	CHEMICAL FORMULA (MW)
Acetylene	Flammable	0.91	C_2H_2 (26.0)
Ammonia	Flammable, corrosive, toxic	0.59	NH_3 (17.0)
Carbon monoxide	Flammable, toxic	0.97	CO (28.0)
Diborane	Flammable, toxic	0.96	B_2H_6 (27.7)
Ethene (ethylene)	Flammable	0.97	C_2H_4 (28.0)
Helium	Asphyxiant	0.14	He (4.0)
Hydrogen	Flammable	0.07	H_2 (2.0)
Hydrogen cyanide	Flammable, corrosive, toxic	0.93	HCN (27.0)
Hydrogen fluoride	Corrosive, toxic	0.69 (theoretical)	HF (20.0)
Methane	Flammable	0.55	CH_4 (16.0)
Natural gas (10% ethane, 90% methane)	Flammable	0.60	Mixture (17.4)
Neon	Asphyxiant	0.70	Ne (20.2)
Nitrogen	Asphyxiant	0.97	N_2 (28.0)

WATER BEHAVIOR

How will a hazardous material behave if it is released from its container and it contacts water (Figure 4-7)? We want to know this information because the material may react violently with water and explode. It is also important to know whether the material floats on the water or sinks to the bottom. Or maybe it mixes? This is what is meant by water behavior. In order to understand how a hazardous material will behave when it contacts water, three things are vital to know: (1) Will it react violently with water? (2) If not, will it dissolve in water? and (3) If it does not mix with water, will it float or sink? Let's see how to find out this information.

Water Reactivity

Some materials are extremely water reactive, such as sodium metal. Sodium metal is stored under oil to prevent contact with even the small amount of moisture in the air. It ignites and will explode on contact with water. A truck transporting sodium metal that has lost its load on a bridge over water presents a serious water reactive hazard. Thus it is important to consult reference materials and check for water reactivity! Chemical reactivity, including water reactivity, is listed in the NIOSH guide, on the MSDS sheet, in WISER, and in many other reference materials.

Water Solubility

After we are confident that the material is not water reactive, the next question to ask is whether the material is water **soluble**. Water solubility describes the extent to which a material will dissolve in water. For example, n-Butanol is 9% soluble in water. Be careful though, because this percentage is not what you are used to. Percent water solubility means that 9 mL of n-Butanol will dissolve in 100 mL water. A counterintuitive example is sodium hydroxide, which is 111% water soluble. This simply means that 111 g of sodium hydroxide will dissolve in 100 mL of water. So how do we convey that something is completely soluble in water in any and all proportions? The term for complete solubility is **miscible**. For example, ethanol is soluble in all proportions in water. That means I can dissolve a drop of ethanol in a bathtub full of water, or I can dissolve a drop of water in a bathtub full of ethanol, or any ratio of ethanol and water in between.

soluble ■ Capable of dissolving in a solvent such as water or alcohol to form a solution.

miscible ■ Completely soluble in all proportions.

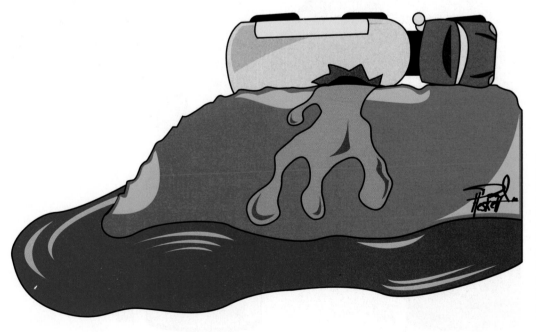

FIGURE 4-7 It is critical to understand how spilled liquids will behave in the presence of water. First determine whether the material is water reactive, find out what the water solubility of the material is, and finally determine the specific gravity of the material if it is insoluble.
Art by David Heskett

TABLE 4-4	Water Solubility of Selected Chemicals
CHEMICAL	**WATER SOLUBILITY**
Gasoline	Insoluble
Tetrachloroethylene (PERC)	0.02%
Chlorine	0.7%
n-Butanol	9%
Methyl ethyl ketone (MEK)	28%
Ammonia	34%
Hydrogen chloride	67% (at 86°F)
Sodium hydroxide	111%
Acetone	Miscible
Ethanol	Miscible
Sulfuric acid	Miscible

Specific Gravity

specific gravity ■ The mass of a given volume of material compared with the mass of an equal volume of water.

Specific gravity, akin to vapor density for gases, tells whether a liquid will float on top of the water or sink to the bottom (Figure 4-8). Specific gravity compares the weight of a volume of liquid to an equal volume of water. The specific gravity of water is 1.0. Thus, an insoluble liquid with a specific gravity of 0.8 will float on water, whereas an insoluble liquid with a specific gravity of 1.8 will sink in water. Note that specific gravity is also a unitless number because we are comparing the density of a liquid to the density of water.

Well, how does knowing all of this information affect a hazardous materials incident? Let's imagine that a railcar full of butanol has derailed and is leaking, and product is

FIGURE 4-8 Insoluble substances with a specific gravity of less than 1 float on water, whereas ones with a specific gravity greater than 1 sink. *Art by David Heskett*

TABLE 4-5	Specific Gravity of Selected Materials

CHEMICAL	SPECIFIC GRAVITY
Ammonia	0.62 (exists as a gas at ambient temperature and pressure)
Gasoline	0.72–0.76 (at 60°F)
Acetone	0.79
Ethanol	0.79
Methyl ethyl ketone (MEK)	0.81
n-Butanol	0.81
Water	1.0
12 M Hydrochloric acid	1.2
Chlorine	1.41 (at 6.86 atm) (exists as a gas at ambient temperature and pressure)
Tetrachloroethylene (PERC)	1.62
Sulfuric acid	1.85

flowing toward a river. One defensive option you might have is to go downstream, ahead of the release, and place a boom across the river (a technique discussed in Chapter 10). Butanol is only 9% soluble, so most of the material should not dissolve. Because the specific gravity of butanol is 0.81, the bulk of the material should float, and you will be able to capture it behind a boom. Right? Not so fast. What happens after the bulk of the butanol is trapped behind the boom, and the river continues to move underneath the material? Because the butanol is 9% soluble, the water will continue to pick up 9 g of butanol for every 100 mL of water that flows by. Therefore, if it takes the environmental company 2 hours to arrive and start pumping off the butanol, there will likely not be much left to pump out. On the other hand, if a completely insoluble material were trapped behind the boom, all of the material would still be there when the environmental company arrived.

FIRE BEHAVIOR

Flammability is one of the most serious dangers a hazardous material can pose. As an operations level first responder, you must quickly recognize and deal with these hazards. Two of the most important properties to consider are the flammable range and flashpoint.

Flammable Range

The **flammable range** is the vapor concentration, usually measured in percent, at which the product will burn. The flammable range is bracketed by the **lower explosive limit (LEL)** and the **upper explosive limit (UEL)** (Figure 4-9). Below the LEL there are not enough vapors to burn, and the mixture is too lean. Above

flammable range ■ The concentration of gas or vapor between the lower and upper explosive limits in ambient air. Also known as the explosive range.

lower explosive limit (LEL) ■ The concentration of a gas or vapor in air below which it will not ignite.

upper explosive limit (UEL) ■ The maximum concentration of a gas or vapor in the ambient air that will burn.

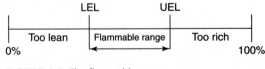

FIGURE 4-9 The flammable range.

SOLVED EXERCISE 4-2

How would the water solubility of ethanol affect the incident commander's decision-making process in the opening scenario?

   Solution: Ethanol is miscible in water, which means ethanol dissolves completely in water in all proportions. The ethanol will mix with any water it encounters, whether that is a river or groundwater, making it more difficult to clean up. Mixing the ethanol with water can reduce the flammability when the temperature is above the freezing point of water (32°F). The freezing point of ethanol is significantly below ambient temperature.

TABLE 4-6 | Flashpoint and Flammable Range of Common Materials

CHEMICAL	FLASHPOINT	FLAMMABLE RANGE (LEL–UEL)
Propane	N/A (gas)	2.1% to 9.5%
Acetylene	N/A (gas)	2.5% to 100%
Hydrogen	N/A (gas)	4% to 74%
Methylamine	N/A (gas)	4.9% to 20.7%
Methane	N/A (gas)	5% to 15%
Ammonia	N/A (gas)	15% to 28%
Ethyl ether	−49°F	1.9% to 36%
Gasoline	−45°F	1.4% to 7.6%
Acetone	0°F	2.5% to 12.8%
Hydrogen cyanide	0°F	5.6% to 40%
Methanol	52°F	6% to 36%
Ethanol	55°F	3.3% to 19%
Styrene	88°F	0.9% to 6.8%
Acetic acid	103°F	4.0% to 19.9% (at 200°F)
Kerosene	100–162°F	0.7% to 5%
Chlorine	N/A (nonflammable)	N/A (nonflammable, oxidizing gas)

N/A—not applicable

the UEL there is too much fuel and not enough oxygen to burn, and the mixture is too rich. Below the LEL and above the UEL, even if a flammable gas or vapor is present, and there is an ignition source, the mixture will not ignite.

Flashpoint

The flashpoint tells us a lot of information about the flammable material. It is determined by a combination of the lower explosive limit and the vapor pressure. The flashpoint is the temperature at which a liquid gives off enough vapors to ignite, but not necessarily to sustain combustion (Figure 4-10). Therefore, the flashpoint gives information about the evaporation rate (temperature at which a liquid gives off vapors) and the LEL (enough vapors to ignite).

chemical property ■ A characteristic of a substance involving a molecular change in that substance. Examples of chemical properties include flashpoint, LEL, and ionization potential.

Knowledge of the flashpoint can make or break a hazmat incident involving flammable materials. For example, an uninsulated tanker rolls over on the interstate and spills approximately 250 gallons (946 L) of methanol. This spill occurs next to a large apartment complex. The temperature is currently 25°F (-4°C). Do we need to evacuate the apartment complex due to the threat of fire? The first **chemical property** we should investigate is the flashpoint, which turns out to be 52°F (11°C). What does this mean? It

SOLVED EXERCISE 4-3

Based upon the flammability characteristics of ethanol in the opening scenario, the incident commander wants to evacuate the area for a half mile (0.8 km) and shut down the interstate in both directions. Are these actions appropriate?

Solution: No. The temperature of the ethanol is 16°F (-9°C), which is significantly below its flashpoint. Even in the presence of an ignition source, the ethanol vapors would not ignite because the pool of liquid ethanol is not producing sufficient amounts of ethanol vapor to reach the lower explosive limit (LEL).

FIGURE 4-10 The effects of flashpoint. *Art by David Heskett*

means the temperature of the methanol needs to be at least 52°F before there are enough vapors to ignite. Therefore, as long as the temperature of the methanol is below 52°F, an ignition source can be present and the vapors will not ignite.

Ignition Temperature

The ignition temperature, also known as the **autoignition temperature**, is the minimum temperature at which a material will initiate self-sustained combustion without an external ignition source. This means that when a material reaches this temperature it will start to burn and release flammable decomposition products. An example of this is a situation where oil-soaked rags burst into flames after being stuffed into a trash can. The oxidation of oils such as linseed oil generates heat. As the heat builds up and reaches the ignition temperature of the linseed oil (650°F/343°C), the oil ignites and catches the contents of the trash can on fire. This is often referred to as spontaneous combustion. Another example of a material reaching its autoignition temperature is the common kitchen fire caused by burning vegetable oil. When the temperature of the cooking oil in the pan reaches the ignition temperature of the vegetable oil, the oil spontaneously combusts and catches fire. Materials with low ignition temperatures, therefore, pose fire hazards. When these materials are reactive, such as linseed oil, which oxidizes in air, the hazard is significantly increased.

autoignition temperature ■ The minimum temperature a substance must possess in order for combustion to occur without an external ignition source. Also known as ignition temperature.

Boiling Liquid Expanding Vapor Explosions (BLEVEs)

Boiling liquid expanding vapor explosions (BLEVEs) occur when a flammable liquid inside a tank or vessel is heated above its boiling point and the tank catastrophically fails due to a structural defect or because it has been pressurized beyond its limitations (Figure 3-31). Historically many people, including first responders, have been killed by BLEVEs caused by brittle metal used in the construction of tanks and the complete lack of pressure-relief devices or inadequate pressure-relief devices. Even now, BLEVEs occur at regular intervals when the pressure inside tanks increases faster than the pressure-relief devices can dissipate that pressure. This pressure increase is typically caused by fires

Based upon the NIOSH guide entry of ethyl alcohol, would it be safe for fire personnel to extricate the critically injured accident victim in the opening scenario using structural firefighters' protective clothing?

Solution: Yes. Ethanol is not particularly toxic, and flammability is of minimal concern at 16°F (−9°C).

underneath partially empty or mostly empty tanks carrying flammable liquids or liquefied compressed gases such as propane.

When a tank catastrophically fails in this manner, the liquid contents inside the tank are instantaneously converted to a gas or vapor and are ignited almost immediately with an ensuing fireball. The catastrophic failure of the tank sends deadly shrapnel up to a mile and a half away (especially in the case of railcars). The fireball can engulf everything that is nearby. Box 4-1 describes one such BLEVE that occurred at Kingman, Arizona.

CHEMICAL REACTIVITY

polymerization ▪ The generally exothermic chemical reaction in which monomer molecules react to form larger molecules (polymers).

catalyst ▪ A substance that increases the rate of a chemical reaction without itself being consumed in the reaction.

Chemical reactivity refers to the ability of substances to interact and combine with one another, forming one or more different substances. This is called a chemical reaction. Some examples of chemical reactions are combustion and **polymerization**. The rate of the chemical reaction depends on the chemical and physical properties of the materials involved; the temperature, amount, and concentration of the materials; the pressure of the materials; and the presence of any catalysts. **Catalysts** increase the rate of a chemical reaction without being consumed in the reaction. An example of a catalyst is the platinum metal that helps remove pollutants in a car's catalytic converter.

Combustion is a very rapid oxidation-reduction reaction. The most familiar oxidation-reduction reactions are fire and rusting. Fire is an extremely rapid rate of oxidation that releases a lot of energy quickly in the form of heat and light. Rusting is a very slow rate of oxidation that produces neither a great amount of heat nor light. Box 4-2 describes some of the toxic products of combustion that firefighters face every day in their jobs. These toxic products of combustion are hazardous materials from which firefighters must protect themselves using respiratory protection such as a self-contained breathing apparatus (SCBA).

BOX 4.1 HAZMAT HISTORY: LPG TANK CAR EXPLOSION IN KINGMAN, ARIZONA

On July 5, 1973, 11 firefighters were killed in Kingman, Arizona, due to the catastrophic failure of a tank containing liquefied petroleum gas (LPG). A railroad tank car carrying the LPG was being unloaded when it caught fire. The fire department was called and responded. Its initial actions were to attempt to extinguish the fire in close proximity using hand lines and master streams. The tank unexpectedly, but not unpredictably, failed due to a catastrophic boiling liquid expanding vapor explosion (BLEVE).

The heat of the impinging fire eventually heated the liquefied propane to a temperature at which the propane's vapor pressure exceeded the bursting pressure of the tank. Once the internal pressure of the heated propane exceeded the capabilities of the tank, it failed at its weakest point. Because the temperature of the still liquefied propane far exceeded the boiling point of the propane when the containment vessel failed, the LPG almost instantly turned to gas. It BLEVEd! The expansion ratio of propane is 1:270. This means that for every gallon of LPG in the tank, there were instantaneously 270 gallons of flammable vapor after the BLEVE. The resulting explosion is what killed the 11 firefighters.

BOX 4.2 TOXIC PRODUCTS OF COMBUSTION

Courtesy of Jeff French, Superior Township (MI) Fire Department

TOXIC PRODUCT OF COMBUSTION	CHEMICAL FORMULA	PHYSIOLOGICAL EFFECT
Acrolein	$CH_2{=}CHCHO$	Respiratory irritant
Ammonia	NH_3	Respiratory tract and lung damage, including pulmonary edema, due to formation of caustic ammonium hydroxide (corrosive)
Carbon monoxide	CO	Chemical asphyxiation by preventing oxygen absorption into the blood
Hydrogen chloride	HCl	Respiratory tract and lung damage, including pulmonary edema (corrosive)
Hydrogen cyanide	HCN	Chemical asphyxiation by preventing oxygen from being used by the tissue and cells
Hydrogen fluoride	HF	Respiratory tract and lung damage, including pulmonary edema (corrosive); fluoride poisoning
Hydrogen sulfide	H_2S	Chemical asphyxiation by preventing oxygen absorption into the blood
Isocyanates	$R\text{-}CN$	Respiratory tract and lung damage, including pulmonary edema. Sensitization leading to acute allergic reaction
Nitrogen oxides	NO, NO_2	Respiratory tract and lung damage, including pulmonary edema, due to formation of nitric acid (corrosive)

(continued)

TOXIC PRODUCT OF COMBUSTION	CHEMICAL FORMULA	PHYSIOLOGICAL EFFECT
Perfluoroisobutene	C_4F_8	Respiratory tract and lung damage, including pulmonary edema, due to formation of fluorophosgene (corrosive); fluoride poisoning
Phosgene	$COCl_2$	Respiratory tract and lung damage, including pulmonary edema (corrosive)
Sulfur dioxide	SO_2	Respiratory tract and lung damage, including pulmonary edema, due to the formation of sulfuric acid (corrosive)

Combustion is the rapid oxidation of fuel, what is commonly called burning or fire. Complete combustion of hydrocarbon fuels should produce only carbon dioxide and water. High-efficiency appliances, such as propane-driven forklifts, are examples of complete combustion. However, when fuels burn rapidly in ambient air, they usually do not burn efficiently or cleanly, and they produce toxic by-products of combustion. Incomplete combustion can occur for several reasons, the two most common of which are insufficient oxygen and complex fuels.

When hydrocarbon fuel burns with too little oxygen, toxic products such as the following are produced: carbon monoxide, partial breakdown products of the hydrocarbon such as benzene and formaldehyde, as well as large particulates or soot. Complex fuels have other atoms in them besides carbon and hydrogen, such as nitrogen, chlorine, fluorine, and sulfur among others. When fuel contains additional elements such as these, additional combustion by-products are produced such as hydrogen cyanide (nitrogen), hydrogen fluoride (fluorine), hydrogen chloride (chlorine), ammonia (nitrogen), oxides of nitrogen, oxides of sulfur, phosgene (chlorine), and many more. Together, these toxic products of combustion make up what is commonly referred to as smoke. The smoke from each fire is unique as it depends on the types of fuel that are burning and the amount of oxygen available to the fire.

---

exothermic ■ A chemical reaction that emits heat into its surroundings.

acid ■ A corrosive material that has a pH below 7. A substance that corrodes steel or destroys tissue at the site of contact. A compound that releases hydrogen ions when dissolved in water. Examples of acids include battery acid, muriatic acid (hydrogen chloride), nitric acid, and vinegar.

base ■ A substance which releases hydroxide ions (OH^-) in water. Bases form solutions that have a pH above 7. Also known as alkali, alkaline material, or caustic.

FIGURE 4-11 A valve corroded by nitric acid. Hazardous materials releases are often caused by using incompatible fittings, transfer hoses, or containers. This was a less-than-30-second exposure to concentrated nitric acid.

Polymerization is the formation of a larger molecule from two smaller ones. It typically occurs when a relatively homogenous material is able to react with itself to form a continuous chain of a longer polymer. The plastics industry is based upon the polymerization reaction. Polymerization is an **exothermic** reaction, which means it liberates heat during the reaction. Runaway polymerization is, therefore, extremely dangerous because large amounts of heat can be generated in an enclosed environment, which may cause a BLEVE. Railcars of styrene, which is the monomer used in the polymerization reaction that produces the common plastic polystyrene, have been involved in lethal BLEVEs during derailments.

As mentioned previously, chemical reactions can be accompanied by a release of energy, typically in the form of heat and sometimes by light, as in fire. When large amounts of chemicals are involved, or the chemicals release a large amount of energy, chemical reactivity can be extremely dangerous. Whenever there is a potential for large amounts of incompatible materials to come in contact with each other, a large isolation distance must be enforced.

CORROSIVES AND THE pH SCALE

Corrosives are materials that cause visible destruction of the skin or significant corrosion of steel (Figure 4-11). There are two types of corrosive materials: **acids** and **bases**. The difference between these two types of corrosives is their **pH**. The pH scale is a way to measure the corrosive properties of a material that contains water (Figure 4-12). A liquid that contains water is called an aqueous

pH

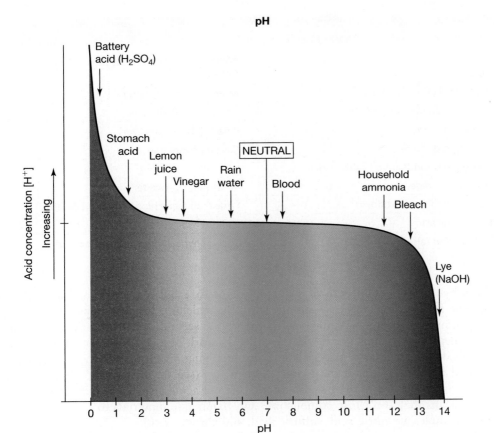

solution. A pH of 7 is considered neutral, a pH below 7 is considered acidic, and a pH above 7 is considered basic.

The pH scale is based upon water (H_2O) dissociating into a proton (H^+) and a hydroxide ion (^-OH):

$$H_2O \leftrightarrow H^+ + {}^-OH$$

pH is actually the concentration of protons in the solution. The more protons there are, the more acidic the solution is. The pH scale is logarithmic; this means every change of one unit of pH means a tenfold change in the acidity of the solution. Thus, a pH of 6 is tenfold more acidic than a pH of seven. Also, pH 1 is a millionfold more acidic than pH 7 because they are separated by six pH units. Figure 4-12 illustrates the logarithmic nature of the pH scale. The farther away from pH 7 a substance's pH is, the faster the corrosive properties of that liquid increase.

Although both high and low pH are corrosive, acids and bases have very different properties. Table 4-7 shows examples of both acids and bases. Acids are materials that contribute protons to a solution. For example, muriatic acid, or hydrogen chloride (HCl), almost completely dissociates into a proton and a chloride ion in aqueous solution. This is why muriatic acid is such a strong acid; it increases the concentration of protons. Bases, on the other hand, contribute hydroxide ions to solution, which causes the hydroxide to combine with protons in the solution and form water, thereby decreasing the proton concentration and increasing the pH. For example, sodium hydroxide (NaOH), or lye, almost completely dissociates into a hydroxide ion and a sodium ion in aqueous solution. This is why lye is such a strong base. Bases are also called alkalis, alkaline materials, or caustics.

Acid burns and caustic burns are not the same. Acid burns are painful at the site of contact and can usually be washed off relatively easily with water. The most dangerous

pH ■ The measure of corrosiveness. Low pH indicates an acid; high pH indicates a base; pH 7 is considered neutral.

TABLE 4-7 | Selected Acids and Bases

ACIDS	BASES (ALSO KNOWN AS ALKALIS OR CAUSTICS)
Acetic acid (also flammable)	Ammonia (also flammable)
Chromic acid (also an oxidizer)	Ammonium hydroxide
Hydrogen bromide	Bleach (also an oxidizer)
Hydrogen chloride	Calcium hydroxide
Hydrogen fluoride (also highly toxic)	Lye
Hydrogen iodide	Potassium hydroxide
Nitric acid (also an oxidizer)	Potassium hypochlorite (also an oxidizer)
Phosphoric acid	Sodium hydroxide
Sulfuric acid (also an oxidizer)	Sodium hypochlorite (also an oxidizer)

acid is hydrofluoric acid, which is not only corrosive but also toxic and skin absorptive. The fluoride ion has a strong attraction for calcium. Because bones have a lot of calcium, hydrofluoric acid is also called the "bone seeker." Hydrofluoric acid can also lead to heart dysrhythmias and was the cause of death of a law enforcement officer who responded to a hazardous materials incident involving a railcar carrying a product that liberated hydrofluoric acid on contact with water (see the Bridgman, Michigan, train derailment incident discussed in Chapter 1). Some acids also have oxidizing potential, which means they can cause combustible and flammable materials to ignite on contact. Some examples are nitric acid, sulfuric acid, and chromic acid.

Conversely, caustic burns may not initially be painful at the site of contact. Caustic burns feel slippery on the skin because the hydroxide ion actually dissolves the skin and creates a soap-like substance. **Saponification** is the process of making soap. Soap has traditionally been made from animal fat and lye. Remember, lye is a strong base, and human skin is a fatty tissue, which is why bases are more difficult to decontaminate from the skin. Bases actually react with skin and penetrate more deeply. As they penetrate, they often inactivate nerve endings as well, deadening the sense of pain. It is very important to thoroughly decontaminate all chemical exposures, especially caustic exposures, using copious amounts of water.

saponification ■ The reaction of triglycerides with alkalis to form soap.

RADIATION

Radiation is all around us, and most of it is harmless and even useful. For example, visible light is a form of nonionizing radiation and is obviously very useful, as is the radiation used in a microwave oven. Figure 4-13 shows the electromagnetic spectrum. When discussing radiation in the context of hazardous materials, it is about ionizing radiation, which contains a lot of energy and is very dangerous to humans because the energy is transferred to our tissues and causes extensive damage. There are several different types of ionizing radiation: alpha, beta, gamma, X-ray, and **neutron** radiation. Not all ionizing radiation

neutron ■ A particle located in the nucleus that has no charge. Neutrons that are violently ejected from the nucleus of radioisotopes are known as neutron radiation.

SOLVED EXERCISE 4-5

What are the pertinent chemical and physical characteristics of ethanol?

Solution: Based upon the DOT hazard class and the NIOSH guide description, ethanol is considered a flammable liquid. Flammability is, therefore, of significant concern. The pertinent chemical property to consult is the flashpoint, which is the temperature at which the ethanol emits enough vapors to ignite (reach the lower explosive limit, or LEL). The NIOSH guide lists the flashpoint of ethanol as 55°F (13°C).

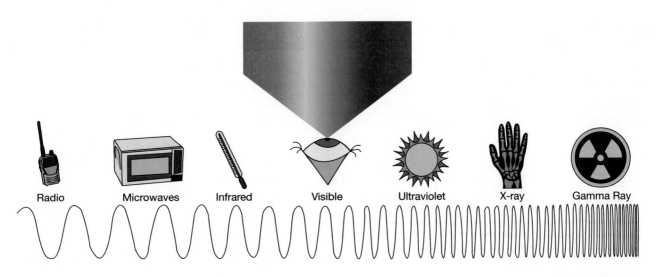

FIGURE 4-13 The electromagnetic spectrum. Ionizing radiation is only a small portion of the electromagnetic spectrum. *Art by David Heskett.*

is pure energy like visible light. Actually, only **gamma (γ) radiation** and X-rays are purely electromagnetic radiation. Table 4-8 lists the attributes of the different types of radiation.

Radioactive materials are also found all around us, in equipment we use as well as naturally occurring radiation in our bodies. Some common uses of radioactive materials are americium in smoke detectors, cesium used in radiography instruments and nuclear density gauges, X-ray generators in medicine, tritium used in gun scopes and emergency exits, cobalt used to sterilize equipment and food, and various radioactive isotopes used in cancer treatments (Table 4-9). Radioactive materials are also the by-products of industry. You are probably familiar with the nuclear waste generated by nuclear power plants, but the oil industry often generates radioactive sludge from drilling operations when naturally occurring radioactive isotopes are brought to the surface.

 gamma (γ) radiation Ionizing electromagnetic energy of very short wavelength and comparatively high energy that is emitted from certain radioisotopes.

TABLE 4-8	Types of Radiation		
RADIATION	**MASS**	**CHARGE**	**SHIELDING**
Alpha	4	+2	Paper
Beta	1/1600	-1	Aluminum or 3/8–inch Plexiglas
Gamma	None	0	Lead
Neutron	1	0	Concrete

TABLE 4-9	Table of Common Radioactive Isotopes		
ISOTOPE	**EMISSION**	**HALF-LIFE**	**COMMERCIAL USES**
Americium-241	Alpha, gamma	432.2 years	Smoke detectors, gauges
Arsenic-73	Gamma	80.3 days	Tracer
Bismuth-213	Alpha, beta, gamma	46 minutes	Cancer therapy
Cadmium-109	Gamma	1.3 years	Mineralogy
Californium-252	Alpha	2.6 years	Cancer therapy, nuclear reactors, density gauges
Carbon-11	Beta, gamma	20.4 minutes	PET imaging
Cesium-137	Beta, gamma	30.1 years	Gauges, tracer
Chromium-51	Gamma	28 days	Medical tracer
Cobalt-57	Gamma	271 days	Medical research
Cobalt-60	Beta, gamma	5.3 years	Cancer therapy, sterilization, radiography, density gauges
Copper-64	Gamma	13 hours	Medical research
Copper-67	Beta, gamma	62 hours	Cancer therapy, medical research
Dysprosium-165	Beta, gamma	2.4 hours	Medical treatment
Erbium-169	Beta	9.4 days	Medical treatment
Fluorine-18	Positron, gamma	1.8 hours	PET imaging
Gallium-67	Gamma	78 hours	Medical imaging
Germanium-68	Gamma	271 days	PET imaging
Gold-198	Beta, gamma	65 hours	Tracer
Holmium-166	Beta, gamma	26 hours	Cancer therapy
Indium-111	Gamma	67 hours	Medical research
Iodine-123	Gamma	13.3 hours	Medical research
Iodine-125	Gamma	60 days	Cancer therapy, medical research
Iodine-131	Beta, gamma	8 days	Cancer therapy
Iridium-192	Beta, gamma	74 days	Cancer therapy, density gauges
Iron-59	Beta, gamma	45 days	Medical diagnosis
Krypton-85	Beta	10.8 years	Density gauges
Lutetium-177	Beta, gamma	6.7 days	Medical treatment
Nickel-63	Beta	100 years	Power source, APD2000 detector
Nitrogen-13	Gamma, positron	10 minutes	PET imaging
Oxygen-15	Gamma, positron	2 minutes	PET imaging
Palladium-103	Beta, gamma	17 days	Cancer therapy
Phosphorus-32	Beta	14 days	Medical research, medical treatment
Potassium-42	Beta, gamma	12.4 hours	Medical diagnosis
Rhenium-186	Beta, gamma	3.8 days	Cancer therapy
Rhenium-188	Beta, gamma	17 hours	Medical treatment
Rubidium-81	Gamma	4.6 hours	Generator (Kr-81m)
Rubidium-82	Gamma, positron	75 seconds	PET imaging
Rubidium-87	Beta	48 billion years	Atomic clocks
Samarium-153	Beta, gamma	47 hours	Cancer therapy
Selenium-75	Gamma	120 days	Medical research, radiography

TABLE 4-9 Table of Common Radioactive Isotopes (*Continued*)

ISOTOPE	EMISSION	HALF-LIFE	COMMERCIAL USES
Silicon-32	Beta	132 years	Tracer
Sodium-22	Gamma	2.6 years	Medical research
Sodium-24	Beta, gamma	15 hours	Medical research
Strontium-89	Beta	50.5 days	Cancer therapy
Strontium-90	Beta	28.8 years	Density gauges
Technetium-99m	Gamma	6 hours	Medical imaging, tracer
Thallium-201	Gamma	73 hours	Medical treatment
Thallium-204	Beta	3.8 years	Density gauges
Tritium	Beta	12.32 years	Medical research, tracer
Tungsten-188	Beta	69.8 days	Generator (Re-188), medical research
Xenon-133	Beta, gamma	5.3 days	Medical research
Ytterbium-169	Gamma	32 days	Medical research
Yttrium-90	Beta	64 hours	Cancer therapy

Overexposure to radioactive materials can be dangerous. It is impossible to reduce your radiation exposure to zero. Therefore, in the workplace, radiation exposure is kept to "as low as reasonably achievable" (ALARA). **ALARA** is the mantra used in workplaces where radiation is routinely used, such as hospitals, research laboratories, nuclear reactors, and doctor and dentist offices that use X-ray machines. Box 4-3 describes a workplace exposure to radioactive radium used to paint watch dials before and during World War II.

Radiation is emitted from the nucleus of radioactive materials, or radionuclides, at a very high rate of speed, which means these particles have a lot of energy. For example, purely electromagnetic radiation travels at the speed of light, whereas heavier radiation, such as **alpha (α) particles**, travels at about 1/20th the speed of light. This is still extremely fast and energetic!

Alpha particles, the largest form of ionizing radiation, are composed of two protons and two neutrons, which is 4 atomic mass units (amu), and have a net charge of +2. This makes them very heavy and very charged. Because of their size and charge, alpha particles do not travel very far in air, only a few inches (Figure 4-14). This is because they tend to interact very rapidly with the molecules in air. Therefore, it is easy to protect ourselves

ALARA ■ Acronymn for "as low as reasonably achievable." This mnemonic is used to minimize exposure to radiation because it cannot be eliminated altogether.

alpha (α) particle ■ A fragment emitted from the nucleus of a radioisotope at a high rate of speed that contains two protons and two neutrons and has a charge of +2 and a mass of 4 amu (a helium nucleus). Alpha particles travel only a short distance in air (a few inches) and can be shielded using a piece of paper.

BOX 4.3 HAZMAT HISTORY: THE RADIUM GIRLS

In 1917 five women who had been working at the U.S. Radium Corporation in Orange, New Jersey, sued their employer after becoming severely ill with radium jaw disease and bone tumors. Radium jaw is a debilitating bone disease caused by the incorporation of radium into the bone. The women painted watch dials for use by fighter pilots in the military using glow-in-the-dark paint. This paint, aptly named "Undark," was radioluminescent due to the effect of a mixture of radioactive radium and zinc sulfide in the paint. The women would point the brushes using their lips to generate a fine tip. During this process, they repeatedly ingested small amounts of radium over long periods of time, leading to significant chronic exposure. The employer had previously told them the paint was harmless. This historic case established a worker's right to sue his or her employer after contracting an occupational illness.

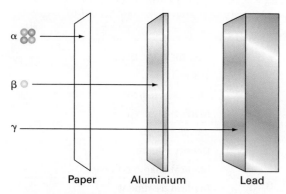

α

β

γ

Paper   Aluminium   Lead

FIGURE 4-14 The relative penetrating power of alpha, beta, and gamma radiation.

alpha (α) emitter ■ Any radioactive substance that emits an alpha particle.

beta (β) particle ■ A fragment emitted from the nucleus of a radioisotope that has a charge of −1 and the mass of an electron. Beta particles travel about 10 feet in air and can be shielded using Plexiglas or a thin sheet of metal.

beta (β) emitter ■ Any radioactive substance that emits a beta particle.

gamma (γ) emitter ■ Any radioactive substance that emits gamma radiation.

from alpha radiation sources. For example, we just need to stay back a foot or use a piece of paper for shielding. In fact, alpha particles will not penetrate a layer of dead skin, but they may damage the eyes. However, if radioactive materials that emit alpha radiation are ingested, inhaled, or otherwise enter the body, they are extremely dangerous. Once the **alpha (α) emitter** (the radioactive material or radionuclide) is inside the body, all of the alpha particles transfer their energy to the body and cause extensive damage.

Beta (β) particles are the smallest form of ionizing radiation, except for gamma and X-rays, which have no mass at all. They have a mass and charge identical to an electron's, but are emitted from the nucleus at nearly the speed of light. They have a mass of 1/1600th of an atomic mass unit and a net charge of −1. Although beta particles are relatively light, they are strongly charged. Due in large part to their charge, beta particles do not travel very far in air, only a few feet. Beta particles can be an external hazard to the skin and eyes, and may lead to surface burns similar to a sunburn, called a beta burn. Aluminum or plastic sheeting is required to effectively shield radioactive materials that emit beta radiation. **Beta emitters** are also an internal hazard if they are ingested or inhaled.

Gamma radiation and X-rays are purely electromagnetic radiation and have no mass or charge. They travel at light speed and can travel several hundred feet in the air. Gamma rays are so energetic that approximately 75% of gamma radiation passes through the body without doing any damage. However, the other 25% makes up for the rest! X-rays have almost identical properties to gamma rays; however, they are generated by different means. For our purposes, they are lumped in together with gamma radiation. Because gamma radiation travels hundreds of feet in the air, shielding must be made of dense material that is several inches thick (such as lead or steel). **Gamma (γ) emitters** are also an internal hazard if they are ingested or inhaled.

Neutron radiation is relatively massive at 1 amu; however, these particles have no charge (as their name implies). Because neutron radiation is uncharged, it can also travel hundreds of feet in the air, like gamma radiation. Neutron radiation requires massive amounts of water or concrete for shielding. Neutron radiation is typically encountered in nuclear processes, such as operating nuclear reactors and detonating nuclear weapons. Box 4-4 describes a Boy Scout who successfully tried to build a small breeder reactor near Detroit, Michigan. Neutron emitters are also an internal hazard if they are ingested or inhaled.

The most effective way to protect ourselves from radiation is time, distance, and shielding. The less time spent near a radioactive source, the less radiation we receive. Thus, minimizing time spent near a source is a very effective way of reducing radiation exposure. The more shielding placed between ourselves and the radioactive source, the less radiation can reach our bodies. However, shielding is often not a practical option with highly penetrating radiation such as gamma and neutron radiation due to the weight of the shielding. The most effective way to protect ourselves from radiation is to keep our distance. This is due to the inverse square law:

$$R_1 \times (D_1)^2 = R_2 \times (D_2)^2$$

This equation means that every time you double the distance you move away from the source, you decrease the amount of radiation received by fourfold (Figure 4-15). So if I move twice the distance away, let's say from 1 foot to 2 feet (30 cm to 60 cm) from the source, the radiation I receive reduces from 100 mrem to 25 mrem (1 mSv to 0.25 mSv). This is important to remember when establishing the isolation distance. If there are indications that a strong radioactive source is present, isolate the area the longest practical distance away.

Health Effects of Hazardous Materials

Hazardous materials may enter the body through four routes of entry: inhalation, skin and eye absorption, ingestion, and injection (see Chapter 2). Inhalation is the most effective route of entry. It is an involuntary act, and the lungs are designed to exchange gases (absorb oxygen and excrete carbon dioxide). Unfortunately, the lungs are not specific for these two gases and may also efficiently and effectively absorb many potentially toxic chemical vapors and gases. The second most common route of entry is skin and eye absorption because humans tend to touch materials with the hands if no hazard is perceived. This is another reason why early recognition of hazardous material releases is so important.

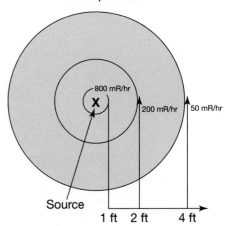

Inverse square law

FIGURE 4-15 The inverse square law and its effects. Doubling your distance away from a radioactive material reduces your exposure by fourfold.

BOX 4.4 HAZMAT HISTORY: THE RADIOACTIVE BOY SCOUT

In the mid-1990s, in a suburb of Detroit, Michigan, a teenager named David Hahn attempted to build a nuclear reactor in the shed of his mother's backyard. He is called the radioactive Boy Scout because he was trying to earn a merit badge to become an Eagle Scout.

The nuclear reactor he was attempting to build was a breeder reactor designed to generate large amounts of energy. He obtained the necessary starting materials through both legal and fraudulent means, including posing as a professor. He started out with radioisotopes such as americium-241 and thorium-232, and ended up with highly radioactive isotopes such as uranium-233. He finally disassembled his breeder reactor when he could detect gamma radiation standing 5 doors down from his mother's house. Now he had the problem of what to do with the highly radioactive remnants of his project. Less than 24 hours after dismantling the reactor, he was questioned by police for loitering in his car late at night in a residential neighborhood. When law enforcement officers searched the car, they found what appeared to be an improvised explosive device and a chemistry lab (chemicals, acids, fireworks, foil-wrapped cubes containing a gray powder, lantern mantles, mercury switches, clock faces, and a sealed toolbox). David warned the officers that some of the material was radioactive. The police impounded the vehicle and its contents and had it towed to the police station where it was inspected in personal protective equipment (PPE) by the Michigan State Police bomb squad and health physicists from the Michigan Department of Community Health.

David had been experimenting with chemistry since he was a boy, had made several explosives, and had at least one accident requiring a trip to the emergency room. His breeder reactor project came to a dramatic end on June 26, 1995, when contractors working for the Environmental Protection Agency (EPA) dismantled the shed in full PPE and disposed of it and its contents as radioactive waste. Although David Hahn did not appear to have a sinister intent, you can imagine how this process could be used by terrorists.

What should the law enforcement officers have initially done differently? This incident dramatically illustrates the importance of hazardous materials awareness and operations level training for law enforcement officers and public health department workers, as well as others that could be expected to encounter hazardous materials. (Ken Silverstein, *The Radioactive Boy Scout* [New York: Random House, 2004])

HAZMAT HANDLE

Reduce your exposure to radiation to as low as reasonably achievable (ALARA) by using time, distance, and shielding (TDS). And remember, increasing your distance from the radioactive material is the most effective way to limit exposure!

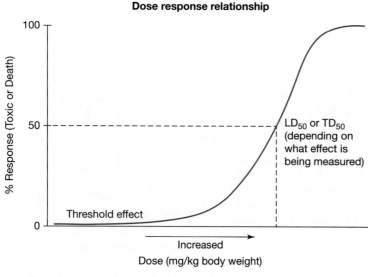

Dose response relationship

(y-axis) % Response (Toxic or Death) — 100, 50, 0

LD_{50} or TD_{50} (depending on what effect is being measured)

Threshold effect

Increased
Dose (mg/kg body weight)

FIGURE 4-16 A dose response curve.

Ingestion can be an effective route of entry because the stomach is designed to absorb nutrients through the gastrointestinal tract; however, eating and drinking are voluntary activities. Therefore as long as you do not eat, drink, or smoke at hazardous materials incidents (or carry contaminants home), you should be just fine regarding this route of entry. Injection is probably the least common route of entry, but if a hazardous material is injected directly into the body, it can enter the bloodstream and be distributed throughout the body very quickly. Victims who have been involved in an explosion or industrial accident may have puncture wounds from contaminated shrapnel or equipment.

Hazardous materials may have acute effects or chronic effects. Acute effects occur quickly upon exposure and may have short-term or long-term effects on the victim. Chronic effects occur over a longer period of time, and signs and symptoms of exposure develop slowly. Similarly, exposure to hazardous materials can be via acute exposures or chronic exposures. An acute exposure to a hazardous material occurs over a short time, whereas a chronic exposure occurs over the long term. For example, a responder to a hazardous materials incident unwittingly gets too close. He smells a pungent, sweet odor; immediately realizes he is too close; and backs away to a safe distance. That responder just received an acute exposure to that hazardous material. On the other hand, someone working in a factory that uses benzene as a solvent without adequate ventilation for 5 years has had a chronic exposure to benzene.

The amount, or dose, of a toxin or medicine that is received will determine the result on the human body. A dose response curve describes how a given material affects the body (Figure 4-16). The dose response curve can describe the dose-dependent healing effects of a medicine, or the dose-dependent damage caused by a hazardous material. Some chemicals have a threshold effect below which no ill effects are seen. Other toxins can act synergistically, or together, meaning the effect of the two chemicals together is worse than that of either chemical separately. Still other chemicals may be harmless by themselves yet cause damage when combined with other materials; this is called a potentiated effect, an example of which is acetaminophen and alcohol. The cumulative effects of some chemicals may not appear for many years after exposure, which is called a **latent effect**. As you can see, it is extremely important to avoid exposure to chemicals whenever possible.

The health effects of hazardous materials may linger if proper decontamination is not performed. An individual is contaminated if the hazardous material is on or in the body. If the material is on the outside of the body, it is called external contamination. If the material has entered the body—through either ingestion, inhalation, or injection—it is called internal contamination. Often, patients become both internally and externally

latent effect ▪ Disease or impairment that is manifested after an extended time period following exposure to the hazardous substance.

HAZMAT HANDLE

Avoid exposure to chemicals first using engineering controls, then good work practices, and finally personal protective equipment as a last line of defense.

contaminated. It is very difficult to remove internal contamination. Secondary contamination may occur if a contaminated person tracks the hazardous material outside the immediate area of release and others become contaminated.

Hazardous materials may harm individuals by several different mechanisms. Asphyxiants interfere with breathing, oxygen transport in the body, or oxygen uptake in the cells. Simple asphyxiants displace oxygen, whereas **chemical asphyxiants** interfere with oxygen transport or uptake in the body. Some materials are convulsants, which cause the body to have convulsions and seizures. Some materials are irritants, which can cause reversible skin, eye, or respiratory inflammation. Skin or eye irritation can be redness, swelling, or a rash at the sight of contact. Respiratory irritation can lead to coughing, sneezing, swelling of the respiratory passages, and difficulty breathing. Some materials are allergens, which lead to allergic reactions. Some materials are **sensitizers**. This means that after repeated exposure, or even a single exposure, they can lead to severe allergic reactions including anaphylactic shock.

Yet other materials are carcinogens, which can cause cancer. Carcinogens are an example of a latent effect, which may occur a significant time after exposure. This could be months, years, or even decades later. Some materials have target organ effects. Target organ effects are damage, and the resulting signs and symptoms, caused by hazardous materials affecting specific organs. Table 4-10 shows some of these target organ effects and examples of chemicals that cause them.

chemical asphyxiant ■ A substance that prevents the body from absorbing, transporting, or using oxygen properly, which usually results in incapacitation or death. Examples of chemical asphyxiants include hydrogen sulfide, carbon monoxide, and hydrogen cyanide.

sensitizer ■ A substance that after the initial exposure causes a more severe allergic reaction on subsequent exposures. Examples of sensitizers are isocyanates.

MEASUREMENT OF HEALTH EFFECTS

It is important to have a way to measure the potential health effects that hazardous materials may have after exposure to them. If someone is exposed to a hazardous material, at what level does the exposure become dangerous? Exposure limits must be understood in

TABLE 4-10	Target Organ Effects		
TARGET ORGAN EFFECT	ORGAN AFFECTED	SIGNS AND SYMPTOMS	SELECTED EXAMPLES OF CAUSATIVE AGENTS
Hepatotoxins	Liver damage	Jaundice, liver enlargement	Carbon tetrachloride, nitrosamines
Nephrotoxins	Kidney damage	Edema, proteinuria	Halogenated hydrocarbons, uranium
Neurotoxins	Nervous system damage	Varied (see CNS and PNS below)	Acrylamide, nerve agents (such as sarin and VX), many venoms
Central nervous system (CNS)	Brain injuries	Drooping of upper eyelids, slurred speech, respiratory difficulty, seizures, unconsciousness	Heavy metals (such as lead, mercury, and thallium)
Peripheral nervous system (PNS)	Nerve injuries	Numbness, tingling, decreased sensation, change in reflexes, decreased motor strength	Arsenic, lead, toluene, styrene
Hematopoietic toxins	Red blood cell damage	Cyanosis, loss of consciousness	Carbon monoxide, benzene
Pulmonary toxins	Lung damage	Cough, chest tightness, shortness of breath	Silica, asbestos, hydrochloric acid
Reproductive toxins	Mutations and teratogenesis	Birth defects, sterility	Lead, DBCP
Cutaneous hazards	Skin damage	Defatting of the skin, rashes, irritation	Ketones, chlorinated compounds
Eye hazards	Eye damage	Conjunctivitis, corneal damage	Organic solvents, acids

order to choose the appropriate engineering controls, decide on appropriate work practices, and be able to choose the appropriate level of PPE at hazardous materials incidents. So how can the health effects that hazardous materials may have be measured? Many different exposure limits have been set by government and industry. Two of the most important ones are the permissible exposure limit (PEL) and the immediately dangerous to life and health (IDLH) level.

Permissible Exposure Limit (PEL)

The permissible exposure limit (PEL) is set by the Occupational Safety and Health Administration (OSHA) and has the force of law. The PEL is the maximum airborne concentration of a chemical that a healthy worker can be exposed to for 8 hours a day, 40 hours a week, throughout his or her career without experiencing any permanent health problems. The PEL is an 8-hour time weighted average (TWA), which means that the PEL may be temporarily exceeded under certain conditions. Generally, if the airborne concentration of the chemical, as determined by air monitoring, is below the PEL, no respiratory protection is required. Above the PEL, respiratory protection is required.

Immediately Dangerous to Life and Health (IDLH)

The immediately dangerous to life and health (IDLH) level is set by the National Institute for Occupational Safety and Health (NIOSH). The IDLH is specifically mentioned in the confined space regulations and thus also carries the force of law. Any exposure above the IDLH requires supplied air respiratory protection. Exposures below the IDLH, but above the PEL, require some type of respiratory protection, either supplied air respiratory protection or an air-purifying respirator. As you can see, understanding the significance of exposure limits, especially the PEL and IDLH, is extremely important for the appropriate selection of personal protective equipment (Figure 4-17). This topic is discussed in greater detail in Chapter 6, Operations Level Responders Mission-Specific Competencies: Personal Protective Equipment.

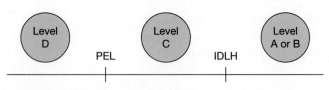

FIGURE 4-17 The PEL and IDLH should be used for the selection of personal protective equipment (PPE).

SOLVED EXERCISE 4-6

What are the toxic exposure level values of ethanol according to OSHA?

Solution: The PEL of ethanol is 1000 ppm and the IDLH is 3300 ppm. In this case, the IDLH is not based upon toxicity, but rather upon flammability because it is set at 10% of the LEL. The LEL of ethanol is 3.3%, which is 33,000 ppm. OSHA considers 10% of the LEL to be at IDLH levels in confined spaces. Therefore, the IDLH has been set at 3300 ppm.

Summary

Operations level responders should be able to gather information about the hazardous material or weapon of mass destruction that has been released. In addition, they must be able to interpret this information and make sound decisions to protect themselves and the public, and to prevent the incident from becoming worse. Understanding what chemical and physical properties indicate about a hazardous material is crucial. It is also vital to understand how the hazardous material will behave on contact with air and with water, as well as its chemical reactivity including flammability. This information can be gathered from such sources as the ERG, MSDS sheets, the NIOSH guide, WISER, CHEMTREC, and other technical experts. Anticipating how a hazardous material will act under the circumstances being faced can make the difference between a safe and successful mitigation and a disaster.

Review Questions

1. What physical property describes how quickly a substance evaporates?
2. What are the three properties you need to know about a hazardous material should it come into contact with water?
3. What is the definition of flashpoint? Will the vapors of a material burn if the temperature of the material is below its flashpoint?
4. What are the four primary types of ionizing radiation?
5. What information can be found in the NIOSH guide?
6. Why are technical experts useful at the scene of a hazardous materials release?

Problem-Solving Activities

1. A material has a flashpoint of 52°F (11°C), and the material is spilled from a cargo tanker onto a roadway on a chilly 14°F (−10°C) evening. Will the vapors of the substance ignite if an ignition source is present?
2. A liquid is flowing toward a pond. The NIOSH guide indicates it is water reactive, is water soluble, and has a specific gravity of 2.3. What will happen when it reaches the water? What is your primary concern? What actions should you take as an operations level responder?
3. A different liquid is flowing toward a river. The NIOSH guide indicates it is not water reactive, is water soluble, and has a specific gravity of 0.3. What will happen when it reaches the water? What is your primary concern? What actions should you take as an operations level responder?
4. A highway cargo tanker that has overturned is carrying fluorosulfonic acid. Which reference materials would you consult? What are the chemical and physical properties of fluorosulfonic acid? What are your primary concerns? What other questions do you need to ask to answer this question completely?
5. A 300-gallon (1135 L) tote is leaking in a chemical storage area. What questions do you need to ask the facility manager? What chemical and physical properties are you most concerned about and why?

References and Further Reading

Centers for Disease Control and Prevention. (2007). *NIOSH Pocket Guide to Chemical Hazards*. Washington, DC: U.S. Government Printing Office.

Meyer, Eugene. (2009). *Chemistry of Hazardous Materials* (5th ed.). Upper Saddle River, NJ: Pearson/Brady.

National Fire Protection Association. (2008). NFPA 472, *Standard for Competence of Responders to Hazardous Materials/Weapons of Mass Destruction Incidents*. Quincy, MA: Author.

Occupational Safety and Health Administration. (1990). 29 CFR 1910.120, *Hazardous Waste Site Operations and Emergency Response (HAZWOPER)*. Washington, DC: U.S. Department of Labor.

Silverstein, Ken. (2004). *The Radioactive Boy Scout*. New York: Random House.

U.S. Department of Transportation. (2008). *2008 Emergency Response Guidebook*. Washington, DC: Pipeline & Hazardous Materials Safety Administration.

Weber, Chris. (2007). *Pocket Reference for Hazardous Materials Response*. Upper Saddle River, NJ: Pearson/Brady.

OBJECTIVES

- Implement the incident command system (ICS) at a hazardous materials or WMD incident.
- Institute scene control measures at a hazardous materials or WMD incident.
- Formulate response objectives at a hazardous materials or WMD incident.
- Describe the difference among life safety, incident stabilization, and property and environmental conservation objectives.
- Describe the advantages and disadvantages of personal protective equipment (PPE) available for hazardous materials response.

- Describe the advantages and disadvantages of available respiratory protection for hazardous materials response.
- List the essential components of a safety briefing.
- Describe the key features of a decontamination line.

You Are on Duty! Crashed Highway Cargo Tanker (Part 3)

Paw Paw, Michigan

On the evening of January 28, 2004, a 911 call is received stating that a tanker truck and a pickup truck just collided on I-94 between Paw Paw and Mattawan in Van Buren County, Michigan. The weather conditions are a temperature of 16°F (−9°C), 10 to 15 mph winds out of the west, and light snow with 11 inches (27.5 cm) already on the ground. The pickup truck driver is critically injured. Based upon the placard and the shipping papers, the tanker contains 8500 gallons (32,000 L) of ethanol. It also appears that the tanker is leaking a small amount of clear liquid.

- How would you handle this emergency incident?
- How would you treat and transport the injured driver?
- How would you protect the public?
- How would you handle interstate traffic?

Let's try to answer these questions in this chapter.

Operations level responders take action at hazardous materials and weapons of mass destruction (WMD) incidents to protect the public, stabilize the incident, and protect property and the environment. In this chapter, we explore ways in which to accomplish these goals. We present the basic components of a hazardous materials response at the operations level.

The current hazardous materials operations standard, NFPA 472, 2008 edition, has been greatly expanded from previous editions. Several roles that were reserved for hazardous materials technicians have been made available as mission-specific competencies to the operations level responder. Operations level personnel have traditionally played primarily a defensive role, especially if they were only trained to the core competency level. However, the new mission-specific competencies allow more aggressive tactics and some hot zone work. For example, law enforcement personnel may enter the hot zone for evidence preservation and sampling missions, if they are properly trained and they have the appropriate personal protective equipment (PPE) available (Chapter 9, Operations Level Responders Mission-Specific Competencies: Evidence Preservation and Sampling). Firefighters may enter the hot zone for victim rescue, if they are properly trained and they have the appropriate PPE available (Chapter 12, Operations Level Responders Mission-Specific Competencies: Victim Rescue and Recovery). EMS personnel can start patient treatment in the warm zone with appropriate training and PPE, and law enforcement personnel may enter the hot zone to investigate illicit laboratories with appropriate training and PPE (Chapter 13, Operations Level Responders Mission-Specific Competencies: Illicit Laboratory Incidents).

Each agency must determine its response level, whether it is operations level with the basic core competencies, or operations level with one or more mission-specific competencies, or forming a hazardous materials response team at the technician level. The primary determinants for which response level to choose are the needs of the agency, what equipment has been or will be purchased, and the extent of training that will be provided, remembering that all of this equipment and training must be maintained to competency from year to year. Agencies have never had more flexibility in being able to choose the level of service and specific functions they will provide at the hazardous materials operations level. As the different levels of response in this chapter are explored, we will reference the appropriate mission-specific competencies in later chapters, in case a higher level response capability is needed.

Dispatch and Response

Most agencies classify the severity of hazardous materials incidents into different levels. These different levels require specific response capabilities, including varied equipment, personnel training levels, and other needed support agencies. Classifying hazardous materials incidents in such a way helps organize and expedite the response. Dispatchers should be trained to ask the correct questions when taking 911 calls, such as the following:

1. What is the nature of the emergency?
2. What is the location of the emergency?
3. What is the best way to access the emergency?
4. Is anybody hurt?
5. Is anyone contaminated?
6. When did the incident start?
7. What type of occupancy or vehicle is involved?
8. What type of container is involved?
9. Is the container damaged?
10. What chemicals are involved? Spell out the chemical name to responding units.
11. What quantity of chemical is involved?
12. Has an explosion or fire started?
13. Is the release ongoing or has it stopped?
14. Is the incident escalating or has it stabilized?
15. Is a knowledgeable individual available? Have him or her meet the first arriving units in a safe location (upwind, uphill, and upstream, at a safe isolation distance).
16. Determine current weather conditions and any changes that are forecast.

Based upon the dispatch information, the scope of the hazardous materials incident must be classified. Here is one way to classify hazardous materials incidents:

Level 1: Incidents that can typically be handled by one or two responding units
Level 2: Incidents that can typically be handled by a trained local hazmat team
Level 3: Incidents that require regional, state, or federal assets to mitigate

The classification system should be useful for informing potential responders of the type of incident they will face and the resources they will likely need.

Once dispatched, hazardous materials incidents should be approached from upwind, uphill, and upstream whenever possible (Figure 5-1). This will minimize exposure to any gases, vapors, or hazardous runoff.

How should you approach the incident described in the opening scenario?

Solution: You should approach hazardous materials from upwind, uphill, and upstream. Because the interstate runs east to west and the winds are out of the west, the best approach would be from the west. If there

GUIDE 127	FLAMMABLE LIQUIDS (POLAR/WATER-MISCIBLE)	ERG2008

POTENTIAL HAZARDS

FIRE OR EXPLOSION

- **HIGHLY FLAMMABLE: Will be easily ignited by heat, sparks or flames.**
- Vapors may form explosive mixtures with air.
- Vapors may travel to source of ignition and flash back.
- Most vapors are heavier than air. They will spread along ground and collect in low or confined areas (sewers, basements, tanks).
- Vapor explosion hazard indoors, outdoors or in sewers.
- Those substances designated with a **"P"** may polymerize explosively when heated or involved in a fire.
- Runoff to sewer may create fire or explosion hazard.
- Containers may explode when heated.
- Many liquids are lighter than water.

HEALTH

- Inhalation or contact with material may irritate or burn skin and eyes.
- Fire may produce irritating, corrosive and/or toxic gases.
- Vapors may cause dizziness or suffocation.
- Runoff from fire control may cause pollution.

PUBLIC SAFETY

- **CALL Emergency Response Telephone Number on Shipping Paper first. If Shipping Paper not available or no answer, refer to appropriate telephone number listed on the inside back cover.**
- As an immediate precautionary measure, isolate spill or leak area for at least 50 meters (150 feet) in all directions.
- Keep unauthorized personnel away.
- Stay upwind.
- Keep out of low areas.
- Ventilate closed spaces before entering.

PROTECTIVE CLOTHING

- Wear positive pressure self-contained breathing apparatus (SCBA).
- Structural firefighters' protective clothing will only provide limited protection.

EVACUATION

Large Spill

- Consider initial downwind evacuation for at least 300 meters (1000 feet).

Fire

- If tank, rail car or tank truck is involved in a fire, ISOLATE for 800 meters (1/2 mile) in all directions; also consider initial evacuation for 800 meters (1/2 mile) in all directions.

Page 200

Courtesy of the U.S. Department of Transportation Pipeline and Hazardous Materials Safety Administration

is a low-lying area immediately west of the incident, you should move the response vehicles to a more westerly position on higher ground. Based upon the 2008 ERG (Guide 127), the immediate area should be isolated for 150 feet (45 m) in all directions. The hot zone or exclusion zone is considered to be the area within a 150-foot radius.

ERG2008	FLAMMABLE LIQUIDS (POLAR/WATER-MISCIBLE)	GUIDE 127

EMERGENCY RESPONSE

FIRE
CAUTION: All these products have a very low flashpoint: Use of water spray when fighting fire may be inefficient.
Small Fire
- Dry chemical, CO_2, water spray or alcohol-resistant foam.

Large Fire
- Water spray, fog or alcohol-resistant foam.
- Use water spray or fog; do not use straight streams.
- Move containers from fire area if you can do it without risk.

Fire involving Tanks or Car/Trailer Loads
- Fight fire from maximum distance or use unmanned hose holders or monitor nozzles.
- Cool containers with flooding quantities of water until well after fire is out.
- Withdraw immediately in case of rising sound from venting safety devices or discoloration of tank.
- ALWAYS stay away from tanks engulfed in fire.
- For massive fire, use unmanned hose holders or monitor nozzles; if this is impossible, withdraw from area and let fire burn.

SPILL OR LEAK
- ELIMINATE all ignition sources (no smoking, flares, sparks or flames in immediate area).
- All equipment used when handling the product must be grounded.
- Do not touch or walk through spilled material.
- Stop leak if you can do it without risk.
- Prevent entry into waterways, sewers, basements or confined areas.
- A vapor suppressing foam may be used to reduce vapors.
- Absorb or cover with dry earth, sand or other non-combustible material and transfer to containers.
- Use clean non-sparking tools to collect absorbed material.

Large Spill
- Dike far ahead of liquid spill for later disposal.
- Water spray may reduce vapor, but may not prevent ignition in closed spaces.

FIRST AID
- Move victim to fresh air.   • Call 911 or emergency medical service.
- Give artificial respiration if victim is not breathing.
- Administer oxygen if breathing is difficult.
- Remove and isolate contaminated clothing and shoes.
- In case of contact with substance, immediately flush skin or eyes with running water for at least 20 minutes   • Wash skin with soap and water.
- In case of burns, immediately cool affected skin for as long as possible with cold water. Do not remove clothing if adhering to skin.
- Keep victim warm and quiet.
- Ensure that medical personnel are aware of the material(s) involved and take precautions to protect themselves.

Page 201

FIGURE 5-1 Always approach a hazardous materials incident from upwind, uphill, and upstream whenever possible. *Art by David Heskett*

The Incident Management Process

It is important to have a system in place to manage hazardous materials incidents. Most **incident commanders (ICs)**, fire company officers, paramedics, and law enforcement officers have their own system prepared to handle routine calls. This system ensures that key steps are not missed and accidents do not happen. The same is true for hazmat incidents, except that, fortunately for most first responders, hazmat incidents are few and far between. Mnemonics can therefore be very helpful.

At the awareness level, it has been found that RAIN—recognize, avoid, isolate, and notify—can be beneficial. Even as an operations level responder, this acronym has merit. Often responders are not dispatched specifically to a hazmat call, but rather to a domestic disturbance that turns out to have an illicit lab in the kitchen, or to an unresponsive patient that has committed suicide using sodium cyanide, or to a garage fire that turns out to be a makeshift chemical storage warehouse. Responders always need to be alert to recognize hazardous materials incidents early, before they become too committed or, worse yet, contaminated, exposed, or injured.

Even when the incident is dispatched as a hazardous materials incident—for example, a gasoline spill at a gas station—it is vital to remember our capabilities, training, and equipment. Responders may have applied absorbent to a 5-gallon (19 L) gasoline spill a dozen times, but this time the nozzle and hose were ripped off the pump and the emergency shutoff failed to engage. Now there are 50 to 100 gallons (189 to 378 L) of gasoline on the ground and the hazmat team is needed there. Make all necessary notifications early! Recognize your limitations.

Several mnemonics are useful, specifically at the operations level. These include the DECIDE process developed by Ludwig Benner in the 1970s; the Eight Step Process© developed by Greg Noll, Mike Hildebrand, and Jim Yvorra; and the AFIRMED process (see Chapter 3) developed by the author. These acronyms are designed to jog your memory to perform critical steps at any hazmat incident.

Ludwig Benner, a renowned accident investigator, retired from the National Transportation Safety Board (NTSB). He formulated the DECIDE process after noticing patterns in the accidents he had investigated. The DECIDE process consists of:

Detect   Detect the presence of hazardous materials.
Estimate   Estimate the likely harm that would occur without intervention.

incident commander (IC) ▪ The person in charge of the emergency incident.

Choose	Choose your response objective(s).
Identify	Identify the action options with available resources and training.
Do	Do the best option.
Evaluate	Evaluate your progress.

The DECIDE process stresses recognition, analysis of the problem in conjunction with the available resources, implementing appropriate actions, and a continuous evaluation of the effectiveness of your actions. The strength of this system is that it stresses basing your decisions on your ability to improve the outcome of the incident. There is no point in acting if your actions will make the incident worse, endanger yourself and other first responders, or put the public at risk.

The Eight Step Process© is a management process that addresses the tactical and strategic goals at hazardous materials incidents. Designed to address the important aspects of hazardous materials response, this process consists of:

Step 1   Site management and control
Step 2   Identification of the problem
Step 3   Hazard and risk assessment
Step 4   Selection of the proper level of PPE
Step 5   Coordination of information and resources
Step 6   Selection and implementation of response objectives
Step 7   Decontamination procedures
Step 8   Termination procedures

This method is designed for use at recognized hazardous materials incidents and is excellent for managing the response to hazmat incidents from beginning to end. The strength of this management process is the level of detail with which the actions specifically needed at a hazmat incident are addressed.

The AFIRMED process consists of:

Assess	Assess the situation prior to committing resources.
Formulate	Formulate an incident action plan (IAP) consistent with your level of training, available resources, and the current situation.
Initiate	Initiate actions that will support the response objectives and IAP.
Reassess	Reassess the situation frequently to ensure a positive outcome.
Modify	Modify the incident action plan based upon the reassessment.
Extended ops	Continue with extended operations to mitigate the incident.
Demob	Demobilize all resources safely and with accountability.

This management process addresses the incident command concepts needed at complex incidents, including hazardous materials incidents and terrorist incidents. The strength of AFIRMED is that specific incident tactical considerations are kept to a minimum in favor of stressing the overall incident management philosophy from beginning to end in a NIMS-compliant way.

Choose one or more of these incident management processes to help guide your response. It is important to have a consistent and complete response to any WMD or hazmat incident.

Establishing an Incident Command System (ICS)

HAZWOPER (29 CFR 1910.120) mandates the use of the incident command system (ICS) at all hazardous materials incidents, which include criminal and terrorist events using weapons of mass destruction. The incident command system, as a method to manage emergency incidents at large events, has been standardized as the National Incident Management

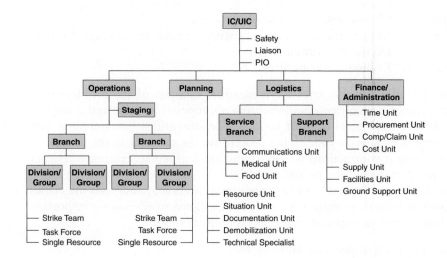

System (NIMS) at the federal, state, tribal, and local level. All agencies that respond to emergency incidents should be NIMS compliant in order to be eligible to receive federal funding.

What does NIMS compliant mean? With regard to the incident command system, it means that participating agencies will use common terminology, interoperable communications, a defined span of control, unity of command, and a unified command approach when multiple agencies or multiple jurisdictions are involved. Figure 5-2 is a sample incident command chart for an incident or event using NIMS standardized terminology. ICS is designed to be able to expand and contract in a modular fashion to accommodate the smallest to the largest incident. Basic ICS concepts will be covered in this book, but detailed ICS concepts will not. As an emergency responder, you should receive further NIMS-compliant ICS training based upon your job duties.

INCIDENT COMMAND

Incident command should be established immediately upon arrival of the first emergency responders, whether they are law enforcement officers, firefighters, EMS personnel, or an industrial emergency response team. An incident commander (IC) should remain on scene until the last emergency response personnel leave the scene. Command is often transferred at incidents of long duration, or those that have multiple operational periods so that ICs have a chance to rest. Transfer of command means appointing a new IC in place of the preceding one. Command is also transferred as more qualified individuals arrive on the scene.

A single incident commander in charge of any emergency incident is responsible for the safe and effective mitigation of the incident, which includes accomplishing the following actions:

Analyzing the incident (scene size-up)
Ensuring the safety of all emergency responders and the public
Determining strategies and tactics
Appointing positions within the ICS framework
Developing an incident action plan (IAP)
Ordering and managing necessary resources (logistics)
Disseminating information to the public
Documenting the incident (for legal and financial purposes)

As you can see, this is a complex set of tasks, which may be difficult for a single person to keep track of at even a small incident, let alone a large incident, without a lot of practice and experience. The incident commander may therefore delegate much of this responsibility to others using the ICS framework described next. As an operations level trained responder, you are qualified to assume the role of IC at a hazardous material incident, assuming you have the requisite incident command training and experience as well.

At larger or more complex incidents, multiple agencies or jurisdictions may form a **unified incident command (UIC)** composed of entities that have significant responsibilities for successful termination of the incident. Terrorist incidents are often managed using the UIC concept due to the overlapping legal responsibilities of law enforcement agencies, fire departments, and EMS agencies. The technical knowledge required to successfully manage and mitigate complex incidents is usually beyond the capabilities of a single agency or person and requires a unified command structure.

unified incident command (UIC) ■ The joint command of an incident by incident commanders from different agencies and/or jurisdictions.

COMMAND STAFF

Command staff consists of the incident commander (IC), safety officer, liaison officer, and public information officer (PIO). The command staff can be viewed as the incident commander's support staff, or office staff, as they report directly to the IC. These positions are appointed by the incident commander when the situation becomes too complex for him or her to handle alone. At smaller incidents, these functions are typically handled by the incident commander directly. Additional command staff may be appointed depending on the complexity of the incident (such as legal counsel or other technical experts).

command staff ■ In an ICS, the safety officer, liaison officer, and public information officer (PIO). These positions report directly to the incident commander.

Safety Officer

The safety officer monitors incident operations and advises the incident commander (IC) on any safety issues. The safety officer must coordinate closely with the IC and his or her general staff to ensure the IC understands incident objectives, and the strategies and tactics that are being used. The safety officer has the authority to overrule the incident commander when life safety is immediately at stake. He or she must notify the incident commander of any corrective actions as soon as possible to ensure that the incident action plan is not compromised.

HAZWOPER mandates the appointment of a safety officer at every hazardous materials incident. This person not only should be an experienced hazardous materials responder but also must be able to generate a site safety plan and evaluate tactics and actions for their safety and effectiveness. The safety officer also usually delivers the **safety briefing** immediately before any personnel enter the hot zone.

safety briefing ■ The oral description of the hot zone safety hazards, instituted safety precautions, and work expectations given to the entry team prior to entry.

At larger incidents, where hazardous materials are one piece of a much larger puzzle, hazardous materials operations will be carried out by the hazmat branch or hazmat group (see "Operations Section"). Due to the dangers and complexity of hazardous materials operations, another safety officer—termed the assistant safety officer, hazmat—will likely be appointed specifically for the hazmat branch.

Liaison Officer

The **liaison officer** is the point of contact for representatives from other agencies that are involved in responding to the incident. These outside agencies may include federal, state, and local government agencies; nongovernmental organizations (NGO) such as the Red Cross; and private companies such as cleanup contractors. The liaison officer can therefore be viewed as the incident receptionist. However, the liaison officer must have the authority to speak for the incident commander after appropriate direction and consultation have been given.

liaison officer ■ In an ICS, the person who serves as the primary contact with supporting agencies and organizations at an emergency incident.

Public Information Officer (PIO)

The **public information officer (PIO)** is responsible for interfacing with anyone requiring information regarding the incident. This includes not only the public and news media but also other agencies that need incident information. For example, the PIO may keep other agencies apprised of road closures, smoke and vapor cloud conditions, or evacuation areas for informational purposes. The PIO gathers pertinent incident information such as size, resources committed, and estimated duration. This information must be current and accurate, and is released only after the IC's approval. At larger incidents, the PIO may have assistants to help gather and process the information.

public information officer (PIO) ■ In an ICS, the person responsible for informing the general public of incident developments as instructed by the incident commander (IC).

The PIO's role of disseminating information quickly and accurately is extremely important at hazardous materials incidents. Evacuation and shelter-in-place orders can be facilitated by using the media to rapidly notify residents. Rumors can be minimized and controlled when the media have access to accurate and timely incident information. Traffic backups and other inconveniences due to road closures can be reduced when other agencies such as road commissions and public works departments have access to information quickly. These agencies often have the ability to adjust traffic patterns rapidly with modern technology.

GENERAL STAFF

General staff consists of the operations section chief, the logistics section chief, the finance/administration section chief, and the planning section chief. As with the command staff, general staff is appointed when the incident becomes too complex for the IC to handle all of these incident management requirements alone. The incident commander is still responsible for all aspects of the incident outcome irrespective of what authority he or she delegates.

Operations Section

The **operations section** is the workhorse of any large incident. The operations section, managed by the operations section chief, is responsible for carrying out the incident objectives by formulating the tactics, and managing and assigning the appropriate resources to accomplish them. That is not to say that the other sections are not important. They are in large part there to support the work of the personnel assigned to operations to bring the situation under control.

As a hazardous materials operations level responder, you will most likely be operating within the operations section, in the hazardous materials branch, in one of the hazmat groups. Branches, divisions, and groups are implemented when the number of personnel and resources at an incident becomes too large for a single individual to manage. This is referred to as exceeding the span of control. Ideally, any single individual is in charge of five subordinates; an acceptable span of control is three to seven subordinates. Span of control is designed to optimize the use of resources by not overtaxing individual managers. Thus, as the number of personnel and resources grows at a scene, the number of managers must be increased; this is accomplished by adding branches, divisions, and groups.

Branches Different functions—such as fire, law enforcement, and EMS—are often grouped into branches at larger incidents. Thus there may be a hazmat branch at complex incidents that involve not only hazardous materials response but also fire suppression, law enforcement, and patient treatment. Branch directors manage branches, and they may have deputy branch directors as necessary.

Divisions and Groups Divisions and groups are another way to organize resources and maintain a viable span of control. Divisions and groups are managed by supervisors who report to their branch director. **Divisions** are entities that cover a geographical area; for example, there could be a chemical storage area division at a hazardous materials incident. **Groups**, on the other hand, cover a functional area, such as a **mass decontamination** group or a research group. There may be both divisions and groups within any particular branch at any given incident.

A hazmat branch director or hazmat group supervisor manages the hazardous materials specific functions at the incident. Let's examine what the incident command system would look like for the hazmat branch or the hazmat group at a larger incident. Figure 5-3 shows a sample incident command chart for a potential hazmat incident. If the incident is large enough to warrant a hazardous materials branch, there will likely be several groups corresponding to the common functions that will be carried out. For example, there may be a research group, a decontamination group, a dressout or PPE group, a product control group, an evidence collection group, a medical group, and so on. If the hazardous materials incident

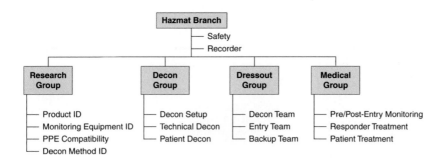

FIGURE 5-3 Sample ICS flowchart for the hazardous materials branch.

is this large and complex, you—as the hazardous materials operations level responder—will likely be supporting a technician level hazardous materials response team (HMRT).

Resources   Resources may be organized into single resources, strike teams, and task forces. Single resources are commonly used at emergency incidents, especially hazardous materials incidents. **Strike teams** are a group of the same kind and type of resource. **Task forces** are a group of different resources put together to gain functionality as a group. Strike teams and task forces must adhere to the normal span of control limits.

There may be times when strike teams and task forces become necessary. Strike teams are often formed when a large number of the same resource is needed at an incident, for example, ground crews at a wildfire, or when the incident is extremely large and resources are grouped to maintain span of control. Task forces are used when different skills are needed repeatedly. For example, during a large snowstorm that renders the roads impassable, task forces consisting of a snow plow, police cruiser, fire engine, and ambulance may be formed. These task forces could then respond to most emergency incidents that no individual resource could tackle alone.

strike team ■ In an ICS, a set of identical resources grouped together for a specific purpose.

task force ■ In an ICS, a set of different resources grouped together for a specific purpose.

Logistics Section

The **logistics section** is in charge of procuring resources, in the form of personnel, equipment, and services. Personnel in this section provide the resources needed to accomplish the response objectives in the incident action plan (IAP). Without the logistics section, the operations section could not carry out the incident objectives. The logistics section ensures that supplies and services needed at the incident are obtained. For example, if the supply of level A suits were running low at a hazmat incident, the logistics section would ensure that more are procured. The logistics section is also tasked with making sure enough water, food, and sanitary facilities are available to accommodate all the responders.

logistics section ■ In an ICS, the personnel responsible for ensuring that the needed service and support resources are ordered and available at the incident.

Finance/Administration Section

The **finance/administration section** is in charge of keeping track of expenditures and labor-hours at complex incidents; in other words, how much the incident costs. The personnel are responsible for ensuring vendors get paid after the incident and any federal or state reimbursements are received. They must therefore keep track of responders' hours and receipts for equipment and services purchased. They are also accountable for keeping the cash flow going when vendors must be paid immediately. Hazardous materials incidents can be very costly due to equipment and service needs, sometimes into the millions of dollars. Clearly this is no job for an amateur! Usually individuals from the finance or accounting department of the agency or jurisdiction in charge assume the role of finance chief and any subordinate positions because they possess the requisite knowledge, skills, and experience.

finance/administration section ■ In an ICS, the personnel in charge of keeping track of expenditures such as salaries and orders, and paying for needed equipment and supplies.

Planning Section

When incidents become complex and last for days, weeks, or months, a **planning section** is appointed to handle preparations for the next operational period. Operational periods are increments of time, which may be hours or days but are typically 12 hours, after which the incident is formally reassessed and the incident action plan (IAP) is modified.

planning section ■ In an ICS, the personnel responsible for planning the next operational period of the incident action plan (IAP).

Draw a possible incident command structure for the incident in the opening scenario.

Solution:

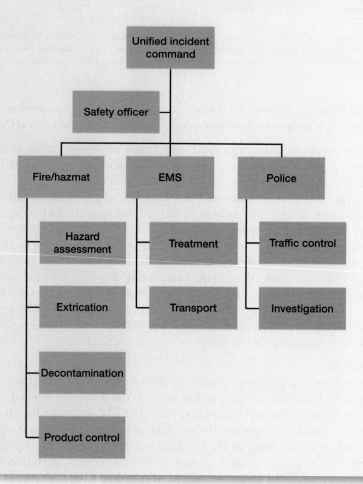

The planning section chief gathers incident information from the operations section chief and determines the necessary resources and how the incident objectives need to be updated or modified for the next operational period. The planning section chief must also notify the logistics section chief of the needed resources for the next operational period.

THE PLANNING PROCESS

An effective plan is the cornerstone of an effective response. The better the plan, the better the response outcome will be. The five phases of the NIMS planning process are:

1. Understanding the situation
2. Establishing incident objectives and strategy
3. Developing the incident action plan (IAP)
4. Preparing and disseminating the IAP
5. Executing, evaluating, and revising the plan

It is essential to understand as much about the situation being faced as quickly as possible, especially for hazardous materials incidents. This involves making an overview of the scene, accessing information from preplans, and contacting other agencies and jurisdictions that may have helpful information. It is also important to know the capabilities and extent of your own resources as well as those available through mutual aid partnerships. This can be accomplished only through effective networking, training, and preplanning before the incident occurs. These are essential activities for hazardous materials responders, especially at the hazmat operations level.

The Planning P

The U.S. Coast Guard developed the **planning P** in response to difficulties it faced at complex incidents. The planning P—a graphical representation of the operational period planning cycle—is illustrated in Figure 5-4. When properly implemented, it accomplishes the goals of the planning process. The incident starts at the foot of the P and then progresses into the loop, where each loop is cycled once per operational period. During each operational period, incident objectives are developed, tactics are determined, a planning meeting is held, the plan is approved by the IC, there is an operational briefing, and the new plan is then executed during the new operational period. Then during the new operational period, the planning section works on the IAP for the following operational period using the latest information provided by the operations section.

Forming an Incident Action Plan

An important component of any incident management system (IMS) is the formation of an incident action plan (IAP). Every incident has an incident action plan, although only complex incidents or incidents of long duration have written incident action plans. Typically, written incident action plans are used to assign resources and describe tactics and objectives for incidents that span one or more operational periods. A written incident action plan facilitates the dissemination of information from one shift to another across operational periods, and to large numbers of first responders from different agencies or geographical areas at large incidents. Incident objectives should be measurable, flexible, and time sensitive so that they can be evaluated for effectiveness and modified if they are not.

Part of the incident action plan is a safety analysis of the incident, often called the site safety plan. Key components of the site safety plan, according to the U.S. Environmental Protection Agency (EPA), are:

- Site description, including control zones (hot, warm, and cold)
- Response objectives and entry team objectives
- On-site incident command system organization
- On-site personnel control (by zones)
- Hazard evaluation, including confined spaces, trenches, and other hazards
- Personal protective equipment (PPE)
- On-site work plan
- Communication procedures
- Decontamination procedures
- Site safety and health plan

Operational Period Planning Cycle

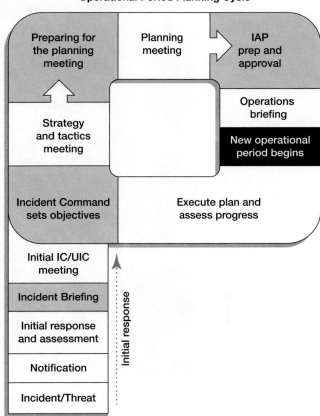

FIGURE 5-4 The planning P illustrates the planning process for long-term incidents. *Courtesy of FEMA*

planning P ■ A process developed by the U.S. Coast Guard by which an incident can be successfully managed in the long term.

Safety briefings should be held for all personnel at hazardous materials incidents and weapons of mass destruction incidents. The safety briefing should give everyone on scene a site safety overview, and let everyone know how to notify the incident commander and the safety officer in the event of an emergency. It specifies what type of communications will be used on the incident, including which frequencies are used by which branches, and how to relay information up the chain of command. It is especially important to hold a detailed safety briefing, called the entry team briefing, for any responders entering the hot zone (Figure 5-5).

Documentation

Proper documentation is an important aspect of any response to hazardous materials or weapons of mass destruction incidents. Especially at large incidents, documentation improves communication, enhances the quality of the information, and provides a rationale for decision making, which improves scene safety, facilitates after-action reviews, and provides a solid foundation if any aspects of the incident go to court. Documentation takes many shapes and forms, including handwritten notes, photographs and video recordings, standardized response forms that are filled out during the response (such as the IAP), the results of air-monitoring and sample identification equipment, electronic data output from instrumentation used on scene, and recorded radio transmissions. After the termination of the incident, incident reports, medical surveillance, and critiques also become part of the incident documentation.

During large and/or complex incidents, document control can become difficult. Specific personnel, such as an incident recorder, should be assigned to keep track of documentation. Document control procedures may include:

- Listing all documents in a document inventory
- Collecting documents at regular intervals during extended operations
- Keeping copies of documents in a centralized, secure location
- Assigning a control number to all documents

Good documentation has many advantages. When response objectives are written down, it is easier for the safety officer to accurately review them. When strategies and tactics are written down, it is easier for other agencies and subordinates to accurately and safely accomplish them. As an added bonus, after the incident is over, it will be much easier to glean lessons learned and improve agency operations when the documentation accurately reflects the actions taken under the given circumstances.

ICS FORMS

NIMS-compliant ICS forms have been developed to help facilitate the seamless coopera-tion of local, state, tribal, and federal agencies (Figure 5-6). The ICS forms have been standardized and follow a logical format. There are at least two dozen different forms,

ICS Form 202

INCIDENT OBJECTIVES	1. INCIDENT NAME	2. DATE	3. TIME
4. OPERATIONAL PERIOD (DATE/TIME)			
5. GENERAL CONTROL OBJECTIVES FOR THE INCIDENT (INCLUDE ALTERNATIVES)			
6. WEATHER FORECAST FOR OPERATIONAL PERIOD			
7. GENERAL SAFETY MESSAGE			

8. Attachments (☐ if attached)

☐ Organization List (ICS 203)   ☐ Medical Plan (ICS 206)   ☐ _____

☐ Assignment List (ICS 204)   ☐ Incident Map   ☐ _____

☐ Communications Plan (ICS 205)   ☐ Traffic Plan   ☐ _____

9. PREPARED BY (PLANNING SECTION CHIEF)	10. PREPARED BY (PLANNING SECTION CHIEF)

FIGURE 5-6 ICS 202, incident objectives form.

and when completed they constitute the basis of the incident action plan (IAP). The following are the core forms of the IAP:

ICS 201	Incident briefing, summary of current actions, current organization, and resource summary
ICS 202	Incident objectives
ICS 203	Organization assignment list
ICS 204	Assignment list
ICS 205	Radio communications plan
ICS 206	Medical plan
ICS 207	Incident organization chart (wall mounted)
ICS 209	Incident status summary
ICS 210	Status change
ICS 211	Incident check-in list
ICS 213	General message
ICS 215	Operational planning worksheet
ICS 215A	Hazard risk analysis

INCIDENT COMMAND FACILITIES

At larger incidents, the sheer number of personnel and resources will require more substantial facilities to effectively operate.

Incident Command Post (ICP)

incident command post (ICP) ▪ The physical location of the command functions and where the incident commander or the unified incident command is located.

Command and control functions take place at the **incident command post (ICP)**. This is where the IC and most likely the command staff and most of the general staff are located when appointed. Thus, the ICP can become crowded at larger incidents if the site has not been chosen wisely. The ICP should be as close to the incident as safety considerations and practical situations dictate. The ICP must be located in a clean area upwind, uphill, and upstream of the hazardous materials release, and should be secured from unauthorized access by the public and other first responders. This is especially important at criminal incidents, especially WMD and terrorist incidents.

At smaller incidents, the incident command post (ICP) is typically the vehicle in which the incident commander arrived. At larger incidents, the ICP may be moved to a mobile command post that has superior lighting, electrical, sanitary and communications capabilities. At extremely large incidents, the ICP may be moved to a building, possibly remote from the site, that has good visual and verbal communication to the incident location via radios and cameras.

Emergency Operations Center (EOC)

emergency operations center (EOC) ▪ A permanent ICS installation that facilitates the deployment of resources by acting as a logistics clearinghouse at larger emergency incidents. Most cities and counties have an EOC that is maintained by the local emergency management agency (EMA).

The **emergency operations center (EOC)** is a fixed facility run and maintained by the emergency management agency having jurisdiction. It is a location with excellent communications resources that can accommodate representatives from all the stakeholding agencies involved in the incident. The members of the EOC respond to resource requests from the incident commander. The EOC and emergency manager provide an important resource to the incident commander but do not assume command of the incident. When the logistics section has been established, it is usually in close communication with the EOC.

Staging Area

staging area ▪ In an ICS, the location where incoming resources wait until they are needed. The staging area is located far enough away from the hot zone so as not to interfere with incident operations, but close enough to permit rapid deployment.

staging area manager ▪ In an ICS, the individual in charge of the staging area.

The **staging area** is where requested resources report to when they arrive at the incident. This keeps them from interfering with ongoing operations and causing traffic jams and other access and egress problems. The staging area is managed by a **staging area manager**, who reports directly to the incident commander or the operations section chief (if appointed). In order to keep the staged resources supplied and ready for mobilization, the staging area manager requests resources from the logistics section.

A well-run incident will have resources arriving before they are actually needed. The staging area is located in a safe location, out of the immediate work zone, at which the responding units can receive an incident briefing and report to the appropriate manager when they are needed.

Joint Information Center (JIC)

At larger incidents, or multi-jurisdictional incidents, a **joint information center (JIC)** may be formed to coordinate the release of information by several different agencies or jurisdictions at an incident. For example, at a hazardous materials incident involving a large plume, environmental agencies such as the EPA may be monitoring the plume, while the public health department conducts swab surveys, and the hazmat team assesses and controls the leak. Information from all three agencies must be coordinated in order to give the public accurate information and to ensure the information advances the incident objectives (such as evacuation versus in-place sheltering advice).

joint information center (JIC) ■ In an ICS, the facility at which PIOs from different agencies and jurisdictions can brief the public with a unified voice.

Collecting and Analyzing Information

Once you have arrived at the hazardous materials incident and established command, gather information as quickly as possible. This includes using binoculars to look at placards, labels, and container shapes and sizes from a safe distance (Figure 5-7). Such information may include:

- Types of containers involved
- Identification markings on containers

FIGURE 5-7 Using binoculars to assess the incident scene and collect information.

FIGURE 5-8 A Knox box used to store building keys and maps for rapid emergency responder access. Local emergency responders have a universal key to all Knox boxes in their jurisdiction.

- Quantity and/or capacity of containers
- Materials involved
- Type of release
- Size of the release

Information may also be obtained from facility preplans located at your station or in a **Knox box** at the facility (Figure 5-8), shipping papers, witnesses, drivers, or workers at the facility, as explained in Chapter 3. Once there is some idea of what we are dealing with, we can consult technical experts such as CHEMTREC, or reference materials such as the DOT *Emergency Response Guidebook*, the NIOSH guide, or MSDS sheets, as explained in Chapter 4. Using the chemical and physical properties gleaned from the reference materials, we then need to understand how the product will behave, estimate the potential harm it may cause, and predict where it might go.

Key questions to ask are:

Are there any victims?
Has a release occurred?
Where is the product going?
Is the situation getting worse?
Could a BLEVE or a catastrophic failure occur?
Do I have the appropriate and necessary resources on scene?

Also note the surrounding conditions (Figure 5-9), such as:

Knox box ■ Storage devices for building keys, maps, and facility emergency plans that emergency responders within a jurisdiction have access to through a common key.

- Topography
- Land use
- Accessibility
- Weather conditions
- Bodies of water
- Public exposure potential
- Overhead and underground wires and pipelines
- Storm sewer drains
- Possible efficient sources
- Adjacent land usage such as rail lines, highways, and airports
- Nature and extent of injuries
- Building information

secondary device ■ An explosive device designed to injure first responders and people evacuating the site of a previous attack.

Consider additional hazards, such as **secondary devices** in the case of a terrorist or criminal incident, armed resistance and use of weapons, booby traps, and secondary contamination from handling victims and patients.

As soon as feasible, verify the information you have received. Witnesses can be mistaken, reference materials have misprints, and information and numbers can be transcribed incorrectly. Verification can be accomplished by using multiple reference sources and interviewing numerous witnesses to corroborate the information. Such errors and misconceptions, if left uncorrected, can lead to catastrophic results.

Information can be verified from a distance using various methods such as a thermal imaging camera to determine the product level in the container. Assuming a worst-case scenario (that the container is full), an estimate of the spilled product would be the full container capacity less the amount of material left in the container. Air-monitoring equipment can be used to determine the extent of any vapor cloud (see Chapter 11,

Operations Level Responders Mission-Specific Competencies: Air Monitoring and Sampling).

DETERMINING RESPONSE OBJECTIVES

It is important to determine the level of response that will be provided at any incident. At the operations level, that response is primarily defensive. Defensive operations are generally performed at a distance from the hazardous materials release. There are some exceptions within the mission-specific competencies outlined in later chapters. Response objectives to any incident, including hazardous materials incidents, should be prioritized by:

1. Life safety
2. Incident stabilization
3. Property and environmental conservation

At most incidents there are limited resources. It is therefore extremely important to correctly prioritize incident objectives. Lives could depend on it!

Some key questions to ask when determining the response objectives are:

How many victims can be saved with the available resources?
Is there a potential for secondary devices or a secondary attack?
How big is the leak?
Can we control the leak safely and effectively with the available resources?

This is part of a hazard/risk assessment process. A risk/benefit analysis is key to making sound strategic and tactical decisions at the hazardous materials incident. It is very important to examine the risks associated with any action, as well as the benefits to be gained from it. If the risks outweigh the benefits, a different response objective should be chosen.

Life Safety

Life safety objectives in hazardous materials incidents include isolating the immediate area, maintaining scene control, removing non-ambulatory victims from the hot zone, and prioritizing emergency medical care of the victims (triage). Depending on the nature of the hazardous materials incident, some of these objectives may be more advanced than your training at the hazardous materials operations level. If you determine that you do not have the resources or training to meet certain objectives, it is very important to notify the appropriate agencies that do.

Incident Stabilization

Incident stabilization objectives in hazardous materials incidents cover a wide variety of actions (Figure 5-10). Any objective that keeps an incident from getting worse is considered an incident stabilization objective. Examples of such goals are product control, including remote valve shutoff, diking and damming, building overflow and underflow dams in rivers and streams, and setting up water curtains for gases and vapors. Such actions require more advanced training than the core competencies covered in this chapter, and are discussed in Chapter 10, Operations Level Responders Mission-Specific Competencies: Product Control.

FIGURE 5-10 Incident stabilization through product control. In this case, entry team members are placing absorbent pads to limit the spread of the hazardous materials and reduce vapor generation.

As the incident commander, what would your response objectives be for the incident described in the opening scenario?

Solution:
1. *Life safety:* Patient extrication, treatment, and transport.
2. *Incident stabilization:* Collect the ethanol that is leaking (confinement). Stop the ethanol from leaking out of the tank (containment).
3. *Property and environmental conservation:* Remove contaminated soil. Reopen the interstate.

If any of these response objectives cannot be met with the resources on scene, the appropriate resources must be called, such as a hazmat team, heavy vehicle towing services, or a cleanup contractor.

Property and Environmental Conservation

Last, but not least, on the list of response objectives are property conservation and environmental conservation. After life safety has been handled and the incident has been stabilized, it is time to consider how to minimize both property damage and the environmental impact of the incident. Property conservation may include preventing runoff from affecting neighboring areas (exposures), decontaminating property and equipment, and transferring residual material inside leaking tanks to undamaged tanks. Environmental conservation may include preventing runoff from affecting lakes and streams, and preventing groundwater contamination by excavating contaminated soil.

MANAGING RESOURCES

After determining the scope of the problem, additional resources may be needed, such as the local hazmat team, cleanup contractors, the state Department of Environmental Quality, the local health department, the department of public works, the bomb squad, the meth team, or other technical experts. Network with representatives from these agencies now, before an incident occurs, and have their contact information handy! Make these notifications as early into the incident as possible.

Hazardous materials incidents can be resource-intensive operations. Most hazardous materials incidents will require the presence of a hazmat unit or a local hazardous materials response team (HMRT). These are typically requested through dispatch. However, on larger hazardous materials incidents that may affect the wider community, it may be a good idea to think about activating the local emergency operations center (EOC). Many different agencies may be needed at a large hazmat incident, including emergency medical services, the fire department, law enforcement, public works, utility companies, public works, and state and federal environmental agencies; and the EOC is an excellent tool for requesting resources.

When releases above the reportable quantity (RQ) occur, a minimum of three notifications must be made at the local, state, and federal level. At the local level, it is usually the local emergency management agency or LEPC that must be notified. If you are a public safety agency and have been dispatched to a hazardous materials incident, chances are your dispatcher has automatically notified the emergency management agency as well. At the state level, it typically means notifying the Department of Environmental Quality or the Department of Natural Resources. At the federal level, the National Response Center (NRC) must be notified at 1-800-424-8802. Technically, it is the responsibility of the spiller to make these notifications, but it is a good idea to double-check that they have been made. Notification of the NRC is especially important because of the many resources, equipment, personnel, and money it can provide.

EVALUATING THE PLANNED RESPONSE

Once the response objectives have been formulated and are starting to be executed, it is important to evaluate the effectiveness of these measures. How well have the needs of the incident been anticipated? Have we taken care of all life safety considerations, including downwind populations? Is the incident being stabilized? Has the release been stopped or slowed? Or is the release increasing in scope or size? If the answers to these questions are not favorable, the plan must be modified. If the situation becomes too dangerous, responders may have to be withdrawn to a safe distance.

Scene Control Procedures

The first step in scene control is to isolate the immediate area (Figure 5-11). If some information is known about the hazardous material or the mode of transportation, the isolation distances can be obtained from the current edition of the *Emergency Response Guidebook*. Generally, the isolation distance for solid materials is 75 feet; for liquid materials it is 150 feet; and for gaseous materials it is 300 feet. If little information is known, the size of the hazard area must be estimated from the size of the release, witness accounts, air-monitoring equipment, or a combination of all three. Once the size of the hazard area has been determined, it is known as the exclusion zone or the hot zone. This hazard area should be increased, especially downwind, if the release cannot be stopped quickly or is expected to get worse.

IMMEDIATE ISOLATION AND THE USE OF CONTROL ZONES

Initially, the incident is divided into two zones, the hot zone and the cold zone (Figure 5-12). The hot zone is any area that currently has contamination or can reasonably be expected to have contamination at any time before the incident is mitigated. Everyone should be evacuated from the hot zone; evacuation can be accomplished by public address systems, alarm activation, and searching room to room. Only personnel properly trained and equipped with personal protective equipment (PPE) should be allowed to enter the hot zone. Use of personal protective equipment is covered in detail in Chapter 6, Operations Level Responders Mission-Specific Competency: Personal Protective Equipment.

FIGURE 5-11 Scene control using law enforcement and the public works department.

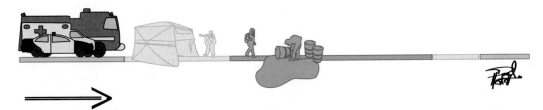

The hot zone/cold zone boundary should be clearly marked. For example, existing control features such as fences can be used to strategically set the boundary. When this is not feasible, barrier tape, cones, or temporary fencing may be used. It is very important that all emergency response personnel on the scene know where the hot zone/cold zone boundary is and what it means. It is equally important that any bystanders and the public understand how dangerous the hot zone is and where the hot zone perimeter is located. It may also be helpful to set up an outer perimeter beyond the hot zone to keep the public and onlookers at a safe distance so they will not interfere with emergency response operations in the cold zone.

The area that is not contaminated is called the cold zone, or support zone. A warm zone, or decontamination zone, is carved out of the cold zone on the upwind side and, when possible, uphill side of the hot zone. Because the warm zone comes from the cold zone, it is initially not contaminated and may be set up without the use of special personal protective equipment. However, once contaminated patients or entry personnel start returning from the hot zone, it is considered contaminated and access points must be controlled. **Emergency decontamination** is also carried out in the warm zone. Techniques for emergency decontamination and mass decontamination will be covered in detail in Chapter 7, Operations Level Responders Mission-Specific Competencies: Mass Decontamination and Chapter 8, Operations Level Responders Mission-Specific Competencies: Technical Decontamination respectively.

DOWNWIND EVACUATION

The decision about whether to evacuate or shelter in place depends upon several factors:

- *Resources:* It may be impractical to completely evacuate the designated area. Targeted evacuation from the most dangerous area may be the most effective solution.
- *Nature of the emergency:* The vapor cloud may have already passed, making evacuation unnecessary, or evacuation may mean the occupants will have to pass through the vapor cloud, which dramatically increases the exposure.
- *Nature of the chemical:* Evacuation is preferred for flammable vapor clouds due to the risk of ignition, although evacuating people through a flammable vapor cloud is more dangerous. If occupants are sheltered in place, advise them to stay away from windows and glass in case of explosion.

When the decision to evacuate has been made, ensure that occupants are evacuated into a safe atmosphere using air-monitoring equipment. If it does become necessary to evacuate occupants through a vapor cloud, ensure that the rescuers and the occupants are properly protected before exiting the structure. Use radio, television, and reverse

emergency decontamination ■ The cursory removal of contaminants using any available means. Usually emergency decon is less effective than technical decon and must be verified and often completed when resources permit. Examples of emergency decon include a fire department hose line in the field and an emergency shower at a fixed site facility.

SOLVED EXERCISE 5-5

How would you control access to the scene described in the opening scenario?

Solution: Based on the 150-foot (45-m) exclusion zone listed in the ERG, both directions of the interstate should be shut down using law enforcement officers. Before sending law enforcement downwind to perform traffic control, air monitoring should be conducted to ensure the area is safe. Law enforcement should remain at least 150 feet (45 m) from the accident scene. Given the cold temperatures, the exclusion zone could most likely be reduced if appropriate air-monitoring readings were taken (CGI and PID) and they are non-detect. The ERG isolation zone values are for guidance only and can be modified using appropriate air-monitoring data or the advice of trusted technical specialists. In this case, you could even consider reopening the opposite lane of traffic if conditions warrant.

911 systems to notify residents efficiently. If first responders are sent into the protective action zone (evacuation zone), ensure that they are properly trained and equipped with personal protective equipment (if necessary).

The downwind evacuation distance can be estimated using the green section of the *Emergency Response Guidebook* (ERG) (Figure 2-27). Table 1 of the green section gives downwind evacuation distances, for large and small spills as well as for daytime and nighttime spills, for any materials that are highlighted in the blue and yellow sections.

Downwind evacuation or in-place sheltering is appropriate for chemicals that are highlighted in the ERG and are *not* on fire. If the product is on fire, most of the vapors are being consumed and will most likely not pose a significant downwind exposure hazard. The values found in Table 1 can be applied in the following way:

1. Identify the chemical by name or UN/NA identification number. If it is highlighted and not on fire, locate the appropriate entry in Table 1 using the identification numbers, which are listed in numerical order.
2. Determine whether the spill is large or small. A small spill is from one or more containers that have an aggregate capacity of approximately 55 gallons (200 L) or less. A small leak from a large container may be classified as a small spill under appropriate conditions. However, keep in mind that if the leak is getting worse, or could be reasonably expected to get worse, estimate accordingly. All other spills are classified as large.
3. Isolate the area in all directions using the appropriate isolation distance listed in Table 1 of the green section of the ERG (for a small or large spill).
4. Determine the appropriate downwind evacuation distance: The first step is to establish the time of day you are planning for. The daytime evacuation distances are shorter than the nighttime evacuation distances. During the daytime, atmospheric mixing disperses the vapor cloud and more quickly lowers the airborne chemical vapor concentration. At night, more stable atmospheres allow concentrated vapor clouds to drift farther downwind. If the incident started at night or is likely to continue into the night, use the nighttime protective action distances. Daytime is classified as the period between sunrise and sunset.
5. The protective action zone extends downwind from the spill the distance indicated in the appropriate column (depending on spill size and time of day), and extends half of the distance to either side. This creates a square whose left side is centered at the release point, and extends downwind (see Figure 5-13). Start the evacuations as close to the spill site as safely possible, and continue downwind as far as resources permit. Call in additional resources if the evacuation or in-place sheltering cannot be carried out in a timely fashion. See the following section for guidance on evacuation versus in-place sheltering.

This method is also described on pages 298–299 of the 2008 ERG.

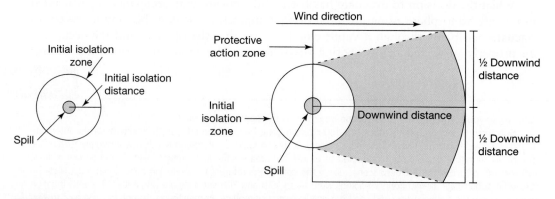

FIGURE 5-13 How to determine the downwind evacuation distance according to the DOT *Emergency Response Guidebook*. *Courtesy of the U.S. Department of Transportation Pipeline and Hazardous Materials Safety Administration*

How far would you evacuate the area in the opening scenario and why? There is an apartment complex 200 feet (61 m) to the north of the accident. Should it be evacuated? Why or why not?

Solution: The entry for ethanol (1170) is not highlighted, so we do not need to consult the green section. The spill appears to be small (less than 55 gallons or 208 L), and therefore people within 150 feet (45 m) should be evacuated. The apartment complex therefore does not need to be evacuated.

IN-PLACE SHELTERING

In-place sheltering may be considered when a gas or vapor cloud will move through the area relatively quickly and not linger, especially when it is not feasible to evacuate due to a lack of resources or time. Sheltering in-place may *not* be appropriate for:

- Flammable vapors
- A lingering vapor cloud that will infiltrate the building
- Buildings that cannot be sealed tightly
- Incidents involving explosives

In-place sheltering, or sheltering in place, is accomplished by notifying the occupants of the building to close all of their windows and doors and to turn off their air-handling system (HVAC) for a specified amount of time. The amount of time should be long enough to allow the vapor cloud to safely pass the sheltered-in-place occupancy. Maintain contact with the occupants if at all possible, keeping them updated of changing conditions and when the in-place sheltering order has been lifted.

In-place sheltering is often used when occupancies are difficult or dangerous to evacuate, such as hospitals and nursing homes. When a large number of occupancies need to be notified, reverse 911 systems or call centers can be helpful. As a last resort, responders may enter the area to notify occupants in person or via vehicle-mounted public address systems. However, this may put those responders at risk, and they should have the appropriate personal protective equipment.

Personal Protective Equipment

Personal protective equipment (PPE) for hazardous materials incidents consists of respiratory protection and chemical protective clothing (CPC) (Figure 5-14). This will be covered in more detail in Chapter 6, Operations Level Responders Mission-Specific Competencies: Personal Protective Equipment. The most important question to ask is whether the available PPE is appropriate for the task at hand.

RESPIRATORY PROTECTION

Respiratory protection can be considered the most important aspect of personal protective equipment because inhalation is the most efficient and dangerous route of entry for hazardous materials into the body. Of course, exceptions to this rule of thumb exist, and the most efficient route of entry is different for specific chemicals. Therefore, it is vital to use the appropriate reference materials when determining the dangers of any particular chemical.

Materials penetrate the respiratory passages at different efficiencies. Large particulates such as dust and soot are trapped by the nasal passages and upper respiratory tract,

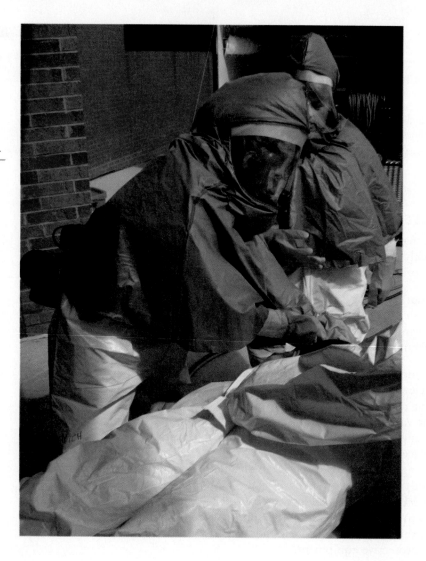

FIGURE 5-14 Personal protective equipment being used in the warm zone for patient triage and treatment. The PPE pictured consists of hooded powered air purifying respirators (PAPR) and chemical protective clothing (CPC).

dust mask ■ A respirator designed to protect the wearer from particulate matter. This type of respirator does not offer any chemical protection or protection from infectious agents.

as any firefighter who has blown his or her nose after a structure fire knows. **Dust masks** and other particulate filters can protect against large particulates and dusts. At the other extreme are gases that penetrate into the alveoli, the smallest pockets in the lungs that exchange gases between the air and the bloodstream. Therefore, there are several levels of respiratory protection available based upon the hazard. Table 5-1 summarizes the advantages, limitations, uses, and operational components of respiratory protection.

The OSHA Respiratory Protection Standard (29 CFR 1910.134) regulates the use of respirators in the workplace. This law was implemented due to the dangers that respirator users face. It requires that respirator users receive a physical exam prior to use (unless the respirator use is voluntary). The wearer must also be physically and mentally capable of using the respirator safely and effectively.

Particulate Respirator (Dust Mask)

Particulate respirators, or dust masks, provide very limited respiratory protection as the simplest and least expensive type of respirator. Maintenance personnel may use this type of mask to protect against dust, or EMS personnel may refer to it as a TB mask

TABLE 5-1 | Types of Respiratory Protection Devices

RESPIRATORY PROTECTION	ADVANTAGES	LIMITATIONS	USES	OPERATIONAL COMPONENTS
Self-contained breathing apparatus (SCBA)	Highest level of respiratory protection	Air bottle size (30 minutes to 1 hour of air)	Most common choice	■ Mask ■ Air tank ■ Harness ■ Regulator
Air-line respirator	Highest level of respiratory protection with unlimited air supply	Hose line is vulnerable to damage	Longer duration incidents Tight spaces	■ Mask ■ Harness ■ Regulator ■ Air hose ■ Compressor or air tank in cold zone ■ Escape bottle
Closed-circuit SCBA	Highest level of respiratory protection with longer entry times than SCBA	Expensive Chemical CO_2 scrubber creates heat	Longer duration incidents	■ Mask ■ Oxygen tank ■ Chemical scrubber ■ Harness ■ Regulator
Powered air-purifying respirator (PAPR)	Less expensive than SCBA Longer entry times with positive pressure at seal	Cannot be used in IDLH atmospheres Must have appropriate cartridge	Decontamination line and situations requiring longer entry times	■ Mask ■ Harness ■ Blower motor ■ Chemical absorptive cartridge
Air-purifying respirator (APR)	Comparatively inexpensive Lightweight Longer entry times Easy to operate tactically	Cannot be used in IDLH atmospheres May leak significantly around seal Must have appropriate cartridge	Decontamination line and situations requiring longer entry times	■ Mask ■ Chemical absorptive cartridge
Particulate respirator (dust mask)	Inexpensive Lightweight	No protection from gases or vapors!	Protects from large particulates	■ Mask

(Figure 5-15). TB masks, officially known as N95 filtering face pieces, are rated to keep out 95% of the particulates. This means that 5% of the contaminants make it into your respiratory tract. Particulate respirators are usually disposable and filter out dusts, fumes, and mists. The mask consists of a filtering type material and must be replaced when it becomes clogged (difficult to breathe through), discolored, or damaged. Because these respirators provide no protection against chemical vapors and gases, they are not considered acceptable respiratory protection against hazardous materials.

Air-Purifying Respirator (APR)

Air-purifying respirators (APR), also commonly referred to as "gas masks," are tight-fitting respirators that filter out chemicals in the air using a filter cartridge system (Figure 5-16). Filter cartridges are designed to remove specific contaminants from the air, usually by binding the contaminant in an absorbent matrix or by neutralizing the chemical via a chemical reaction. Thus APRs clean the ambient air that passes through the cartridge. The APR mask forms a tight-fitting seal against the face and forehead, and straps that tighten against the back of the head maintain this seal. When the wearer inhales, the negative pressure (or vacuum) forces ambient air from the outside through the filtering cartridge. Depending on the type of cartridge and the physical fitness of the user, this can strain the lungs.

air-purifying respirator (APR) ■ Respiratory protection that filters the ambient air using a filter cartridge containing a material that removes or neutralizes the contaminant. The identity of the contaminant and its concentration must be known before an APR may be used. The filter cartridge must be appropriately selected to remove the contaminant(s). APRs may not be worn in IDLH atmospheres. Also known as cartridge respirator.

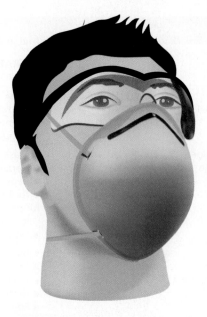

FIGURE 5-15 A dust mask. Dust masks do not provide protection against gases and vapors.

FIGURE 5-16 An air-purifying respirator (APR).

Because APRs filter the ambient air using a chemical-specific process, the following requirements must be met before use:

- Sufficient oxygen must be present (19.5% to 23.5% per OSHA).
- The contaminant must be known (to be able to select the appropriate filtering cartridge).
- The concentration of the contaminant must be known (to remain below IDLH levels and avoid oversaturating the cartridge filter).
- IDLH atmospheres have been ruled out (toxicity, flammability, oxygen deficiency, etc.).

APRs are effective only when used properly. Because inhaling creates a vacuum on the inside of the mask, the mask's seal to the face must be airtight. Facial hair or a poor fit can cause the mask to leak, which means contaminated ambient air is blowing by the seal and entering the user's respiratory system—obviously not a good thing! The APR is also only as good as the selected cartridges, and many types of cartridges are available. If an organic vapor cartridge is selected for an acid gas incident, there is essentially no protection, which, again, is obviously not a good situation. Using APRs requires quite a bit of training and is beyond the scope of the hazmat operations level first responder trained only to the core competency. If your agency is using APRs, you should also be trained to the mission-specific competency of personal protective equipment (see Chapter 6).

Powered Air-Purifying Respirator (PAPR)

Powered air-purifying respirators (PAPRs) are APRs with a blower motor that forces air through the filter cartridge. Thus they essentially provide a positive-pressure environment (although it is not considered to be a true positive-pressure device), which makes them safer and easier to use. They are safer because a negative pressure situation never exists inside the mask. For this reason, a tight-fitting mask is not necessary, and

powered air-purifying respirator (PAPR) ■ Respiratory protection that forces ambient air through a filter cartridge containing a material that removes or neutralizes the contaminant using a belt-mounted pump. The identity of the contaminant and its concentration must be known before a PAPR may be used. The filter cartridge must be appropriately selected to remove the contaminant(s). PAPRs may not be worn in IDLH atmospheres.

hooded PAPRs are very popular in agencies and industries where facial hair is common. Hospitals typically use hooded PAPRs to perform emergency decontamination and preliminary patient treatment on contaminated patients (Figure 5-14). They are easier to use because the lungs do not need to work as hard in forcing the ambient air through the filter cartridges. The drawbacks to PAPRs are that they are heavier (due to the blower motor and battery pack), and the motor may unexpectedly fail in the contaminated area.

Positive-Pressure Self-Contained Breathing Apparatus (SCBA)

The self-contained breathing apparatus (SCBA), which is a type of supplied air respirator (SAR), offers the highest level of respiratory protection when used in positive-pressure mode. Positive-pressure mode refers to a constant air pressure delivery to the mask. If the seal of a positive-pressure SCBA is broken, air will rush out due to the positive pressure delivered to the inside of the mask by the regulator. SCBAs consist of a compressed air bottle, a tight-fitting mask, and one or more regulators to reduce the air pressure. This is the respiratory protection firefighters use (Figure 5-17). As with APRs, the mask must be used properly and fit tightly against the skin. This is why fire departments have strict facial hair grooming standards.

The main advantage of SCBA respiratory protection is that it does not rely on ambient air, but rather its own supply carried in the compressed air tank. The drawback is that SCBA bottles typically come in only 30-minute to 60-minute varieties, which limits the amount of time and work that can be accomplished while wearing an SCBA, a distinct disadvantage compared to APRs. SCBAs are also significantly heavier than APRs, weighing 15 to 25 pounds depending on design and air capacity. Nevertheless, SCBAs are the standard respiratory protection when hazmat personnel must enter unknown or IDLH atmospheres.

FIGURE 5-17 An SCBA is used as part of structural firefighter protective gear.

Positive-Pressure Air-Line Respirator with Escape Unit

Air-line respirators are a type of supplied air respirator in which the air supply is not carried on the back, but rather supplied through an air hose from a remote location. This remote location is usually one or more large compressed air tanks or an air compressor located outside the hot zone. This type of respirator is similar to an SCBA, with the advantage that it is much lighter in weight. Its disadvantage is that the air line may become compromised due to pinching, severing, or bursting. Therefore, a responder must carry a small emergency escape bottle into the hot zone when using an air-line respirator. These respirators are popular when there is little room available, such as in confined spaces, and when longer operation times are required, such as in chemical laboratories, during cleanup operations, and at decontamination lines.

Closed-Circuit SCBA

closed-circuit SCBA ■ A form of supplied air respiratory protection in which the air supply is recycled and used again. The carbon dioxide is removed via a chemical reaction (scrubbed), and oxygen is added from a concentrated source. Closed-circuit SCBAs can be worn for extended periods of time, although the heat generated by the scrubber makes them more uncomfortable and prone to cause hyperthermia in the wearer.

Closed-circuit SCBAs, also known as "rebreathers," are a type of supplied air respirator designed to be worn on the back like an SCBA, but they have a longer work time. The work time can be from 1 to 6 hours depending on the configuration. A closed-circuit SCBA consists of a tight-fitting mask, two or more regulators, a compressed air cylinder, a compressed oxygen cylinder, and a chemical scrubbing system. The chemical scrubbing system removes the exhaled carbon dioxide from the breathing air. This chemical reaction is quite exothermic, or heat releasing, which causes the breathing air to heat up as the unit is used. The unit also feels warm on the back. Both of these factors lead to increased heat stress with closed-circuit SCBA use.

These respirators were developed many years ago for the mining industry and are used today in hazardous materials response when air-line respirators are not practical. The advantage of the closed-circuit SCBA is an extended work period without the inconvenience and distance limitations of an air-line respirator. The disadvantages are that the breathing air is warm to downright hot due to the exothermic reaction of the scrubber and that the units are even heavier than a standard SCBA unit. Today, rebreathers are often used in mines and tunnels, during both rescue operations and hazardous materials incidents.

PROTECTIVE CLOTHING

Protective clothing is used to protect the wearer's body from one or more hazards. As with respiratory protection, different types of protective clothing are available. Leather gloves protect hands from abrasion, insulated clothing protects from temperature extremes, and chemical protective clothing (CPC) protects from hazardous materials. Table 5-2 summarizes their properties.

TABLE 5-2	Protective Clothing		
PROTECTIVE CLOTHING	**PURPOSE**	**ADVANTAGES**	**LIMITATIONS**
Thermal protective clothing	High-temperature flammable liquid firefighting	Protects from radiant heat	Little, if any, chemical protection
Structural firefighter protective clothing	Structural firefighting	Protects from moderate radiant heat with some flash fire protection	Little, if any chemical protection
Chemical protective clothing	Hazardous materials response	When appropriately selected, chemical resistant	No thermal protection Heat stress

Thermal Protective Clothing

Thermal protective clothing is designed to protect the wearer from extremes in temperature. This may be a pair of cryogenic gloves used to handle liquid nitrogen, it may be structural firefighter protective gear used to fight fires, or it may be aluminized high-temperature thermal protective clothing. High-temperature thermal clothing, also called a proximity suit, is used by refinery fire brigade personnel and airport fire personnel due to the extreme heat produced by large quantities of burning hydrocarbons. Thermal protective clothing is *not* considered chemical protective clothing.

Structural Firefighter Protective Clothing

Structural firefighter protective clothing, or turnout gear, is a type of thermal protective clothing designed to protect the wearer from extremes in heat. Yet even this is limited. Turnout gear consists of a helmet, SCBA with a personal alert safety system (PASS) device, a thermal protective hood, a thermal protective coat (bunker coat), thermal protective pants (bunker pants), and steel-toed boots.

Structural firefighter protective clothing has very limited to no chemical protective qualities and should generally not be used in the hot zone at hazardous materials incidents. However, certain exceptions apply: Turnout gear may be appropriate when the fire risk outweighs the other health and safety risks. An example would be a methane or propane leak in which a valve must be turned off. The primary hazard is flammability, the biggest risk is inhalation and not skin absorption, and a positive-pressure SCBA is part of the turnout gear. Turnout gear is clearly much safer than chemical protective clothing in this case. Other examples are immediate lifesaving rescues (consult the 3/30 rule discussed in Chapter 12, Operations Level Responders Mission-Specific Competencies: Victim Rescue and Recovery) and hazardous materials releases that require thermal protection with minimal to no chemical protection (such as some cryogenic gases).

Chemical Protective Clothing (CPC)

Chemical protective clothing (CPC) is designed to protect the wearer from chemicals. The different types of CPC are intended to keep the chemical from coming into contact with the body. Some CPC protects from liquid exposures (splash protection), whereas other CPC protects from gaseous and vapor exposure as well (vapor-protective suits). CPC comes in a variety of protective qualities, sizes, shapes, colors, thicknesses, flexibilities, shelf lives, and costs. Some CPC is disposable, whereas other CPC is reusable. Some is complicated to use and maintain, wheras other types are simple to use and maintain. It is vital to become familiar with the type of CPC your agency uses. CPC is an extremely important component that keeps you safe at hazardous materials incidents.

Chemical protective clothing comes in a variety of configurations. The more effective, and expensive, CPC consists of several chemical-resistant materials that are laminated together. The different types of materials have different resistance to chemicals and must be appropriately selected using manufacturer-supplied information (compatibility charts) (Figure 5-18). The next chapter covers the types of chemical protective clothing and selection criteria in greater detail.

Chemicals can breach CPC in three ways: penetration, degradation, and permeation. **Penetration** refers to the passage of material through a macromolecular (relatively large) opening in CPC, such as a zipper, seam, or tear. **Degradation** refers to the chemical breakdown or dissolution of the CPC material. An example of degradation is Styrofoam dissolving in acetone. **Permeation** refers to the passage of material through a microscopic passage in the CPC. CPC is composed of materials, whether textiles or plastics, that are weaved from strands of material. There are spaces between the weave that some chemicals can get through, or permeate. All chemicals permeate through all materials at some rate. An example of permeation is how helium leaks out of a rubber balloon in a day or two so that the balloon no longer rises. For a compatible piece of CPC, this will be several hours or more; for an incompatible material, it will be seconds or minutes.

penetration ■ The process of a chemical moving through a sizable opening in a material, such as through a zipper, seam, or tear.

degradation ■ The breakdown of clothing, material, or equipment through the action of a chemical.

permeation ■ The process of a chemical moving through a material at the molecular level (through the natural pore spaces of a polymer).

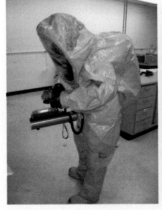

FIGURE 5-18 Manufacturer's compatibility chart for selection of chemical protective gloves. *Courtesy of Showa Best Glove Inc.*

Key To Degradation Rating

E = Excellent	
G = Good	
F = Fair	
P = Poor	
NR = Not Recommended	

Key To CPC Index Number

0, 1, 2, 3, 4, 5

NEOPRENE 6780   ULTRAFLEX NEOPRENE 32

CHEMICAL	CAS NUMBER	EUROPEAN TOXICITY RISK CODE	NEOPRENE 6780 DEGRADATION RATING TIME IN MIN 5	30	60	240	PERMEATION BREAKTHROUGH MDL PPM	BDT MIN	RATE ug/cm²/min	CPC INDEX RATING (0-5)	NEOPRENE 32 DEGRADATION RATING TIME IN MIN 5	30	60	240	PERMEATION BREAKTHROUGH MDL PPM	BDT MIN	RATE ug/cm²/min	CPC INDEX RATING (0-5)
171. PROPYLENE GLYCOL MONOMETHYL ETHER	107-98-2	X	E	E	E	P	0.02	ND	ND	0	E	E	E	E	0.02	ND	ND	0
172. PROPYLENE GLYCOL MONOPROPYL ETHER	1569-01-3	X	E	E	E	E	0.02	ND	ND	0	E	E	E	E	0.02	177	52	3
173. PROPYLENE OXIDE	75-56-9	X, CANCER	G	G	F	F	0.02	11	204	5	G	G	G	G	0.02	43	47	3
174. PROPYL PROPASOL SOLVENT	1569-01-3	X	E	E	E	E	0.02	ND	ND	0	E	E	E	E	0.02	177	52	3
175. PYRIDINE	7291-22-7	X	E	F	P	P	0.02	NR	NR	5	G	P	NR	NR	0.02	NR	NR	5
176. REFRIGERANT 123A	306-83-2	V	E	E	E	E	0.02	73	62	3	E	E	G	G	0.02	85	180	4
177. REFRIGERANT 141B	1717-00-6	V	E	G	F	F	0.02	68	2688	5	E	G	F	F	0.02	21	502	4
178. RUBBER SOLVENT	8032-32-4	X	E	E	E	E	0.02	99	10	2	E	E	E	E	0.02	47	110	4
179. SAFROTIN (50% in ROH)	31218-83-4	X	E	E	E	E	0.02	ND	ND	0	E	E	E	E	0.02	ND	ND	0
180. *SODIUM HYDROXIDE 50%*	1310-73-2	Cx	E	E	E	E	0.02	ND	ND	0	E	E	E	E	0.02	ND	ND	0
181. SODIUM HYPOCHLORITE 4–6%	7681-52-9	C	E	E	E	E	45.0	ND	ND	0	E	E	E	E	45.0	ND	ND	0

EPA LEVELS OF PROTECTION

The EPA uses four levels of protection based upon a combination of chemical protective clothing (CPC) and respiratory protection offered. The lowest level of protection is designated level D, which consists of nothing more than a uniform or other clothing. There is limited (dust mask) or no respiratory protection and little to no chemical protective clothing. Level C consists of minimal respiratory protection, such as an air-purifying respirator (APR), and chemical splash protection. Level B consists of supplied air respiratory protection and chemical splash protection. Level A, the highest level of protection, consists of supplied air respiratory protection and a vapor-tight suit. Figure 5-19 shows examples of all four EPA levels of protection.

Level D

Level D protection essentially consists of no chemical or respiratory protection at all (Figure 5-19, far right). Street clothes or a daily work uniform can be considered level D protection. However, even the universal precautions used by paramedics—latex or nitrile gloves, and eye and face protection—are also merely level D protection. Even with the addition of chemical protective clothing, a dust mask, or a TB mask, it would still be considered level D protection. Level D protection may be used only in environments with airborne contaminant levels below the permissible exposure limit (PEL).

FIGURE 5-19 The four levels of PPE according to the EPA (from left to right): Level A, Level B, Level C, and Level D.

Level C

Level C protection consists of chemical protective clothing that provides splash protection and an air-purifying respirator (APR) (Figure 5-19, second from right). This level of protection is typically used when some splash protection and some level of respiratory protection are required. The EPA considers any level of respiratory protection below an APR to be level D.

An air-purifying respirator, or cartridge respirator, filters the ambient air and removes contaminants. Figure 6-6 shows some common examples of APR cartridges and masks.

Several criteria must be met for the use of level C protection:

1. Sufficient ambient oxygen exists.
2. All dangerous contaminants have been identified.
3. Contaminant concentrations are known.
4. Contaminant concentrations are below IDLH levels.
5. Appropriate cartridges are available that are able to filter out contaminants.
6. No other IDLH conditions exist.

Because APRs filter only the ambient air, we must first determine that ambient oxygen levels are normal, or at least above 19.5% and below 23.5%. Atmospheres are considered immediately dangerous to life and health (IDLH) when oxygen concentrations fall below 19.5% or rise above 23.5%. We must know what the contaminant is in order to choose the appropriate air-purifying cartridge or canister. The absorbent or reactive material inside the canister must effectively exclude or neutralize the contaminant of interest. Furthermore, the cartridge or canister must have an end of service life indicator (ESLI), or we must determine an appropriate change out schedule based upon the manufacturer's recommendations. The contaminant level must be known before using an APR to ensure ambient contaminant levels are, and remain, below IDLH levels. Contaminant levels can be determined using air-monitoring instruments, the use of which is covered in Chapter 11, Operations Level Responders Mission-Specific Competencies: Air Monitoring and Sampling. Finally, we must complete a safety analysis of the hot zone and determine that no other IDLH conditions are present. Other IDLH conditions include flammable atmospheres, permit-required confined spaces, and other situations that pose an immediate danger to life and health.

Level B

Level B protection is used when some splash protection and the highest level of respiratory protection are needed (Figure 5-19, second from left). It consists of the same types of chemical protective clothing and offers the same splash protection seen with level C. However, instead of using an APR, we now use supplied air respiratory protection. Supplied air respiratory protection may consist of self-contained breathing apparatus (SCBA) or a supplied air respirator (SAR). Because supplied air is being used, level B suits may also be fully encapsulating.

Level A

Level A protection is used when the highest level of skin and respiratory protection is needed. The chemical protective clothing that is used is vapor tight. Level A suits have sealed seams, integrated gloves and booties, an airtight zipper, and one-way exhalation valves (Figure 5-19, far left). Because these suits are vapor tight, they are also thicker and heavier than level B or C suits, and wearers are more prone to heat stress. Level A is the highest level of personal protective equipment ensemble in the EPA's system.

SELECTING PERSONAL PROTECTIVE EQUIPMENT

Agencies have different types of personal protective equipment they use on a daily basis. For example, firefighters have structural firefighter protective ensembles with self-contained breathing apparatus (SCBA); law enforcement officers have bulletproof vests; EMS providers have latex gloves, TB masks, and protective eyewear. Although this personal protective equipment works well for its intended use, none of this equipment is designed for hazardous materials responses. The PPE designed for hazardous materials releases must be chemically resistant and adequately protect responders as needed from the four routes of entry: inhalation, ingestion, skin absorption, and injection. Some of the previously mentioned personal protective equipment may offer limited chemical protection in an emergency, but unless it has been specifically designed and tested for chemical resistance, it is not adequate for hazardous materials responses.

Most of the personal protective equipment used today is disposable. Generally, agencies have cost recovery ordinances in place to replace the equipment they have used. If PPE will be reused, it must be thoroughly decontaminated, cleaned, inspected, and placed back into service. All PPE should be maintained and stored per the manufacturer's recommendation and inspected and tested at reasonable intervals.

Because personal protective equipment designed for hazardous materials response must be impervious to chemicals, it is often heavy and hot to wear. Operations level responders must be aware of their physical, mental, and medical limitations when deciding to use hazardous materials–rated personal protective equipment. Medical conditions such as heart disease, obesity, respiratory disease, and poor physical conditioning may limit the use of PPE. Mental factors, such as claustrophobia, may play a role, too.

Thermal Stress

Thermal stress can be pronounced while wearing PPE. Fully encapsulating suits are most prone to producing heat stress because they trap moisture and heat close to the wearer. Heat cramps are the first indication of heat stress. Heat exhaustion, which is marked by profuse sweating and exhaustion, is a more serious sign of a heat emergency. Heat exhaustion can be quickly followed by heatstroke. Heatstroke, which is marked by red, dry skin and elevated body temperature, is a true medical emergency that leads to many fatalities every year.

At the opposite extreme, cold stress can be a significant problem in colder climates, especially in the winter. Frostbite injuries to the fingers can occur rapidly when working in subzero environments and handling cold metals or cryogenic liquids. Hypothermia is also a concern as it can occur when personnel work in cold environments for a long period of time. Hypothermia occurs when the core body temperature drops because the body cannot produce enough heat to maintain 98.6°F (37°C). Symptoms of hypothermia may include altered mental status, shivering, and cold and pale-looking extremities. Victims of hypothermia must be removed from the cold environment and gradually rewarmed. Hypothermic patients are more susceptible to heart attacks and must be handled gently.

Entry personnel should be rotated in and out of the hot zone at reasonable intervals. It is very important to monitor entry personnel for signs and symptoms of heat and cold stress before and after hot zone entry.

USING PERSONAL PROTECTIVE EQUIPMENT

When using PPE several safety factors should be kept in mind. HAZWOPER (29 CFR 1910.120) and the respiratory protection standard (29 CFR 1910.134) both require that the buddy system be used in immediately dangerous to life and health (IDLH)

SOLVED EXERCISE 5-7

What would be the appropriate personal protective equipment to use for patient extrication?

Solution: Full firefighter structural protective gear including SCBA. The greatest risk of the ethanol is flammability. Turnout gear and SCBA are excellent protection against fire. Chemical protective clothing (CPC) would not be indicated due to the flammability hazard and the low toxicity of ethanol, although, as previously indicated, due to the weather conditions (low temperature), the ethanol does not pose a high flammability risk at this incident.

atmospheres. In addition, a backup team or rescue team should be on standby and available for immediate deployment in the event of an accident.

Personal protective equipment is designed to protect the wearer from coming into contact with the hazardous material. However, it will not shield the wearer from hazards such as explosions, fire, and puncture or abrasion. Personnel wearing PPE should safely approach the work area. Responders should enter the hot zone from upwind, uphill, and upstream of the designated control points near the decontamination line. PPE is bulky and unwieldy, making slip, trip, and fall hazards more likely.

Medical monitoring, both pre-entry and post-entry, is required when using PPE. Medical screening consists of monitoring the vital signs of the PPE wearers before they go in the hot zone (pre-entry screening) and after they return and have been decontaminated (post-entry screening). These values should be compared, and if the vital signs do not return to normal within 10 to 15 minutes, medical monitoring should be continued and medical treatment started. Along with medical screening, proper rehabilitation should be part of the PPE plan. Rehabilitation includes medical monitoring, rest, rehydration, and food. PPE use is very physically taxing, and the wearer should have a chance to recover before being assigned to perform other strenuous duties.

The Buddy System

The buddy system refers to at least two personnel always entering the hot zone together (Figure 5-20). HAZWOPER mandates the buddy system for any personnel entering the hot zone. Personnel should *never* enter the hot zone alone. The incident commander and others may be held criminally liable if an injury or death occurs due to a violation of this rule. The buddy system is also mandated by the respiratory protection standard and specifically refers to personnel using breathing apparatus. Using the buddy system increases the safety of personnel entering the hot zone. If one person suffers a medical emergency, has equipment failure, or is overcome by the chemical, the other person may be able to effect a rescue and summon outside help in the form of the backup team.

Resource **Central**

See Hazardous Waste Operations and Emergency Response (HAZWOPER) and OSHA Respiratory Standard from the Code of Federal Regulations for more information.

FIGURE 5-20 The buddy system in use as an entry team enters the hot zone.

FIGURE 5-21 A backup team standing by in case of an emergency in the hot zone. Note that team members are dressed in PPE and only need to go on air to make a rescue.

Backup Teams

A backup team, or rapid intervention team (RIT), is required by HAZWOPER when personnel enter the hot zone. In the respiratory protection standard, this is referred to as the two-in/two-out rule. The backup team serves as a rescue team should anything happen to the entry team in the hot zone (Figure 5-21). The backup team must be ready to enter the hot zone at a moment's notice, which means team members should have their PPE donned and be ready to go on air. They should be wearing at least the same level of PPE as the entry team.

As with a fire department rapid intervention team, backup team members should have appropriate tools at their disposal, such as air-monitoring equipment and rescue tools. This includes tools to remove the entry team in case they become incapacitated, such as Sked boards or stretchers (see Chapter 12, Operations Level Responders Mission-Specific Competencies: Victim Rescue and Recovery). Remember, it is difficult to remove people dressed in full PPE while you yourself are in PPE. This requires practice and proficiency with the rescue equipment. Also, the backup team must be vigilant to avoid the same hazard that caused the entry team to become incapacitated.

Safety Briefing

Before anyone enters the hot zone, a thorough hazard assessment must be conducted. The hazards must be evaluated and appropriate countermeasures instituted. After the written site safety plan has been formed, the entry team, backup team, and decontamination team must be briefed before anyone enters the hot zone (Figure 5-5). This briefing will alert the team members to the dangers present in the hot zone and the actions they are expected to perform. When complicated incident objectives are to be carried out, such as air monitoring, valve repositioning, or involved leak control procedures, the safety briefing may be longer than if a rescue must be performed. In either case, the briefing should cover only

Write a safety briefing for the personnel entering the hot zone to perform patient extrication in the incident described in the opening scenario.

Solution: You will be entering the hot zone from the west to perform a rapid extrication of the pickup truck driver. Use full structural firefighting protective gear with SCBA. The chemical hazard is posed by ethanol, a nontoxic but flammable liquid. At the current temperature ethanol will not generate enough vapors to pose a flammability hazard. Nonetheless, stay out of the product. In case you do come into contact with the material, a dry decontamination line has been set up at the west side of the hot zone. Contamination will initially be detected, and decontamination will be verified, using a photoionization detector (PID). Wet decon will not be used due to the low temperatures. While you are performing extrication, a second crew will be monitoring the leak and confining the spill. Crew members will perform continuous air monitoring for oxygen and combustible gases using the RKI Eagle monitor. Calibration has been verified through quantitative bump checking, and the monitor has been fresh air zeroed. A backup team will be standing by with a charged hose line and rescue equipment. Have you looked at the site map, and do you understand the area you will be working in? We will have radio communications on tactical frequency #3 and line-of-sight visual contact. Do you have any questions?

the needed information at a level appropriate to the task at hand. A 20-minute briefing cannot, and will not, be remembered. Cover the important stuff, and do it well!

At the minimum, the safety briefing should address the following:

- Health hazards
- PPE use and limitations
- Site layout
- Incident objectives
- Equipment use and limitations
- Radio frequencies
- Contact frequency and stay times
- Emergency signals
- Decontamination line location and procedures

The safety briefing should be carried out in the presence of all personnel entering the hot zone, the decontamination line personnel, and the group leaders. Typically the safety briefing is conducted by the incident commander, hazmat branch director, research group leader, or the safety officer. Who carries out the briefing will depend on the size of the incident, the complexity of the incident, and the technical expertise of the relevant personnel.

Victim Recovery and Triage

Victim recovery, or hot zone rescue, has been a controversial subject at the hazardous materials operations level (Figure 5-22). Traditionally, operations level responders have not entered the hot zone for any reason, including victim rescue and recovery. However, it has been recognized that if hazardous materials response teams are not immediately dispatched, victim rescue and patient treatment may be needlessly and catastrophically delayed. In the new NFPA 472 (2008) standard, operations level responders are allowed to operate in the hot zone under certain circumstances with the proper training and personal protective equipment. Hot zone rescue is covered in greater detail in Chapter 12, Operations Level Responders Mission-Specific Competencies: Victim Rescue and Recovery.

If your agency chooses not to train to the hot zone rescue mission-specific competency, you can still take certain actions to help ambulatory victims exit the hot zone:

1. Verbally call to ambulatory victims to leave the immediate area of release.
2. Have them stop at the edge of the warm zone.

FIGURE 5-22 Victim rescue and recovery.

3. Instruct them to remove their contaminated clothing.
4. Provide emergency decontamination using available means.
5. Give the victims disposable clothing to re-dress.
6. Quarantine the victims in the warm zone.
7. Do not come into contact with the victims to prevent cross contamination.
8. Prioritize emergency medical care using a triage system such as START.
9. If patient treatment is started, use appropriate personal protective equipment.

Keep in mind always to follow your agency's standard operating procedures or guidelines (SOPs/SOGs). For example, this technique could be used at a plating facility with a cyanide release: Operations level responders could use a bullhorn or the facility's public address system to instruct people to exit the building and meet on the upwind side at a specified location. At this point, from a distance, the victims are told to remove their clothing and could be decontaminated using a hose line. Then, after the arrival of the hazardous materials response team and the EMS agency, the patients can receive a more advanced level of care.

Formulating a Decontamination Plan

HAZWOPER requires that a decontamination plan be in place before personnel enter the hot zone. A decontamination line can be as simple as pulling a hose line off a fire engine or using an emergency shower at a facility. Typically, **technical decontamination** lines are more intricate than that. Figure 5-23 shows several examples of decontamination lines, which will typically be set up by personnel with training in the mission-specific competency in technical decontamination covered in Chapter 8 and/or in mass decontamination covered in Chapter 7.

technical decontamination ■ The organized and systematic removal of contaminants from entry team members and victims.

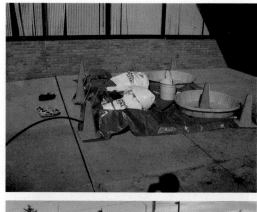

FIGURE 5-23 Decontamination lines may be small or large depending upon the needs of the incident.

Why do we set up a decontamination line? There are two primary reasons: (1) to prevent the hazardous material from causing further injury to people or property, and (2) to prevent cross contamination and the spread of contamination to uncontaminated areas (secondary contamination). However, it can be very difficult to decontaminate both people and property. Some chemicals are soluble in water; others are not. Still others are soluble in soap and water, yet others are not. How do we know that we have successfully decontaminated the patient? We must verify that decontamination is complete using the appropriate detection equipment. For a corrosive material, pH paper may be an appropriate detection method. Until we can verify that decontamination has been successful, we must quarantine the patients and any equipment we have decontaminated in the warm zone. Air monitoring is discussed in greater detail in Chapter 11, Operations Level Responders Mission-Specific Competencies: Air Monitoring and Sampling.

EMERGENCY DECONTAMINATION

Operations level responders may set up emergency decontamination for victims at hazardous materials incidents and weapons of mass destruction incidents (Figure 5-24). This is one of the most important functions operations level responders can perform to ensure public safety. The purpose of emergency decontamination is to prevent further injury to the patient from the hazardous material.

SOLVED EXERCISE 5-9

Would you need a backup team and decontamination plan in the incident described in the opening scenario?
Solution: Yes. The accident scene is considered the hot zone of a hazardous materials incident, and HAZWOPER mandates a backup team and decontamination plan before any personnel enter the hot zone.

Emergency decontamination is very simple and fast to set up. It should be located in the warm zone, upwind, uphill, and upstream of the incident. How emergency decontamination is carried out is up to your imagination and may include:

- A charged hose line
- Water or foam and water fire extinguisher
- Shower facilities in a nearby building
- Emergency shower and eyewash at industrial or research facilities

When life safety is involved, the EPA has stated that runoff does not have to be collected. However, as soon as resources permit or the life safety priorities have been taken care of, all runoff from decontamination efforts must be collected and treated as hazardous waste until proven otherwise.

The faster we can get the hazardous material off the patient, the better his or her prognosis will be. All of a patient's clothing must be removed for decontamination to be effective. This includes underwear and jewelry. If these articles are not removed, chemicals can be trapped underneath. Visible contamination can be brushed off or padded off, and the remaining contamination should be washed off.

Some chemicals can be removed with plain water, whereas others will require soap and water. However, be aware that not all chemicals can be removed even with soap and water. Yet if water is the only decontamination solution available, use it! Emergency decontamination should be followed with technical decontamination or patient decontamination unless complete decontamination can be verified. Emergency decontamination is covered in more detail in Chapter 7, Operations Level Responders Mission-Specific Competencies: Mass Decontamination.

TECHNICAL DECONTAMINATION OF ENTRY TEAM MEMBERS

In many jurisdictions, operations level responders are used to setting up technical decontamination lines for the responding hazardous materials response team. A technical decontamination line has several different stations and is designed to decontaminate entry personnel and allow them to quickly reenter if needed. As such, they may have several different stations, including a primary wash, a secondary wash, an SCBA doffing station, and a re-dress area. Figure 5-25 shows a schematic of a technical decontamination line layout. If you are

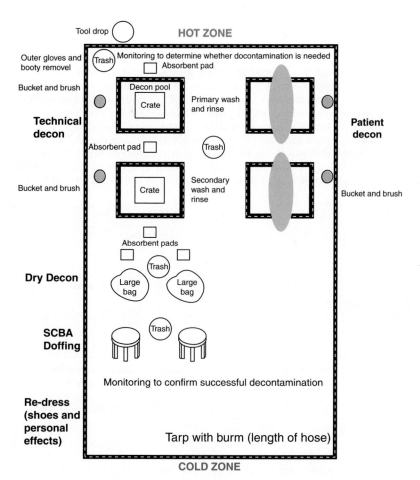

Within the figure:

HOT ZONE

Tool drop

Outer gloves and booty removel

Bucket and brush

Technical decon

Trash

Monitoring to determine whether docontamination is needed

Absorbent pad

Decon pool

Crate

Primary wash and rinse

Patient decon

Absorbent pad

Trash

Crate

Secondary wash and rinse

Bucket and brush

Bucket and brush

Absorbent pads

Trash

Large bag

Large bag

Dry Decon

SCBA Doffing

Trash

Monitoring to confirm successful decontamination

Re-dress (shoes and personal effects)

Tarp with burm (length of hose)

COLD ZONE

expected to set up technical decon, your standard operating guidelines should explain how to set up the technical decontamination line and have a diagram of it. Technical decontamination is covered in much more detail in Chapter 8, Operations Level Responders Mission-Specific Competencies: Technical Decontamination.

Terminating the Incident

The final task that must be completed at a hazardous materials incident is proper termination of the incident. This is especially important for several reasons: Hazardous materials incidents tend to generate hazardous waste, including contaminated equipment, PPE, and runoff that must be disposed of properly. The EPA requires that runoff from decontamination activities be properly disposed of as hazardous waste. It is also important that the incident be properly documented, especially if there have been exposures to hazardous materials. Properly documented exposures will lead to proper medical monitoring in the future, as well as possible worker's compensation claims being honored.

A thorough debriefing and critique should be carried out at the end of hazardous materials incidents. Hazardous materials incidents are uncommon occurrences for most agencies. As such, there are usually many lessons to be learned from the response that can make future responses more effective. If many different departments or agencies were involved in the hazardous materials incident, it is also prudent to have an after-action review with members of these different entities.

Operating at Criminal Incidents and WMD Incidents

Criminal incidents and weapons of mass destruction incidents can be significantly more dangerous than hazardous materials incidents. It is not only that the chemical or weapon

How would decontamination be accomplished in the incident described in the opening scenario? Draw the decon line, and state the decon solution you would use. What are some special considerations for this incident?

Solution: Due to the weather conditions, dry decon would be the best option. Wet decon, such as washing the patient down with a hose line, would greatly increase the risk of hypothermia and severe complications. Ethanol is not particularly toxic and dry decon could be safely performed. Any ethanol contamination found on the skin could be removed using a wet cloth because ethanol is completely water soluble. The patient should be assessed for ethanol contamination at the decon line using at least a combustible gas indicator (CGI) because ethanol is flammable, and preferably using a photoionization detector (PID) because it can detect ethanol at much lower levels. After dry decon is complete, decontamination should be verified using detection equipment (CGI or PID).

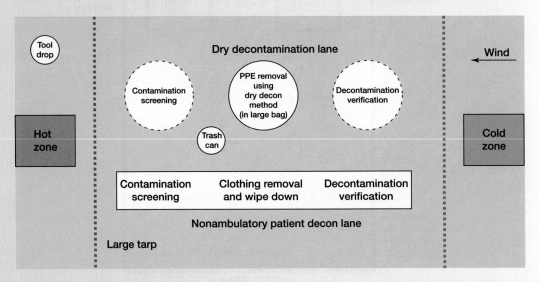

of mass destruction may be that much more dangerous than a normal hazardous material, but also that the intent of the release is to harm people and that secondary devices may be placed to harm you, the operations level responder.

If a terrorist incident is suspected, the National Terrorism Advisory System (NTAS) should be consulted for the latest intelligence information by contacting dispatch and local law enforcement. The NTAS replaces the older color-coded system in use after 9/11. NTAS alerts are only issued when credible and actionable information is available. There are two threat levels according to the U.S. Department of Homeland Security:

- *Elevated Threat Alert*—Warns of a credible terrorist threat against the United States.
- *Imminent Threat Alert*—Warns of a credible, specific, and impending terrorist threat against the United States.

These alerts provide useful information to first responders, including a summation of the possible threat, actions being implemented to maximize public safety, and concrete actions that the public, small and large businesses, and local, state, and tribal governments can take in response to the threat. The NTAS alerts are tailored to the threat. Alerts may be sent only to law enforcement, to targeted areas in the private sector, to the entire nation—or a combination of the aforementioned. Therefore it is important to contact law enforcement in case a limited NTAS alert was issued. NTAS alerts will automatically expire on a certain date and time, eliminating a constant state of alert. Therefore, the latest NTAS alert may contain valuable current information that will point first responders in the right direction at suspected terrorist incidents involving hazardous materials or WMD.

BOOBY TRAPS AND SECONDARY DEVICES

Booby traps and secondary devices are designed to harm the public and first responders. Booby traps are located at locations such as illicit laboratories to protect the contents from both competitors and law enforcement. These booby traps are designed to keep others out. Secondary devices, on the other hand, are designed to kill and maim primarily first responders called to the scene of the original emergency. This is to cause further chaos and fear and to prevent first responders from treating the victims of the initial attack.

Booby traps come in many forms, including explosives, chemicals, firearms, and mechanical devices. Common booby traps include containers full of hazardous chemicals propped above doorways that fall on anyone entering, fishhooks strung at eye level both indoors and outdoors, shotguns rigged with a pull string tied to doorknobs, lightbulbs partially filled with gasoline inserted into light sockets, and improvised explosive devices (IEDs) constructed from scavenged or homemade explosives. It is therefore very important to watch where you walk, be careful opening doors, and not to turn on light switches.

Secondary devices are typically improvised explosive devices (IEDs) designed and placed to cause maximum damage and casualties. The IEDs may contain shrapnel to cause death and injuries and/or force directional devices to maximize the impact of the explosive device in a certain area. Secondary devices were first used against first responders in the United States at abortion clinics and at a nightclub in Atlanta, Georgia, by Eric Robert Rudolph in 1997.

Secondary devices are especially insidious because they are located to maximize the number of casualties at probable staging sites for incoming first responders. It is therefore important to vary response patterns when confronted with routine calls. A terrorist may have triggered the false alarm or minor incident in order to watch your response. If he or she notices a pattern, it may be exploited in the future.

At all incidents, especially suspicious ones, scan the immediate areas in which operations are to be conducted, resources are to be staged, patients are treated and triaged, or the command post is located. Look for areas that could contain secondary devices, such as shrubbery, trash cans, vehicles, or large objects that may redirect the force of an explosion. Try to avoid these areas whenever possible. Visually screen the areas you choose to set up operations in carefully. If you do find something suspicious, do not touch it, isolate the area immediately, and contact the bomb squad. Secure your operational and command areas quickly by isolating nonessential personnel and the public. Enforce these security control zones and conduct periodic security checks at incidents of longer duration.

EVIDENCE PRESERVATION

Evidence preservation is extremely important in criminal incidents and weapons of mass destruction incidents. If proper evidence preservation techniques are not used, it will be almost impossible to find and prosecute the perpetrator. For more detail on evidence preservation, see Chapter 9, Operations Level Responders Mission-Specific Competencies: Evidence Preservation and Sampling.

The core philosophy of evidence preservation is to send as few personnel into the crime scene (hot zone) as possible and have them disturb only what is absolutely necessary to do their job. However, life safety should not be sacrificed in the name of evidence preservation. During patient decontamination, for example, ensure that the clothing is bagged and tagged with the patient's name so it can be examined later by evidence technicians. At the bombing of the Alfred P. Murrah Federal Building in Oklahoma City, Oklahoma, it was an alert law enforcement officer that spotted the axle of the rental van and guarded it until it could be bagged and tagged as evidence. The axle was critical in prosecuting the perpetrators. Evidence preservation should be addressed in your agency's standard operating guidelines and emergency response plans.

Summary

Responding to hazardous materials incidents and weapons of mass destruction incidents can be dangerous, and safety should always be a top priority. Using the incident command system and forming a sound incident action plan that takes safety into consideration is essential. Life safety, incident stabilization, and property and environmental conservation should be the mainstays of your response objectives. Isolating the immediate area and instituting solid scene control procedures will minimize any further life safety hazards. People who have already been contaminated and are injured should be quickly decontaminated by setting up an emergency decontamination line. Minimize cross contamination by using the appropriate personal protective equipment and sound decontamination procedures. Operations level responders should primarily take a defensive role at hazardous materials incidents unless they have been trained to one or more of the mission-specific competencies in the following chapters.

Review Questions

1. What are the key components of an incident action plan (IAP) at a hazardous materials incident?
2. What are three factors to consider when prioritizing response objectives?
3. What are the three control zones called at hazardous materials incidents, and what are the differences among them?
4. List the important criteria for personal protective equipment (PPE) selection.
5. What is the importance of decontamination at hazardous materials incidents?

Problem-Solving Activities

1. In your opinion, what information is essential to know when responding to hazardous materials incidents? Justify your response.
2. Write down an incident command system flowchart for a response to a hazardous materials incident within your agency. What role would you likely play at this incident given your normal job duties?
3. Review a recent response to a hazardous materials incident within your department or jurisdiction. Analyze the response and the incident conditions, and write a site safety plan for that incident.

References and Further Reading

Forsberg, Krister, and S. Z. Mansdorf. (2007). *Quick Selection Guide to Chemical Protective Clothing* (5th ed.). New York: John Wiley.

National Fire Protection Association. (2008). NFPA 472, *Standard for Competence of Responders to Hazardous Materials/Weapons of Mass Destruction Incidents.* Quincy, MA: Author.

National Fire Protection Association. (2008). NFPA 473, *Standard for Competencies for EMS Personnel Responding to Hazardous Materials/Weapons of Mass Destruction Incidents.* Quincy, MA: Author.

Noll, Gregory G., Michael S. Hildebrand, and James Yvorra. (2005) *Hazardous Materials: Managing the Incident.* Chester, MD: Red Hat Publishing.

Occupational Safety and Health Administration. (1990). 29 CFR 1910.120, *Hazardous Waste Site Operations and Emergency Response (HAZWOPER).* Washington, DC: U.S. Department of Labor.

Occupational Safety and Health Administration. (1984). 29 CFR 1910.134, *Respiratory Protection Standard.* Washington, DC: U.S. Department of Labor.

U.S. Department of Homeland Security. (2004, March). *National Incident Management System.* http://www.fema.gov/emergency/nims.

U.S. Department of Homeland Security. (2004, December). *National Response Plan.* http://www.dhs.gov/nrp.

Weber, Chris. (2007). *Pocket Reference for Hazardous Materials Response.* Upper Saddle River, NJ: Pearson/Brady.

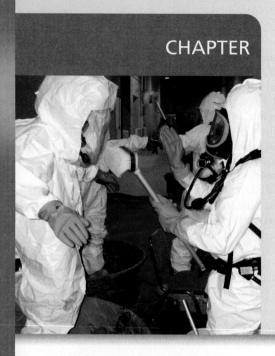

OBJECTIVES

- Select the appropriate respiratory protection for a given hazardous materials or WMD incident.
- Select the appropriate chemical protective clothing (CPC) for a given hazardous materials or WMD incident.
- Describe the types of thermal stress while wearing PPE and the dangers they pose.
- Don and doff personal protective equipment (PPE).
- Describe how PPE should be stored and maintained.
- Describe how PPE should be tested.

You Are on Duty! Train Derailment at the Edge of Town

Minot, North Dakota

In the middle of the night on January 18, 2002, a major train derailment occurred just outside Minot, North Dakota. Tank cars carrying anhydrous ammonia ruptured and released a caustic vapor cloud that drifted into Minot. Thirty-one railcars derailed due to metal fatigue in track components. Public notification was not efficiently accomplished due to the time of the derailment and the remote location of the incident. Local television stations were unstaffed and did not interrupt their programming to notify the public of the emergency. One man died and almost 100 more were injured.

- How could the public be notified to evacuate or shelter in place?
- What type of personal protective equipment (PPE) would be needed to enter the ammonia cloud to notify the public and/or effect rescues?
- What other factors would affect your PPE decision?

Let's try to answer these questions in this chapter.

For most agencies, training at the operations level will include the mission-specific competency in personal protective equipment. Personal protective equipment (PPE) is an essential component of hazardous materials emergency response. It is of utmost importance to understand what PPE is, how it is used, how to maintain it, and how to select the appropriate PPE for the particular circumstances.

In what situations might we expect to use the mission-specific competency in personal protective equipment? Firefighters may be asked to perform victim rescue and recovery at a production plant accident as they initially arrive on the scene. Law enforcement officers may be asked to gather evidence at the scene of a clandestine methamphetamine lab. Paramedics may be asked to treat a patient in the warm zone. Public health professionals may be asked to gather laboratory samples in the wake of a disease outbreak. In all of these cases, personal protective equipment is an essential component necessary for safety.

In this chapter we discuss the selection and use of personal protective equipment in much greater detail than in the last chapter. It is crucial to understand that there are different types of personal protective equipment for different hazards; for example, chemical protective clothing, **respiratory protection**, thermal protection, explosive protection, radiation shielding, hearing protection, and eye protection. You will have to make choices among the different varieties of personal protective equipment your employer offers. Each type of PPE has its particular advantages and disadvantages, safety considerations, chemical compatibility, procedures for donning and doffing, and shelf life and storage criteria.

respiratory protection ◼ PPE that prevents airborne contaminants from being inhaled.

Types of Personal Protective Equipment (PPE)

Personal protective equipment can be classified in many different ways. The NFPA uses ensembles that are based on a self-consistent system of chemical protective clothing. The U.S. Environmental Protection Agency (EPA), on the other hand, uses four distinct levels of protection based primarily on respiratory protection. It is important to understand that personal protective equipment consists of a number of different separate components, including both respiratory protection and thermal and chemical protective clothing.

RESPIRATORY PROTECTION

Respiratory protection can be considered the most important aspect of personal protective equipment because inhalation is the most efficient and dangerous route of entry for hazardous materials into the body. Of course, exceptions to this rule of thumb exist, and the most efficient route of entry is different for specific chemicals. Therefore, it is vital to use the appropriate reference materials when determining the dangers of any particular chemical.

Materials penetrate the respiratory passages at different efficiencies. Large particulates such as dust and soot are trapped by the nasal passages and upper respiratory tract, as any firefighter knows that has blown his or her nose after a structure fire knows. Dust masks and other particulate filters can protect against large particulates and dusts.

protection factor ▪ A unitless number denoting the effectiveness of respiratory protection. A protection factor of 100 indicates that up to 1 in 100 contaminant molecules can be expected to enter the respiratory tract.

OSHA assigns **protection factors** to respirators based upon how effectively they prevent contaminants from entering the body. The higher the protection factor, the more effective the respirator is at protecting you, the wearer. OSHA has assigned the following protection factors:

Half-face cartridge air-purifying respirator (APR)	10
Full-face cartridge air-purifying respirator (APR)	50
Self-contained breathing apparatus (SCBA)	10,000

The protection factor indicates the expected amount of contaminants that may enter the respiratory tract. For example, a full-face APR mask with a protection factor of 50 indicates that 1 in 50 contaminants can be expected to enter the breathing zone of the wearer. Another way of saying this is that 98% of the contaminants are filtered out, while 2% of the contaminants enter the wearer's body. In contrast, a positive-pressure SCBA with a protection factor of 10,000 indicates it filters out 9,999 contaminants and only lets 1 in for every 10,000 contaminant molecules it encounters. That statistic produces a much more confident feeling upon entering a dangerous atmosphere! Table 6-1 summarizes the advantages, limitations, uses, and operational components of respiratory protection.

Air-Purifying Respirator (APR)

filter cartridge ▪ The part of the respirator that removes contaminants from the ambient air. Cartridges must be chosen based upon the specific contaminant and must be replaced at regular intervals.

Air-purifying respirators (APR), also commonly referred to as "gas masks," are tight-fitting respirators that filter out chemicals in the air using a **filter cartridge** system. Filter cartridges are designed to remove specific contaminants from the air, usually by binding the contaminant in an absorbent matrix or by neutralizing the chemical via a chemical reaction. Thus APRs clean the ambient air that passes through the cartridge. The APR mask forms a tight-fitting seal against the face and forehead, and straps that tighten against the back of the head maintain this seal. When the wearer inhales, the negative pressure (or vacuum) forces ambient air from the outside through the filtering cartridge. Depending on the type of cartridge and the physical fitness of the user, this can strain the lungs.

Because APRs filter the ambient air using a chemical-specific process, the following requirements must be met before use:

- Sufficient oxygen must be present (19.5% to 23.5% per OSHA).
- The contaminant must be known (to be able to select the appropriate filtering cartridge).
- The concentration of the contaminant must be known (to remain below IDLH levels and avoid oversaturating the cartridge filter).
- IDLH atmospheres have been ruled out (toxicity, flammability, oxygen deficiency, etc.).

TABLE 6-1	Types of Respiratory Protection Devices			
RESPIRATORY PROTECTION	**ADVANTAGES**	**LIMITATIONS**	**USES**	**OPERATIONAL COMPONENTS**
Self-contained breathing apparatus (SCBA)	Highest level of respiratory protection	Air bottle size (30 minutes to one hour of air)	Most common choice	Mask Air tank Harness Regulator
Air-line respirator	Highest level of respiratory protection with unlimited air supply	Hose line is vulnerable to damage	Longer duration incidents Tight spaces	Mask Harness Regulator Air hose Compressor or air tank in cold zone Escape bottle
Closed-circuit SCBA	Highest level of respiratory protection with longer entry times than SCBA	Expensive Chemical CO_2 scrubber creates heat	Longer duration incidents	Mask Oxygen tank Chemical scrubber Harness Regulator
Powered air-purifying respirator (PAPR)	Less expensive than SCBA Longer entry times with positive pressure at seal	Cannot be used in IDLH atmospheres Must have appropriate cartridge	Decontamination line and situations requiring longer entry times	Mask Harness Blower motor Chemical absorptive cartridge
Air-purifying respirator (APR)	Comparatively inexpensive Lightweight Longer entry times Easy to operate tactically	Cannot be used in IDLH atmospheres May leak significantly around seal Must have appropriate cartridge	Decontamination line and situations requiring longer entry times	Mask Chemical absorptive cartridge

Because some chemicals have very poor warning properties, there is no way to know when the filter becomes saturated, which may lead to serious injury or death due to over-exposure. Therefore, **end of service life indicators (ESLIs)** are often built into these types of cartridges. However, for those many cartridges that are not equipped with an ESLI, a qualified professional, such as a certified industrial hygienist (CIH), should develop an appropriate cartridge change out schedule. Another option is not to rely on air-purifying respirators at all but instead to use supplied air respiratory protection. The following chemicals are known to have poor warning properties [adapted from Mine Safety Appliances Co. (MSA)]:

end of service life indicator (ESLI) ■ A device on a filter cartridge that indicates to the wearer when contaminants are no longer being removed from the ambient air.

Acrolein	Methanol	Phosgene
Aniline	Methyl bromide	Phosphine
Arsine	Methyl chloride	Phosphorus trichloride
Bromine	Methylene chloride	Stibine
Carbon monoxide	Nickel carbonyl	Sulfur chloride
Diisocyanates	Nitric acid	Urethane
Dimethyl sulfate	Nitrogen oxides	Vinyl chloride
Hydrogen cyanide	Nitroglycerin	
Hydrogen selenide	Nitromethane	

APRs are effective only when used properly. Because inhaling creates a vacuum on the inside of the mask, the mask's seal to the face must be airtight. Facial hair or a poor fit can cause the mask to leak, which means contaminated ambient air is blowing by the seal and entering the user's respiratory system—obviously not a good thing! The APR is also only as good as the selected cartridges, and many types of cartridges are available (see Table 6-2). If an organic vapor cartridge is selected for an acid gas incident, there is essentially no protection, which again, is obviously not a good situation.

TABLE 6-2	Types of Cartridges Available for APRs
CARTRIDGE COLOR	**CONTAMINANT FILTERED**
Purple (magenta)	Particulates (P100)
	HE (HEPA) for PAPRs
	Radioactive materials (except tritium and noble gases such as radon)
Orange	Particulates (P95, P99, R95, R99, R100)
Teal	Oil-free particulates (N95, N99, N100)
White	Acid gases
White with 1/2-inch green stripe completely around the cartridge near the bottom	Hydrocyanic acid gas
White with 1/2-inch yellow stripe completely around the cartridge near the bottom	Chlorine gas
Black	Organic vapors
Black/purple (magenta)	Organic vapors and P100
Yellow	Acid gas and organic vapors
Yellow/purple (magenta)	Acid gas, organic vapor, and P100

TABLE 6-2	Types of Cartridges Available for APRs (Continued)
CARTRIDGE COLOR	**CONTAMINANT FILTERED**
Yellow with 1/2-inch blue stripe completely around the cartridge near the bottom	Hydrocyanic acid gas and chloropicrin vapor
Green	Ammonia gas and methyl amine gas
Green with 1/2-inch white stripe completely around the canister near the bottom	Acid gases and ammonia gas
Blue	Carbon monoxide
Brown	Acid gases, organic vapors, and ammonia
Red	Acid gases, organic vapors, ammonia, and carbon monoxide
Olive	Multi-contaminant combinations not listed above CBRNE certified

Because cartridge respirators merely filter the contaminated, ambient air, there is a **maximum use concentration (MUC)** based upon the specifications of the APR. The MUC can be calculated based upon the following formula:

$$MUC = \text{Protection factor} \times \text{Worker exposure limit}$$

For example, if we want to use an APR, with a protection factor of 50, in an atmosphere of hydrogen cyanide, which has a PEL of 10 ppm, then the MUC = 50 × 10 = 500 ppm.

In addition, an APR may never be used in an IDLH atmosphere. Therefore, if the calculated MUC exceeds the IDLH, it should be lowered to the IDLH value for the chemical indicated in the NIOSH guide or other reliable reference source. The IDLH for hydrogen cyanide is only 50 ppm. Therefore, an APR should never be used to protect against hydrogen cyanide concentrations at or above 50 ppm.

maximum use concentration (MUC) ■ The maximum airborne concentration of a contaminant that a respirator may be used in based upon the permissible exposure limit (PEL) and the protection factor of the respirator.

Powered Air-Purifying Respirator (PAPR)

Powered air-purifying respirators (PAPRs) are APRs with a blower motor that forces air through the filter cartridge. Thus they essentially provide a positive-pressure environment (although it is not considered to be a true positive-pressure device), which makes them safer and easier to use. They are safer because a negative pressure situation never exists inside the mask. For this reason, a tight-fitting mask is not even necessary, and hooded

PAPRs are very popular in agencies and industries where facial hair is common. Hospitals typically use hooded PAPRs to perform emergency decontamination and preliminary patient treatment on contaminated patients (Figure 5-14). They are easier to use because the lungs do not need to work as hard in forcing the ambient air through the filter cartridges. The drawbacks to PAPRs are that they are heavier (due to the blower motor and battery pack), and the motor may unexpectedly fail in the contaminated area.

Positive-Pressure Self-Contained Breathing Apparatus (SCBA)

supplied air respirator (SAR) ■ A face mask and source of breathable air required to be used in IDLH atmospheres.

Self-contained breathing apparatus (SCBA), which are a type of **supplied air respirator (SAR)**, offer the highest level of respiratory protection when used in positive-pressure mode (Figure 5-17). Positive-pressure mode refers to a constant air pressure delivery to the mask. If the seal of a positive-pressure SCBA is broken, air will rush out due to the positive pressure delivered to the inside of the mask by the regulator. SCBAs consist of a compressed air bottle, a tight-fitting mask, and one or more regulators to reduce the air pressure. This is the respiratory protection that firefighters use. As with APRs, the mask must be used properly and fit tightly against the skin. This is why fire departments have strict facial hair grooming standards.

The main advantage of SCBA respiratory protection is that it does not rely on ambient air, but rather its own supply carried in the compressed air tank. The drawback is that SCBA bottles typically come in only 30-minute to 60-minute varieties, which limits the amount of time and the amount of work that can be accomplished while wearing the SCBA, a distinct disadvantage compared to APRs. SCBAs are also significantly heavier than APRs, weighing 15 to 25 pounds depending on design and air capacity. Nevertheless, SCBAs are the standard respiratory protection when hazmat personnel must enter unknown or IDLH atmospheres.

Positive-Pressure Air-Line Respirator with Escape Unit

Air-line respirators are a type of supplied air respirator in which the air supply is not carried on the back, but rather supplied through an air hose from a remote location (Figure 6-1). This remote location is usually one or more large compressed air tanks or an air compressor located outside the hot zone. This type of respirator is similar to an SCBA, with the advantage that it is much lighter in weight. Its disadvantage is that the air line may become compromised due to pinching, severing, or bursting. Therefore, a responder must carry a small escape bottle into the hot zone when using an air-line respirator. Another disadvantage is that the maximum length of the hose is limited to 300 feet. These respirators are popular when there is little room available, such as in confined spaces, and when longer operation times are required, such as in chemical laboratories, during cleanup operations, and at decontamination lines.

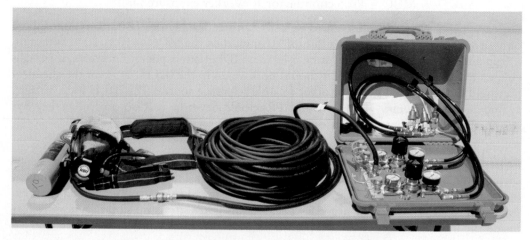

FIGURE 6-1 A supplied air respirator (SAR) often used in confined space operations. The small emergency escape bottle is shown on the left next to the face mask.

Closed-Circuit SCBA

Closed-circuit SCBAs, also known as rebreathers, are a type of supplied air respirator designed to be worn on the back like an SCBA, but have a longer work time. The work time can be from 1 to 6 hours depending on the configuration. A closed-circuit SCBA consists of a tight-fitting mask, two or more regulators, a compressed air cylinder, a compressed oxygen cylinder, and a chemical scrubbing system. The chemical scrubbing system removes the exhaled carbon dioxide from the breathing air. This chemical reaction is very exothermic, or heat releasing, which causes the breathing air to heat up as the unit is used. The unit also feels warm on the back. Both of these factors lead to increased heat stress with closed-circuit SCBA use.

These respirators were developed many years ago for the mining industry and are used today in hazardous materials response when air-line respirators are not practical. The advantage of the closed-circuit SCBA is an extended work period without the inconvenience and distance limitations of an air-line respirator. The disadvantages are that the breathing air is warm to downright hot due to the exothermic reaction of the scrubber and that the units are even heavier than a standard SCBA unit. Today, rebreathers are often used in mines and tunnels, during both rescue operations and hazardous materials incidents.

THERMAL PROTECTIVE CLOTHING

We discussed thermal protective clothing in the last chapter. It is designed to protect the wearer from extremes in temperature (Table 6-3). This may be a pair of cryogenic gloves used to handle liquid nitrogen, it may be structural firefighter protective gear used to fight fires, or it may be aluminized high-temperature thermal protective clothing. High-temperature thermal clothing called a proximity suit is used by refinery fire brigade personnel and airport fire personnel due to the extreme heat produced by large quantities of burning hydrocarbons. Thermal protective clothing is *not* considered chemical protective clothing.

Structural firefighter protective clothing, or turnout gear, is a type of thermal protective clothing designed to protect the wearer from extremes in heat. Yet even this is limited. Turnout gear consists of a helmet, SCBA with a personal alert safety system (PASS) device, a thermal protective hood, a thermal protective coat (bunker coat), thermal protective pants (bunker pants), and steel-toed boots.

Structural firefighter protective clothing has very limited to no chemical protective qualities and should generally not be used in the hot zone at hazardous materials incidents. However, certain exceptions apply: Turnout gear may be appropriate when the fire risk outweighs the other health and safety risks. An example would be a methane or

TABLE 6-3	Types of Protective Clothing		
PROTECTIVE CLOTHING	**PURPOSE**	**ADVANTAGES**	**LIMITATIONS**
Thermal protective clothing	High-temperature flammable liquid firefighting	Protects from radiant heat	Little, if any, chemical protection
Structural fire-fighter protective clothing	Structural firefighting	Protects from moderate radiant heat with some flash fire protection	Little, if any, chemical protection
Chemical protective clothing	Hazardous materials response	When appropriately selected, chemical resistant	No thermal protection Heat stress

propane leak in which a valve must be turned off. The primary hazard is flammability, the biggest health risk is inhalation and not skin absorption, and a positive-pressure SCBA is part of the turnout gear. Turnout gear is clearly much safer than chemical protective clothing in this case. Other examples are immediate lifesaving rescues (consult the 3/30 rule discussed in Chapter 12, Operations Level Responders Mission-Specific Competencies: Victim Rescue and Recovery) and hazardous materials releases that require thermal protection with minimal to no chemical protection (such as some cryogenic gases).

CHEMICAL PROTECTIVE CLOTHING (CPC)

Chemical protective clothing (CPC) is designed to protect the wearer from chemicals. The different types of CPC are intended to keep the chemical from coming into contact with the body. Some CPC protects from liquid exposures (**splash protection**), whereas other CPC protects from gaseous and vapor exposure as well (vapor-protective suits). CPC comes in a variety of protective qualities, sizes, shapes, colors, thicknesses, flexibilities, shelf lives, and costs. Some CPC is disposable, whereas other CPC is reusable. Some is complicated to use and maintain, whereas other types are simple to use and maintain. It is vital to become familiar with the type of CPC your agency uses. CPC is an extremely important component that keeps you safe at hazardous materials incidents.

Chemical protective clothing comes in a variety of configurations. The more effective, and expensive, CPC consists of several chemical-resistant materials that are laminated together. The different types of materials have different resistance to chemicals and must be appropriately selected using manufacturer-supplied information (compatibility charts) (Figure 5-18).

CPC may fail in three different ways: penetration, degradation, and permeation. It is extremely important to consult the manufacturer's compatibility tables before selecting chemical protective clothing for hot zone entries. In addition, the equipment itself can fail due to improper storage and use, or due to a manufacturer's defect. Many features of the CPC may fail, such as zippers, one-way valves, and seams; or the respiratory protection, such as the SCBA or APR, may fail. It is therefore crucial to inspect the PPE before every use.

Degradation

Degradation is a process by which the chemical protective clothing is compromised by being effectively dissolved by the hazardous material with which it comes into contact. Degradation reduces the integrity of one or more materials used in chemical protective clothing, rendering them less effective or completely ineffective. A good example of degradation is what happens when acetone comes into contact with polystyrene: The acetone literally dissolves the polystyrene lid on contact (Figure 6-2).

splash protection ■ PPE designed to protect the wearer from liquid contact but not from gases and vapors.

FIGURE 6-2 An example of a degradation process is acetone dissolving polystyrene.

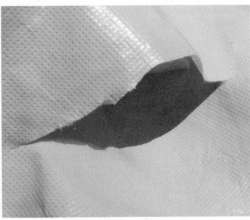

FIGURE 6-3 Liquids and gases can penetrate PPE through functional features such as zippers and unsealed seams (left) or through damage to PPE such as a puncture or tear (right).

Three signs of degradation are discoloration of the material, change in flexibility of the material, and thinning or melting of the suit material. Any one of these signs indicates that the chemical protective clothing may fail due to chemical degradation. It is very important to immediately exit the contaminated area and proceed to decontamination. Let the decon personnel know that you suspect suit failure due to degradation.

Penetration

Chemicals can enter chemical protective clothing via penetration of natural openings in the suit such as zippers, seams, and hand and foot openings (Figure 6-3). Splash protective ensembles have several natural openings, depending upon their design criteria, and are not necessarily vapor and liquid tight. Penetration may also result from mechanical damage to the chemical protective clothing or from defective design openings in the suit. It is crucial to inspect the suit before entry for any manufacturing defects that may be present, and to protect your chemical protective clothing from mechanical damage.

Mechanical damage may occur while navigating through the hot zone. There may be sharp edges due to an explosion that can cause tears or rips in the suit fabric structural features and equipment found normally at the facility, or jagged edges at a transportation accident. Suits may also be punctured from the inside if sharp objects such as rings are worn in the hot zone. Suit breaches may also occur if the chemical protective clothing is too small for the wearer. Zippers and seams, especially in the groin region, may split or tear when the wearer moves around during normal work-related tasks. Proper sizing of PPE is critical for entry team safety.

Permeation

Permeation is the process of chemical movement through the suit at the molecular level (Figure 6-4). Suit integrity remains intact during this process. The suit material is neither degraded nor does the chemical move across larger imperfections of the suit material. The permeation rate is a function of a combination of the chemical properties and the CPC properties. Imagine an insect screen on your window. The effectiveness of the screen depends on the mesh size and the size of the insect you are trying to keep out. If you decide to use chicken wire as a window screen, it will not be able to keep out flies and mosquitoes. However, by using a

FIGURE 6-4 The permeation process involves the movement of a substance through a protective material at the molecular level. The PPE remains completely intact during permeation.

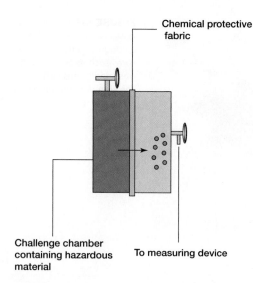

Chemical protective fabric

Challenge chamber containing hazardous material

To measuring device

FIGURE 6-5 Schematic of a permeation test cell.

breakthrough time ■ The time it takes for a hazardous substance to pass through chemical protective clothing.

smaller size mesh, it will keep out those flies and mosquitoes, but gnats may still make it in. You need to further reduce the mesh size of your screen to keep out those pesky gnats! Effectively the insects are permeating through the screen because the screen itself has not been compromised (which would be penetration or degradation). Similarly, chemicals permeate through chemical protective clothing.

Chemical properties that affect permeation rate include the molecular size of the contaminants, the state of matter, and viscosity. Environmental factors such as temperature and pressure also affect the permeation rate. Chemical protective clothing properties that affect the permeation rate include the type of polymer, the thickness of the material, and the density of the material.

So how can we make the proper choices? The most reliable method to determine appropriate work times in chemical protective clothing is to consult the manufacturer's chemical compatibility charts that are based on **breakthrough times** using the ASTM F903 test method (Figure 6-5). A swatch of material to be tested is placed in a permeation chamber. One side of the chamber contains the challenge chemical, while the other side of the chamber is monitored using detection equipment. Breakthrough times are listed for up to 8 hours. Depending on how the chemical protective clothing ensemble is used, and how long it will be used, shorter breakthrough times may be acceptable under certain circumstances.

SOLVED EXERCISE 6-1

Based upon the DuPont chemical protective clothing guideline shown below, what is the breakthrough time of ammonia for the Tychem® TK suit? Would this suit be appropriate to use at the incident described in the opening scenario. Why or why not?

Sub-Class	Chemical Name	CAS Number	Phase	Standardized Breakthrough Time (Minutes)										
				Tychem® CPF1	Tychem® QC	Tychem® CPF 2	Tychem® SL	Tychem® CPF 3	Tychem® F	Tychem® CPF 4	Tychem® BR/ Tychem® LV	Tychem® Responder®	Tychem® TK	Tychem® Reflector®
391	Acetone	67-64-1	L	imm.	imm.	12	12	>480	>480	>480	>480	>480	>480	>480
431	Acetonitrile	75-05-8	L	imm.	imm.	12	12	imm.	157	>480	>480	>480	>480	>480
350	Ammonia gas	7664-41-7	G		imm.	32	32	12	79	>480	46	>480	>480	>480
296	1,3-Butadiene	106-99-0	G		imm.	>480	>480	>480	>480	>480	>480	>480	>480	>480
502	Carbon disulfide	75-15-0	L	imm.	imm.	imm.	imm.	16	>480	>480	>480	>480	>480	>480
330/350	Chlorine gas	7782-50-5	G		imm.	>480	>480	>480	>480*	>480	>480	>480	>480	>480

Courtesy of DuPont

Solution: The breakthrough time listed by DuPont is > 480 minutes. It would be appropriate to use this CPC as long as continuous air monitoring is performed and the ammonia concentration remains below the LEL.

COMMUNICATIONS

Communications between entry team members in the hot zone is essential, yet communications while wearing personal protective equipment can be a challenge. How do we rise to this challenge? The most common type of communication between entry team members will be verbal, whereas the most common type of communication between entry team members and the cold zone will be through radio communications. Verbal communication can be a challenge, especially when using self-contained breathing apparatus and level A suits. An alternative to verbal communication between entry team members may be written communication or hand signals. Hand signals generally have the advantage of being visible over longer distances, yet only relatively simple information can be transmitted.

It is essential that the radio communications in the hot zone do not rely on fine-motor skills. Typically, large press-to-talk buttons are used that can be strategically located and operated using an elbow or hand. For example, the press-to-talk button may be located on the SCBA belt or an SCBA shoulder strap. Because coherent speech is often difficult in PPE, throat microphones or bone microphones are often utilized to increase the accuracy of speech. Even with both of these technologies in use, radio communication is still a challenge at most hazardous materials incidents.

EPA LEVELS OF PROTECTION

The EPA uses four levels of protection based upon the combination of skin protection and respiratory protection offered (a PPE ensemble). The lowest level of protection is designated **level D PPE**, which consists of nothing more than a uniform or other clothing. There is no respiratory protection and little to no chemical protective clothing. **Level C PPE** consists of minimal respiratory protection, such as an air-purifying respirator (APR), and chemical splash protection. **Level B PPE** consists of supplied air respiratory protection and chemical splash protection. **Level A PPE**, the highest level of protection, consists of supplied air respiratory protection and a vapor-tight suit. Figure 5-19 shows examples of all four EPA levels of protection.

Level D

Level D protection essentially consists of no chemical or respiratory protection at all. Street clothes or a daily work uniform may be considered level D protection. However, even the universal precautions used by paramedics—latex or nitrile gloves, and eye and face protection—are also merely level D protection. Even with the addition of chemical protective clothing, a dust mask, or a TB mask, it would still be considered level D protection. Level D protection may be used only in environments with airborne contaminant levels below the permissible exposure limit (PEL).

Level C

Level C protection consists of chemical protective clothing that provides splash protection and an air-purifying respirator (APR). This level of protection is typically used when some splash protection and some level of respiratory protection are required. The EPA considers any level of respiratory protection below an APR to be level D.

An air-purifying respirator, or cartridge respirator, filters the ambient air and removes contaminants. Figure 6-6 shows some common examples of APR cartridges and masks.

Several criteria must be met for the use of level C protection:

1. Sufficient ambient oxygen exists.
2. All dangerous contaminants have been identified.
3. Contaminant concentrations are known.
4. Contaminant concentrations are below IDLH levels.
5. Appropriate cartridges are available that are able to filter out contaminants.
6. No other IDLH conditions exists.

level D PPE ■ Some chemical protective clothing combined with no or very limited respiratory protection (such as a dust mask).

level C PPE ■ Liquid splash protective chemical protective clothing combined with an air-purifying respirator.

level B PPE ■ Liquid splash protective chemical protective clothing combined with supplied air respiratory protection.

level A PPE ■ Vapor-tight chemical protective clothing combined with supplied air respiratory protection.

(a)

(b)

(c)

Because APRs filter only the ambient air, we must first determine that ambient oxygen levels are normal, or at least above 19.5% and below 23.5%. Atmospheres are considered immediately dangerous to life and health (IDLH) when oxygen concentrations fall below 19.5% or rise above 23.5%. We must know what the contaminant is in order to choose the appropriate air-purifying cartridge or canister. The absorbent or reactive material inside the canister must effectively exclude or neutralize the contaminant of interest. Furthermore, the cartridge or canister must have an end of service life indicator (ESLI), or we must determine an appropriate change out schedule based upon the manufacturer's recommendations. The contaminant level must be known before using an APR to ensure ambient contaminant levels are, and remain, below IDLH levels. Contaminant levels can be determined using air-monitoring instruments, the use of which is covered in Chapter 11, Operations Level Responders Mission-Specific Competencies: Air Monitoring and Sampling. Finally, we must complete a safety analysis of the hot zone and determine that no other IDLH conditions are present. Other IDLH conditions include flammable atmospheres, permit-required confined spaces, and other situations that pose an immediate danger to life and health.

Level B

Level B protection is used when some splash protection and the highest level of respiratory protection are needed. It consists of the same types of chemical protective clothing and offers the same splash protection seen with level C. However, instead of using an APR, we now use supplied air respiratory protection. Supplied air respiratory protection may consist of self-contained breathing apparatus (SCBA) or a supplied air respirator (SAR). Because supplied air is being used, level B suits may also be fully encapsulating. Figure 6-7 shows examples of nonencapsulating and encapsulating level B suits.

Level A

Level A protection is used when the highest level of skin and respiratory protection is needed. The chemical protective clothing that is used is vapor tight. Level A suits have sealed seams, integrated gloves and booties, an airtight zipper, and one-way exhalation valves (Figure 6-8). Because these suits are vapor tight, they are also thicker and heavier than level B or C suits, and wearers are more prone to heat stress. Level A is the highest level of personal protective equipment ensemble in the EPA's system.

NFPA CHEMICAL PROTECTIVE CLOTHING STANDARDS

The National Fire Protection Association (NFPA) defines personal protective equipment in a slightly different manner than the EPA does. NFPA 1991, *Standard on Vapor-Protective Ensembles for Hazardous Materials Emergencies*, describes the chemical protective clothing for the EPA's level A protection. NFPA 1994, *Standard on Protective Ensembles for First Responders to CBRN Terrorism Incidents*, describes the specific chemical protective

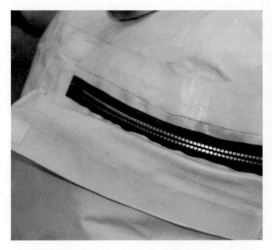

FIGURE 6-8 A sealed seam (left) and a one-way exhalation valve (right) found on a level A suit.

clothing for chemical, biological, and radiological weapons of mass destruction agents. Typically these NFPA standards include the suit, gloves, and footwear but exclude respiratory protection equipment.

NFPA 1991

The NFPA 1991 standard deals with fully encapsulating chemical protective clothing that is gas and vapor tight. Thus, it covers the EPA level A chemical protective clothing but makes no mention of respiratory protection. Specifically, the standard requires permeation testing against 19 different chemicals, six gases, and two WMD agents. The CPC must also pass burst strength, seam strength, tear resistance, abrasion resistance, flammability resistance, and cold temperature performance tests. NFPA 1991 therefore quantifies the fully encapsulating level A suit requirements and sets the bar for manufacturers to meet.

NFPA 1992

NFPA 1992, *Standard on Liquid Splash-Protective Ensembles and Clothing for Hazardous Materials Emergencies*, covers chemical protective clothing designed for liquid splash protection. These suits may be either fully encapsulating or nonencapsulating. Similarly, the NFPA 1992 standard covers what is commonly considered EPA level B chemical protective clothing and, again, makes no mention of respiratory protection requirements. Because respiratory level is not defined, the NFPA 1992 suit may be used in an EPA level B, level C, or even level D ensemble, depending on the level of respiratory protection provided. The key is that the NFPA 1992 CPC passes the burst strength, seam strength, tear resistance, abrasion resistance, flammability resistance, cold temperature performance, and flexural fatigue tests. The strength of the NFPA 1992 standard is that it quantifies the requirements for the complete system of liquid splash-resistant chemical protective clothing.

NFPA 1994

NFPA 1994 describes chemical protective clothing to be used for chemical, biological, radiological, or nuclear (CBRN) terrorism incidents. These garments are designed for one-time use. The NFPA 1994 standard contains four classes: Class 1, also covered by NFPA 1991, describes gas- and vapor-tight suits with supplied air respiratory protection, otherwise known as a level A ensemble. Class 2 describes CPC designed to protect the user from IDLH atmospheres and corresponds to the EPA level B ensemble. Class 3 describes chemical protective clothing designed to protect the user from chemical hazards above the permissible exposure limit (PEL) but below the IDLH, corresponding to level C protection. Class 4 is essentially level D protection (dust masks and HEPA filters).

NFPA 1999

NFPA 1999, *Standard on Protective Clothing for Emergency Medical Operations*, sets forth the requirements for chemical protective clothing for emergency medical operations in the warm zone. The CPC must be liquid tight. Once again, NFPA 1999 CPC must pass certain tests, including burst strength, seam strength, tear resistance, abrasion resistance, flammability resistance, cold temperature performance, and flexural fatigue.

SPECIALTY SUITS

Certain other types of personal protective equipment are used under special circumstances, such as bomb suits, structural firefighter protective gear, proximity suits, and LANX or MOPP suits (Figure 6-9). Bomb suits provide limited protection from the effects of an explosion, such as from shrapnel, thermal effects, and overpressure effects. These suits actually have special air filtering systems to minimize the overpressure effect from an explosion but, but as you can imagine, they are effective for only relatively small detonations. Structural firefighter protective gear is very successful in providing thermal protection from the heat of fires typically encountered during structural firefighting. Yet this type of personal protective equipment provides little to no protection from chemicals

FIGURE 6-9 Specialty suits include aluminized proximity suits (left) and bomb suits (right). *Photo on left © 2011 Lakeland Industries, Inc.*

or from intense heat and flash fires. For this reason, structural firefighter protective gear is not particularly effective for fires involving significant amounts of petroleum products.

LANX and mission oriented protective posture (MOPP) gear are primarily designed for protection from chemical warfare agents. The lining of these suits contain activated charcoal, which absorbs most chemical warfare agents. The suits also offer limited protection against many toxic industrial chemicals (TICs).

HAZARDS AND PPE OPTIONS

The last chapter covered some of the hazards and personal protective equipment options available to first responders for uses unrelated to hazardous materials and WMD responses. Let's now take a look at some of the hazards possible when responding to hazardous materials and WMD incidents. Among these, of course, are chemical and asphyxiating hazards, and we must be able to ensure that the chemical protective clothing and the level of respiratory protection chosen will protect us from the chemical properties of the hazardous materials. In Chapter 4 we learned to interpret some of the chemical and physical properties in order to be able to predict the hazardous materials' behavior, Thermal, mechanical, radiological, and biological/etiological hazards may also be involved.

Simple Asphyxiating Hazards

The human body requires oxygen to survive, and simple asphyxiating hazards prevent the body from getting oxygen by displacing the ambient oxygen in the air. Simple asphyxiants—such as nitrogen, helium, neon, argon, and carbon dioxide—are typically not chemically reactive and so are otherwise harmless to the body. Nitrogen even makes up 78% of the air we breathe! Thus, the most important personal protective equipment used to prevent asphyxiation is the highest level of respiratory protection. It is vital to use supplied air respiratory protection when responding to asphyxiating gases and vapors. Chemical protective clothing is a secondary consideration in this case. For example, if a tank of liquid nitrogen leaks inside an enclosed space such as a laboratory, supplied air

What types of hazards does anhydrous ammonia pose?

Ammonia	Formula: NH_3	CAS#: 7664-41-7	RTECS#: BO0875000	IDLH: 300 ppm
Conversion: 1 ppm = 0.70 mg/m³	DOT: 1005 125 (anhydrous); 2672 154 (10-35% solution); 2073 125 (>35-50% solution); 1005 125 (>50% solution)			

Synonyms/Trade Names: Anhydrous ammonia, Aqua ammonia, Aqueous ammonia
[**Note:** Often used in an aqueous solution.]

Exposure Limits:	Measurement Methods
NIOSH REL: TWA 25 ppm (18 mg/m³) ST 35 ppm (27 mg/m³) **OSHA PEL †:** TWA 50 ppm (35 mg/m³)	(see Table 1): **NIOSH** 3800, 6015, 6016 **OSHA** ID 188

Physical Description: Colorless gas with a Pungent, Suffocating odor.
[**Note:** Shipped as a liquefied compressed gas. Easily liquefied under pressure.]

Chemical and Physical Properties:	Personal Protection/Sanitation (see Table 2):	Respirator Recommendations (see Tables 3 and 4):
MW: 17.0 **BP:** −28°F **Sol:** 34% **FI.P:** NA (Gas) **IP:** 10.18 eV **RGasD:** 0.60 **VP:** 8.5 atm **FRZ:** −108°F **UEL:** 28% **LEL:** 15%	**Skin:** Prevent skin contact **Eyes:** Prevent eye contact **Wash skin:** When contam (solution) **Remove:** When wet or contam (solution) **Change:** N.R. **Provide:** Eyewash (>10%) Quick drench (>10%)	**NIOSH** **250 ppm:** CcrS*/Sa* **300 ppm:** Sa:Cf*/PaprS*/CcrFS/ GmFS/ScbaF/SaF **§:** ScbaF: Pd, Pp/SaF: Pd, Pp:AScba **Escape:** GmFS/ScbaE

[**Note:** Although NH_3 does not meet the DOT definition of a Flammable Gas (for labeling purposes), it should be treated as one.]

Incompatibilities and Reactivities: Strong oxidizers, acids, halogens, salts of silver and zinc
[**Note:** Corrosive to copper and galvanized surfaces.]

Exposure Routes, Symptoms, Target Organs (see Table 5):	First Aid (see Tabel 6):
ER: Inh, Ing (solution), Con (solution/liquid) **SY:** Irrit eyes, nose, throat, dysp, wheez, chest pain; pulm edema; pink frothy sputum; skin burns, vesic; liquid: frostbite **TO**: Eyes, skin, resp sys	**Eye:** Irr immed (solution/liquid) **Skin:** Water flush immed (solution/liquid) **Breath**: Resp support **Swallow:** Medical attention immed (solution)

Courtesy of National Institute for Occupational Safety and Health

Solution: Ammonia is a corrosive, toxic, and flammable gas. According to the NIOSH guide, the PEL = 50 ppm, the IDLH = 300 ppm, and the LEL = 15%. Note that this is in stark contrast to how the U.S. DOT classifies and placards this material.

respiratory protection is essential to perform a rescue (Figure 4-2), whereas chemical protective clothing is not. Therefore, a firefighter in structural firefighter protective clothing using an SCBA is well protected to perform this rescue.

Chemical Hazards

Unlike the simple asphyxiants, most chemicals have further hazards associated with them. Chemical hazards may be damaging via all four routes of entry (inhalation, absorption, ingestion, and injection). The choice of the appropriate level of respiratory protection and chemical protective clothing depends on the route of entry. If the chemical has a low vapor pressure, meaning that it does not evaporate readily but is readily absorbed through the skin, we may be able to choose a lower level of respiratory protection with a high level of chemical protective clothing that offers splash protection to that particular chemical.

Level C may be appropriate in this case. On the other hand, if the chemical has a high vapor pressure and is readily absorbed through the skin, we must choose the highest level of respiratory protection and chemical protective clothing that is impermeable to that particular chemical. Level A protection would be required in this case.

Chemical hazards include corrosives, poisons, chemical asphyxiants, and oxidizers. Examples of corrosives include acids such as sulfuric acid, HCl, and nitric acid, and bases such as ammonia, bleach, and sodium hydroxide. Examples of common poisonous substances include insecticides; isocyanates; and gases such as phosphine, arsine, hydrogen sulfide, and carbon monoxide. Chemical asphyxiants prevent the body from either transporting or using oxygen; two common ones are hydrogen cyanide and carbon monoxide. Oxidizers are a common type of chemical hazard that may increase the propensity of fuels to burn and may directly damage the human body; examples include hypochlorite, hydrogen peroxide, and perchlorates. It is very important to choose the appropriate respiratory protection and chemical protective clothing ensemble when responding to incidents involving chemical hazards.

Thermal Hazards

Thermal hazards may be exhibited by a wide range of hazardous materials, including flammable materials, combustible materials, cryogenics, and explosives. Flammable materials are a prime example of thermal hazards because they are very commonly encountered. Examples include gasoline, methane, propane, ethanol, and many other common industrial and household hazardous materials. It is extremely important not to overlook a thermal hazard potential of hazardous materials. In 1984 in Shreveport, Louisiana, a fire department hazmat team member was killed while responding to an ammonia leak in a cold storage warehouse. He and his partner were wearing level A suits to protect themselves from the corrosive properties of ammonia but neglected to protect themselves from the flammable properties of ammonia. During mitigation activities, they unknowingly found themselves in a flammable atmosphere in the presence of an ignition source. They both suffered severe burns to which one ultimately succumbed (Box 6-1).

Cryogenic materials also pose thermal hazards, although they are the direct opposite of the thermal hazards flammable materials pose. Cryogenic materials are defined as having a boiling point below −130°F (−90°C) by the U.S. DOT. Cryogenic materials are stored and transported as liquids. Often the greatest danger with cryogenic materials is skin contact and the resulting cryogenic burn, which is a type of severe frostbite.

Thermal hazards require special protective clothing. One example of thermal hazard protection is structural firefighter protective gear. This type of PPE is designed to limit the amount of heat that reaches a firefighter's body. Another example of thermal hazard protection is cryogenic gloves that allow employees to handle containers holding cryogenic materials without receiving cryogenic burns. Such gloves are designed to insulate the wearer from the extreme cold of the cryogenic liquid. Neither type of PPE offers a significant amount of chemical protection. Certain chemical protective clothing is designed to be resistant to flash fires; however, this offers little thermal protection. Aluminized suits are also available as overgarments for level A protection and offer limited protection against radiant heat. This type of PPE is often used by airport and industrial firefighters that must face the intense heat of petroleum-based fires.

Explosives are materials that detonate. Detonation involves a very rapid release of energy. In contrast, deflagration is the very rapid burning of material. An example of a material that deflagrates is gunpowder. Unconfined, it burns rapidly but does not lead to an explosion. Confined, on the other hand, it leads to an explosion and may propel a bullet or cause a pipe bomb to explode. Explosives pose a thermal hazard, an overpressure hazard, and a mechanical hazard.

Mechanical Hazards

Mechanical hazards can be found across the hazardous materials incident, especially in the hot zone (Figure 3-35). When we think of mechanical hazards, we often think of

BOX 6.1 COLD STORAGE WAREHOUSE EXPLOSION

In the afternoon of Monday, September 17, 1984, an anhydrous ammonia explosion occurred at the Dixie Cold Storage Company in Shreveport, Louisiana. Anhydrous ammonia is a toxic, flammable, and corrosive gas that is often used as a refrigerant in cold storage warehouses. Two hazardous materials team members from the Shreveport Fire Department Hazardous Materials Unit were caught in the explosion, one of which died. He is the only member of an organized hazardous materials team to have died in the hot zone at a known hazardous materials incident to date.

Workers had discovered a small ammonia leak several days before the explosion, and they attempted to isolate the leak and repair it. On Monday morning, they determined there was a faulty valve at the evaporator in the room, removed the perishable contents from the freezer room, attempted to close an isolation valve, and started repairs. The workers wore only APRs to perform the work and noticed that the ammonia levels were rising steadily. They attempted to neutralize the ammonia with a carbon dioxide fire extinguisher, which created water vapor and reduced the visibility in the room to near zero. Eventually they exited the room and called the fire department for assistance.

Upon arrival, the first due engine company assessed the situation and immediately called the hazmat unit due to the presence of the anhydrous ammonia. After an initial assessment by the hazmat unit, the hazmat team members entered the warehouse and attempted to isolate and repair the faulty valve in the evaporator 17 feet above the floor. The forklift truck they were operating skidded on the wet cement floor and caused a spark that ignited the anhydrous ammonia, which had been able to accumulate to levels in the flammable range due to poor ventilation. The explosion ignited the chemical protective clothing the hazmat team members were wearing, fatally burning the forklift operator and critically injuring the second team member. The fatally injured hazmat team member died approximately 36 hours later in the hospital after receiving burns to over 97% of his body.

According to investigators, three contributing factors to the loss of life at this incident were:

1. "The ignition of a flammable mixture of anhydrous ammonia gas during the emergency scene operation."
2. "The lack of proper precautions by workers to reduce the possibility of a hazardous accumulation of anhydrous ammonia gas."
3. "The lack of awareness by firefighters that the conditions for a hazardous accumulation of flammable anhydrous ammonia gas were present."

Source: Thomas J. Klem, *Fire Investigations: Cold Storage Warehouse, Shreveport, Louisiana.* (Quincy, MA: National Fire Protection Association, September 17, 1984). Reproduced with permission from the National Fire Protection Association, www.nfpa.org. Copyright © 1984, NFPA.

---

explosions where fragmentation and shrapnel are deadly mechanical hazards. In the hot zone, much more mundane mechanical hazards are more commonly encountered, such as jagged edges on metal drums, pneumatic and hydraulic equipment, and other sharp and jagged hazards.

Personal protective equipment for mechanical hazards include thicker chemical protective clothing, leather gloves, and bomb suits for explosive ordnance disposal (EOD) personnel.

Radiological Hazards

Radiological hazards are posed by radioactive materials and other processes that can emit radiation. Examples of common radioactive materials include americium in smoke detectors, tritium in gun sights and exit signs, and medical and research radiopharmaceuticals. Examples of equipment and processes that generate radiation are the nuclear power industry and X-ray generators used for medical diagnostics. The most dangerous forms of radiation are gamma, X-ray, and neutron radiation because they travel long distances in the air. Chemical protective clothing is not effective in preventing exposure to gamma, X-ray, or neutron radiation; however, it will prevent exposure to the radioisotope itself, the material that is emitting radiation. Chemical protective clothing will shield the wearer from alpha radiation and is partially effective in shielding the wearer from beta radiation.

In the opening scenario, how would the ammonia concentration affect your choice of PPE?

Solution: At concentrations between the PEL and approximately 10% of the LEL, the PPE should primarily protect the wearer against the toxic and corrosive effects of ammonia. As the ammonia concentration approaches the LEL, generally above 10% of the LEL, flash protection becomes increasingly important. When possible, engineering controls such as ventilation should be used to reduce the ammonia concentration, and workplace practices should be modified such as by using reverse 911 systems to notify residents rather than entering the hazardous atmosphere.

Biological/Etiological Hazards

Biological and etiological hazards can arise from typical sources such as contamination with human waste or human blood, or stem from properties of the hazardous materials themselves. For example, research laboratories use biological agents in their research. Medical and laboratory waste shipments may also contain etiological hazards. Universal precautions worn in the EMS field are an example of PPE that offers limited protection from biological and etiological hazards.

Limitations of Personal Protective Equipment

Personal protective equipment has its limitations. No one type of PPE protects against all hazards. Therefore, it is essential to properly research not only the properties of the released hazardous material but also the advantages and disadvantages, including chemical compatibility, of the available personal protective equipment ensembles as a whole. As we have just seen, hazardous materials and WMD may exhibit many different hazards. It is very important not only to accurately assess the specific types of hazards posed by the situation as a whole but also to accurately research the capabilities of the available PPE.

Physiological and Psychological Stresses

Working in chemical protective clothing and respiratory protection can be physically and mentally demanding. Encapsulating suits may cause feelings of claustrophobia in some users due to limited sight and hearing. Chemical protective clothing, which is a barrier to chemicals, also prevents evaporative cooling by keeping perspiration and heat inside the suit.

THERMAL STRESS

One of the biggest dangers of using PPE is heat, or thermal, stress. Because chemical protective clothing traps heat inside the suit and near your body, **hyperthermia** is a very real risk. The four indications of suffering from heat stress are heat cramps, heat rash, heat exhaustion, and most seriously heatstroke. The protective action that chemical protective clothing provides must be carefully weighed against the thermal stress it may induce, especially in warmer climates and under difficult working conditions. In colder climates, hypothermia and frostbite may also be issues while using personal protective equipment. How real is heat stress? Here are some statistics on the average number of deaths per year in the United States due to various causes:

hyperthermia ■ The excessive heating of the body above normal body temperature.

Hurricanes   17
Avalanches   28
Lightning   55
Tornadoes   80
Heat stress   334 (annual average from 1979 to 2003)

Let's take a look at some of the signs and symptoms of heat stress, its causes, and some ways to avoid heat stress when wearing PPE.

Heat Cramps and Heat Rash

The mildest forms of heat stress are **heat cramps**. Heat cramps occur when there is a rise in body temperature that induces uncontrollable, painful muscle contractions most commonly in the legs, stomach area, and arms. In fact, you have probably suffered from heat cramps yourself on a hot summer day if you have become slightly dehydrated and overheated.

Some people may also suffer from **heat rash** when they become overheated. Heat rash is a red, mottled appearance of the skin similar to a mild allergic reaction. Rest and fluid replenishment, including electrolytes, are the best treatments for muscle cramps and heat rash. It is extremely important to recognize the early signs of heat stress in order to avoid much more dire consequences down the road!

Heat Exhaustion

Heat exhaustion is a more serious reaction to heat stress. At this point, the body is significantly stressed and has kicked into high gear in order to shed heat. Symptoms of heat exhaustion include weakness and fatigue, dizziness, nausea, headache, and (most visibly) cool and moist skin. A person suffering from heat exhaustion will have wet and clammy skin and feel cool to the touch because the body has overcompensated in its cooling attempts. Although the skin will feel cool to the touch, the body core is still hyperthermic. It is extremely important to quickly get the victim out of the hot environment; rehydrate him or her; and possibly apply cold packs to the armpits, forehead, and groin region.

Heatstroke

Heatstroke, the most serious reaction to heat stress, is a potentially life-threatening situation. At this point the body has failed to cope with the heat. Someone suffering from heatstroke will have dry and red skin, altered mental status, and possibly lose consciousness and/or have seizures. Each year many people die from heatstroke. Heatstroke victims must be rapidly cooled using cold packs, rehydrated, and quickly transported to the hospital for further medical care.

Hypothermia

Hypothermia may be caused by weather conditions or by the hazardous materials themselves. Across large parts of the United States and Canada, the winter months bring intense cold, snow, and ice. These conditions make it difficult to operate at hazardous materials and WMD incidents. The extremities, especially the fingers and toes, are most susceptible to frostbite, which occurs when the skin and underlying tissue is frozen. The ice crystals that form destroy the cells and cause tissue death. If frostbite is severe and not treated, gangrene may set in and amputation may be necessary. Hypothermia is also a concern. It occurs when the core body temperature drops because the body cannot produce enough heat to maintain 98.6°F (37°C). Symptoms of hypothermia may include altered mental status, shivering, and cold and pale-looking extremities. Victims of hypothermia must be removed from the cold environment and gradually rewarmed. Hypothermic patients are more susceptible to heart attacks and must be handled gently.

HEAT EXCHANGE UNITS

Heat exchange units are very effective in regulating body temperature and limiting the consequences of heat stress. When personnel will be working in protective clothing under extreme work conditions or in extreme climates, heat exchangers may mean the difference between life and death.

Air Cooled

Air-cooled heat exchange units are fairly uncommon due to their cost and maintenance. The primary disadvantages of air-cooled units are that they can be heavy, bulky, and

heat cramp ▪ A condition caused by exposure to excessive heat marked by severe involuntary muscle contractions.

heat rash ▪ A condition caused by exposure to excessive heat marked by discolored or mottled skin. Heat rash usually appears first on the legs and arms.

heat exhaustion ▪ A condition caused by exposure to excessive heat marked by pale and sweaty skin, weakness, and nausea.

heatstroke ▪ A condition caused by exposure to excessive heat marked by dry and red skin. This is a true medical emergency and requires immediate advanced medical care. Many people die of heatstroke each year.

hypothermia ▪ The excessive cooling of the body below normal body temperature.

noisy due to mechanical equipment such as fans and compressors; or there is an attached cooling line that may be a trip hazard. The advantages of these units are that they cool in a consistent, predictable manner and can usually be precisely controlled and adjusted. Air-cooled units are most effective during long-term operations.

Ice Cooled

Ice-cooled heat exchange units are typically vests filled with water that are frozen and stored cold until they are needed. The benefits of ice-cooled vests are their simplicity and low cost. The disadvantages of ice vests are the requirement for a freezer, a significant time needed to refreeze the ice vest after use, and an uneven cooling effect. Initially the ice vest provides a significant amount of cooling, possibly too much, which then dissipates over a relatively short amount of time. Ice vests should never be worn directly over the skin.

Water Cooled

Water-cooled heat exchange units are also typically vests filled with water; however, they have an inlet and outlet that can be connected to a water chiller. This technique has the advantage of providing a more consistent cooling effect and being able to be reused, or recharged, rapidly at the scene. The disadvantages of water-cooled vests are the acquisition cost and the need for a water chiller. A more economical version of ice vests uses a relatively inexpensive pump and ice water to cool the vest.

Phase Change Coolers

Phase change vests maintain a constant temperature, typically 59°F (15°C). This is a big advantage over the much lower temperatures of ice vests (32°F or 0°C). The phase change material inside the vest changes from a liquid to a semisolid gel, which maintains a temperature of 59°F (15°C) when electricity is applied, keeping the body cool by absorbing its excess body heat. The disadvantages of this type of vest are its cost and higher weight due to the battery pack.

PSYCHOLOGICAL STRESS

The cramped and confined space inside chemical protective clothing may lead individuals to experience psychological stress. Claustrophobia is experienced by many first-time users of PPE. They may have a very intense desire to get out of the CPC. If you are claustrophobic, or feel uncomfortable in small and confined spaces, please let your instructor know before you **don** personal protective equipment. The good news is that psychological stress, especially due to claustrophobia, can often be managed with the appropriate training.

don ▪ To put on (as in chemical protective clothing).

Technical Decontamination

If you become accidentally exposed to hazardous materials while wearing chemical protective clothing, the CPC will be removed during the process of decontamination (Figure 6-10). HAZWOPER requires a decontamination plan be in place before any personnel enter the hot zone. An operations level responder may don PPE for a number

SOLVED EXERCISE 6-4

What types of physical and psychological stresses could responders expect to encounter when operating in full level A PPE at the incident described in the opening scenario?

Solution: Heat stress and cold stress are both concerns. Hypothermia and frostbite are obvious concerns in North Dakota in January, but heat stress is still a concern during periods of exertion in a level A suit. Following periods of exertion and sweating, cold stress becomes of heightened concern due to rapid heat loss from perspiration.

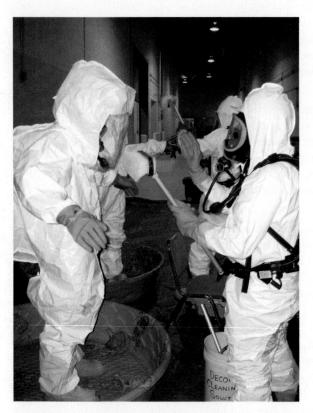

FIGURE 6-10 Wet decontamination being performed in a technical decontamination line.

of different reasons. For example, a firefighter may be asked to enter the hot zone as part of victim rescue and recovery; a law enforcement officer may be asked to enter the hot zone for evidence preservation and collection; a paramedic may be asked to perform decontamination activities and patient treatment in the warm zone. While performing any of these tasks, these personnel may accidentally become contaminated. Technical decontamination is the process by which contaminants are removed from chemical protective clothing and equipment used by first responders. This topic will be covered in much greater detail in Chapter 8, Operations Level Responders Mission-Specific Competencies: Technical Decontamination.

Before entering the hot zone or warm zone—areas of potential contamination—it is critical to understand what type of decontamination will be performed for that particular incident, and the location of the decontamination line. Decontamination may be wet or dry and may contain one or more rinse and wash stations, depending on the contaminant in question. Figure 5-25 illustrates a common decontamination line layout that may be used by first responders, and Box 6-2 describes a common procedure for technical decontamination. For a complete understanding of technical decontamination, it may be beneficial to read Chapter 8, Operations Level Responders Mission-Specific Competencies: Technical Decontamination.

Using PPE

As has been noted already, using personal protective equipment has its risks. Heatstroke can be a life-threatening complication of using PPE. Some agencies have the philosophy of "level A all the way." However, sometimes using the highest level of protection is actually more dangerous due to the environmental or work conditions than the potential chemical exposure itself. For example, a 5-gallon (19 L) spill of acetone in a small warehouse in Arizona, without air-conditioning where 20 fully loaded 55-gallon (208 L) drums need to be moved to get to the spill, is not a good candidate for level A protection (Figure 6-11). Acetone is not particularly toxic, with a PEL of 1000 ppm, and it could pose a significant flammability hazard. In addition, the working conditions in this scenario would be brutal! Likely a much safer alternative to a fully encapsulated level A suit would be a Nomex® undergarment and CPC with an air-purifying respirator. This PPE ensemble would greatly reduce the heat stress to the wearer.

SELECTION OF PPE

Personal protective equipment must be carefully selected based upon the hazard the chemical poses, chemical compatibility, the quantity spilled, the airborne concentration, the environmental conditions, and the type of work that will be performed. The PPE selection process must be a hazard and risk–based approach that accurately assesses the situation.

Choosing a Respirator

The airborne concentration of the contaminant will play the major role in the selection of respiratory protection. At airborne concentrations below the permissible exposure limit (PEL), OSHA does not require respiratory protection. However, carefully consider the long-term implications of low-level exposures, especially to suspected cancer-causing

BOX 6.2 THE TECHNICAL DECONTAMINATION PROCESS

There are two general categories of technical decontamination: **dry decontamination** and **wet docontamination**. Dry decontamination is very effective with gaseous exposures, materials that are not very volatile (generally a vapor pressure below 50 mmHg), and materials that are very volatile (generally a vapor pressure above 300 mmHg). Wet decontamination should be performed for liquids with an intermediate volatility, such as 50 to 300 mmHg, and those that are skin absorptive.

Wet Decontamination Procedure

1. Have entry team drop tools in bucket at edge of hot zone.
2. Decon team members visually inspect the entry team member being decontaminated by having the entry team member rotate 360 degrees and lift his or her feet.
3. If possible, evaluate the extent of contamination using monitoring instruments. If the personnel are not contaminated, they do not need wet decon and they can go straight to dry decon to **doff** their PPE.
4. Remove outermost gloves (third layer) and boot covers, and place in first recovery drum.
5. Wash/scrub any contaminated areas in the first pool and then rinse.
6. Have the entry member step into the second pool, and wash and rinse.
7. Remove outer clothing as in the dry decon procedure.

Dry Decontamination Procedure

1. Remove outermost gloves (third layer) and place in bag (if not already done).
2. Step into bag.
3. Nonencapsulating PPE: Remove SCBA from back while staying on air.
4. Have partner unzip the suit.
5. Carefully peel suit back from zipper. DO NOT allow outside of suit to contact clothing or skin. DO NOT allow outer gloves to touch the inside of the suit or the body.
6. Peel suit down to the boots.
7. Step out of bag toward the cold zone.
8. Go off air and remove face piece (mask) with inner gloves (typically latex or nitrile).
9. Remove inner gloves.

dry decontamination ■ The removal of contaminants without using water or other liquids. Dry decon is necessary in cold climates or when water is scarce. Dry decon is often performed by removing the PPE directly without a wash and rinse when the chemical and physical properties of the contaminant permit.

wet decontamination ■ The removal of contaminants through one or more cycles of washing and rinsing using water or soap and water.

doff ■ To take off (as in chemical protective clothing).

FIGURE 6-11 Use of PPE at a hazmat incident in a warehouse.

materials. Above the immediately dangerous to life and health (IDLH) level, OSHA does require supplied air respiratory protection.

With all of the options that have been discussed, it may at first seem difficult to choose the appropriate respirator. Often this choice is limited by the respirators provided by your employer. Most fire departments that operate exclusively at the operations level will have only SCBA available because this is used to fight structure fires. On the other hand, ambulance services and law enforcement agencies operating at the operations level will typically have only APRs available. These rules of thumb are changing rapidly, especially with the new mission-specific competencies of NPFA 472 (2008).

If an individual is training to the Mission-Specific Competency: Personal Protective Equipment level, your agency most likely has a variety of respirators available for use. The Department of Health and Human Services (DHHS) has developed a respirator decision logic that makes this decision much simpler (Figure 6-12). The respirator decision logic sequence covers key concepts such as IDLH atmosphere, oxygen deficiency, whether the contaminant name and level are known, as well as whether the material is a carcinogen. By following the respirator decision logic, all the key bases are covered. The flammability hazard is not covered in the metric because OSHA has set the IDLH value to a maximum of 10% of the LEL for flammable materials (assuming the material is not toxic at a much lower level). Therefore, determining the IDLH value covers both toxicity and flammability.

Choosing Chemical Protective Clothing (CPC)

Chemical compatibility is one of the most important factors to consider when choosing chemical protective clothing. As has been seen, chemicals may degrade chemical protective clothing and/or permeate through the material. The breakthrough time of chemicals may be as short as seconds to minutes. If the chemical degrades or permeates through the material quickly, you might as well not be wearing any chemical protective clothing. Manufacturers provide breakthrough timetables for their chemical protective ensembles, as illustrated in Figure 5-18. These tables may be color-coded or list minutes until breakthrough. Based upon the chemicals involved and the time that will be spent in proximity to the materials, an appropriate breakthrough time must be chosen.

The quantity of the material that is spilled and the work function will play a major role in the type of CPC that is required. As discussed earlier, heat stress and other environmental conditions should play a major role in CPC selection as well. The risk of potentially deadly heatstroke should be weighed against the additional protection offered by more substantial chemical protective clothing.

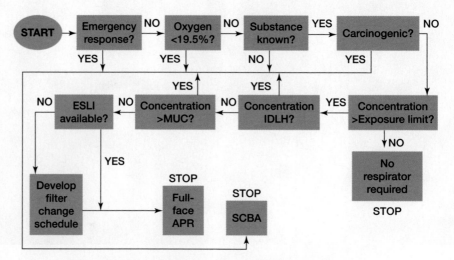

FIGURE 6-12 Respirator decision logic based upon the EPA method. *Courtesy of National Institute for Occupational Safety and Health*

Choosing a PPE Ensemble (EPA Level of Protection)

The combination of respiratory protection and the level of chemical protective clothing selected will determine the level of PPE. If an APR is chosen, the EPA level of protection is automatically level C. If an SCBA is chosen, the EPA level of protection may be level A or level B based upon the CPC chosen.

OSHA requires supplied air respiratory protection above the immediately dangerous to life and health (IDLH) level, which means that a minimum of level B protection is required. Between the permissible exposure limit and IDLH concentrations, air-purifying respirators and powered air-purifying respirators, otherwise known as level C protection, may be used. Under unknown conditions, a minimum of level B protection should be used. In confined areas, level A protection should be strongly considered.

The type of the hazard or risk the chemical poses will indicate the category of personal protective equipment required. For example, if a chemical is highly flammable, a level A suit will not be appropriate in a confined area due to the risk of fire or explosion. In this case, PPE that offers thermal protection and flash protection will be required if no other engineering controls are available, such as ventilation or remote shutoff of the product. Similarly, a chemical that is highly skin absorptive will require a vapor-tight suit—in other words, level A protection.

Many of the same factors will come into play when purchasing PPE. In this case, an agency should examine the SARA Title III sites in its jurisdiction, the transportation routes, and what types of materials are commonly carried. Even with a good hazard-risk analysis based upon the known hazards, much uncertainty always exists for public safety agencies that cover large geographical areas and large population densities. Other factors that should be considered when purchasing PPE are how resistant the material is to common insults such as abrasion, tears, and thermal stress (cold and heat); how easy the material is to clean and decontaminate; the acquisition and maintenance costs; the dexterity while in the suit; and even the color of the suit material. These factors should be evaluated in conjunction with the expected duties of the wearers. For example, cold resistance is probably not the biggest consideration for responders in southern Florida.

DONNING AND DOFFING PPE

Donning and doffing personal protective equipment should be an organized process for several reasons: Each step is very important, and following a step-by-step checklist ensures that no essential step is forgotten. OSHA requires that the buddy system be used on any hazardous materials incident. When all members of both the entry team and a backup team get dressed at the same time, it ensures that all members will be dressed at the same time and ready for their duties at the appropriate time. It is also vital that team members

SOLVED EXERCISES 6-5

What is the appropriate PPE for the incident described in the opening scenario, including chemical protective clothing and respiratory protection?

Solution: It depends on the ammonia level. At lower ammonia concentrations (below the LEL), the primary hazards are corrosiveness and toxicity, and level A protection would provide optimal protection for longer stay times. At ammonia concentrations above the LEL, the flammability hazard is of primary concern and must be addressed. There are two options: The first, and best, option is chemical protective clothing with flash protection. The second option is structural firefighter protective gear with SCBA, which should be used only when life safety is at stake, the stay time will be short, and wet decontamination is immediately available. In this case (January in North Dakota), outdoor wet decontamination is not a viable option, and a heated tent or other specialized decon unit must be available. Even under optimal conditions, the rescue team should expect some skin irritation or burns when performing rescue in turnout gear.

spend the least amount of time waiting for their job assignments in chemical protective clothing to minimize the effects of thermal stress.

Several key procedures should always be performed irrespective of the type of personal protective equipment used:

1. A medical evaluation should be performed prior to donning personal protective equipment to determine whether the user is physically able to perform in PPE. A medical evaluation should also be performed after the PPE is doffed to ensure that the user has not suffered injuries, thermal stress, or overexertion.
2. The personal protective equipment needs to be examined thoroughly for damage and other problems that may interfere with its proper function.
3. All personal effects must be removed before donning chemical protective clothing for two reasons: First, sharp personal effects such as jewelry may puncture or otherwise damage the PPE. For example, rings with sharp edges may puncture gloves in the hot zone and cause penetration of the hazardous material. Second, wallets, cell phones, and jewelry may become contaminated and become hazardous waste if they cannot be properly decontaminated. For example, mercury will amalgamate with gold and silver jewelry and cannot be decontaminated. Leather and paper are porous materials and are difficult to impossible to decontaminate. A good rule of thumb is not to take anything into the hot zone that you are not willing to give up!
4. The on-air and entry times should be recorded for all personnel entering the hot zone.
5. After doffing and decontaminating the PPE, it is essential that decontamination be verified using appropriate monitoring equipment, such as a photoionization detector, pH paper, Geiger counter, or other appropriate detection equipment.

Boxes 6-3 through 6-8 list examples of donning and doffing procedures for common personal protective equipment. These procedures are suggestions only, and each agency may have slight variations.

Resource Central

View an illustrated step-by-step guide on Donning and Doffing Fully Encapsulating Suits

BOX 6.3 DONNING AND DOFFING FULLY ENCAPSULATING SUITS

DONNING

1. Receive medical check.
2. Remove watches, jewelry, leather, shoes, and other personal items, and place in a plastic bag in a secure location.
3. Inspect suit (seams and zippers) and self-contained breathing apparatus (SCBA). Don fire-resistant suit scrubs (if required).
4. Don encapsulating suit to waist.
5. Don chemical-resistant boots with boot covers. For suits with splash guard, position splash guard over the outer boot.
6. Don a cooling vest.
7. Don SCBA and record the bottle pressure.
8. Don communications gear (radio and connections should be secured with tape).
9. Don innermost cotton gloves (if desired) and nitrile gloves.
10. Continue donning suit over arms.
11. Don outermost chemical-resistant gloves.
12. Review hand signals:
    *Thumbs-up signal—OK
    *Hands clutching throat—SCBA malfunction
    *Tapping top of head—Emergency! Get me out of suit
    *Audible high–low siren—Leave Hot Zone Now!

13. Don SCBA face mask: Check seal, perform function test (don't stay on air), and test communications.
14. Don head gear (if required).
15. Assign each person a suit number. Advise operations officer when ready to go on air. Ensure that the entry, backup, and decon teams are properly briefed. Connect regulator to face mask and record time. Ensure wearer is breathing air and ready with thumbs-up signal. Close suit. Record on-air time, entry time, egress time, and off-air time.

DOFFING

1. **NOTE: Remain on air until the last step!** Set out a large decontamination bag. This procedure may be performed as a stand-alone method (dry decontamination), or it may follow wet decontamination.
2. Step into a large bag (dry decon bag).
3. Unzip the suit and carefully peel the suit down to the boots. The outside (dirty) gloves should only touch the outside of the suit, and the inner (clean) gloves should only touch the inside of the suit.
4. Step out of the boots toward the cold zone. You may need the assistance of a tender or a long-handled tool for balance.
5. While remaining on air, remove head gear and communications equipment, loosen the backpack straps, and take off the pack and have a tender hold it or lay it on a chair. Remain on air! Do NOT remove the face piece at this time.
6. Using the inner gloves, remove the face piece and go off air (when in a clean area of the decon zone).
7. Remove the inner gloves.
8. Verify that decontamination was successful.
9. Perform post-entry medical monitoring and rehabilitation.

 ## BOX 6.4 DONNING AND DOFFING NONENCAPSULATING SUITS

Resource
Central

View an illustrated step-by-step guide on Donning and Doffing Nonencapsulating Suits

DONNING

1. Receive medical check.
2. Remove watches, jewelry, leather, shoes, and other personal items, and place in a plastic bag in a secure location.
3. Inspect suit (seams and zippers) and the self-contained breathing apparatus (SCBA) or the air-purifying respirator (APR).
4. Don fire-resistant suit/scrubs (if required). Don nonencapsulating suit to waist.
5. Don chemical-resistant boots with boot covers (if required). For suits with splash guard, position splash guard over the outer boot. For suits without splash guard, fold excess pant leg over outer boot, and tape seam between outer boot and suit (blousing). Leave tab on tape for easy removal.
6. Don a cooling vest (if required). Don suit and zip-up front zipper (if there is no splash flap for zipper, tape leaving tab for easy removal).
7. Don innermost nitrile gloves and then the appropriate chemical-resistant gloves. Tape gloves using appropriate method. Tape should be applied over widest part of hand and leave tab for easy removal. If using an outer chemical- or abrasion-resistant glove, don and secure with tape and tab.
8. Don the appropriate chemical-resistant gloves. Tape gloves using appropriate method. Tape should be applied over widest part of hand and leave tab for easy removal. If using an outer chemical- or abrasion-resistant glove, don and secure with tape and tab.
9. Don SCBA and record tank pressure. Don communications gear (radio and connections should be secured with tape).
10. Review hand signals:
    *Thumbs-up signal—OK
    *Hands clutching throat—SCBA malfunction
    *Tapping top of head—Emergency! Get me out of suit
    *Audible high–low siren—Leave Hot Zone Now!

11. Don respirator face mask: Check seal, perform function test (don't stay on air), and test communications. Don attached hood of suit (if taping seems necessary, consider going to level A protection). Don head gear (if required).

12. Assign each person a suit number. Advise operations officer when ready to go on air. Ensure that the entry, backup, and decon teams are properly briefed. Connect regulator to face mask and record time. Ensure wearer is breathing air and ready with thumbs-up signal. Close suit. Record on-air time, entry time, egress time, and off-air time.

DOFFING

1. **NOTE: Remain on air until Step 6!** Set out a large decontamination bag. This procedure may be performed as a stand-alone method (dry decontamination), or it may follow wet decontamination.

2. Step into the large bag (dry decon bag).

3. While remaining on air, remove head gear and communications equipment. If using an SCBA: While remaining on air, loosen the backpack straps, and take off the pack and have a tender hold it or lay it on a chair. Remain on air! Do NOT remove the face piece at this time.

4. Unzip the suit and carefully peel the suit down to the boots. The outside (dirty) gloves should only touch the outside of the suit, and the inner (clean) gloves should only touch the inside of the suit. The gloves should be taped loosely enough to be able to remove the hands without removing the tape at the wrists.

5. Step out of the boots toward the cold zone. You may need the assistance of a tender or a long-handled tool for balance.

6. Using the inner gloves, remove the face piece and go off air (when in a clean area of the decon zone).

7. Remove the inner gloves.

8. Verify that decontamination was successful. Check the hands and feet first because typically 85% of the contamination is found there.

9. Perform post-entry medical monitoring and rehabilitation.

View an illustrated step-by-step guide on the Double Seam Glove Taping Method

BOX 6.5 DOUBLE SEAM GLOVE TAPING METHOD

1. Don glove and fold the glove back at the wrist, leaving approximately 1 to 2 inches folded over. Place the center of the folded-over flap at the widest part of the hand. Splay the thumb and fingers out to create maximum width.

2. Pull the suit cuff to the center of the folded-over flap. Make sure the seam is located at the widest part of the hand. This minimizes the number of wrinkles and allows the hand to be easily removed for dry decontamination.

3. Tape the seam using a chemical-resistant tape (rated). Minimize the number of wrinkles (less liquid penetration), and make sure the hand can be removed from the glove without removing the tape (for dry decontamination).

4. One or two wraps of tape are sufficient.

5. Fold the flap back. Make sure the seam has been taped well, at the widest part of the hand, and with few wrinkles.

6. Tape the newly created seam in the same way: over the widest part of the hand, with as few wrinkles as possible.

7. Two seams are now taped. The double taping method with few wrinkles gives maximum protection against liquid intrusion. The glove–suit ensemble is taped loosely enough to allow removal of the hand during dry decontamination.

8. Pull the glove back over the hand and repeat for the other side.

BOX 6.6 USING AIR-PURIFYING RESPIRATORS

Resource Central

View an illustrated step-by-step guide on Using Air-Purifying Respirators

1. Before wearing an APR in the hot zone, ensure that a fit test (qualitative or quantitative) has been performed, passed, and recorded with the exact same type and size of mask.
2. Carefully inspect the mask: Inspect the face piece head straps. Look for abnormal wear, degradation, or tears. Make sure all of the straps and fasteners are in place. Check face piece for dirt, scratches, cracks, discoloration, tears, or holes. If present, inspect the voice emitter or other electronics for signs of damage.
3. Check the filter canister to ensure that it is appropriate for use with the contaminant in the hot zone.
4. Check the end of service life indicator on the canister(s) (or determine the appropriate service life time), and make sure the canister(s) is not damaged.
5. Thread the canister(s) into the appropriate face piece port. Ensure that all ports have a canister! Some masks require up to three canisters. Remove both the tab covering the air inlet and the outlet holes of the canister.
6. Loosen the harness head straps. Place your chin in the face piece first, and pull the harness over your head.
7. Tighten the head straps according to the manufacturer, but typically the order is from the bottom (neck area) to the top (temple region).
8. Perform a seal check on the mask. Cover the air inlet with the palm of your hand. Breathe in and hold for 10 seconds. The mask should collapse into your face and remain there. If it leaks or does not stay collapsed, readjust or tighten the straps. **Do NOT use the APR if you cannot get a good seal!**

BOX 6.7 DONNING AND DOFFING LEVEL C WITH HOODED POWERED AIR-PURIFYING RESPIRATORS (PAPRS)

Resource Central

View an illustrated step-by-step guide on Donning and Doffing Level C with Hooded PAPRS

DONNING

1. Organize and inspect all of the personal protective equipment. Is the equipment damaged or worn? Is it the correct level of PPE for the situation? Is it the proper size? Examine the face piece, blower, canister, seams, zippers, and materials for damage.
2. Perform pre-entry medical monitoring. At a minimum, check vital signs. Ensure personnel are hydrated.
3. Remove watches, jewelry, leather, shoes, and other personal items (place in plastic bag and store in a secure location).
4. Check the filter canister to ensure that it is appropriate for use with the contaminant in the hot zone.
5. Check the end of service life indicator on the canister(s) (or determine the appropriate service life time), and make sure the canister(s) is not damaged.
6. Thread all three canisters into the appropriate blower ports. Ensure that all ports have a canister and that the protective caps and tabs are removed (two per canister).
7. Check to make sure the blower flow rate is adequate (6 cfm) using the manufacturer-supplied tester.
8. If not already completed, attach the breathing hose to the blower by tightening the thumb screw.
9. Don innermost nitrile gloves.
10. Don nonencapsulating suit.
11. Don chemical-resistant boots with boot covers (if required). For suits with splash guard, position splash guard over the outer boot. For suits without splash guard, fold excess pant leg over outer boot and tape seam between outer boot and suit (blousing). Leave tab on tape for easy removal.
12. The hood of the suit is not used with a hooded PAPR. Tuck it under so it is under the neck on your back.

13. Don the outer pair of gloves and tape in an appropriate fashion (ensure that your hands can easily come out of the gloves for dry decon).
14. Don blower unit, and tuck the straps away to prevent trip and snag hazards.
15. Turn on blower unit battery.
16. Don the hood. Tuck the inner shroud into the suit, and let the outer shroud hang over the suit.

DOFFING

1. **NOTE: Remain on air until the last step!** Set out a large decontamination bag. This procedure may be performed as a stand-alone method (dry decontamination), or it may follow wet decontamination.
2. Step into the large bag (dry decon bag).
3. Remove tape at the boots and zipper seam, if they were used.
4. If wet decontamination was performed, wipe any excess liquid from the shroud using a towel.
5. Roll the outer shroud up and tape. Do not obscure the vision!
6. Unzip the suit and carefully undress down to the boots. The outside (dirty) gloves should only touch the outside of the suit, and the inner (clean) gloves should only touch the inside of the suit. The gloves should be taped loosely enough to be able to remove the hands without removing the tape at the wrists.
7. Step out of the boots toward the cold zone. You may need the assistance of a tender or a long-handled tool for balance.
8. Using the inner gloves, remove the hood touching only the inner shroud (when in a clean area of the decon zone). Go off air.
9. Remove the inner gloves.
10. **Verify that decontamination was successful.**
11. Perform post-entry medical monitoring and rehabilitation.

Resource
Central

View an illustrated step-by-step guide on Donning and Doffing LANX Suits and MOPP Gear

 BOX 6.8 DONNING AND DOFFING LANX SUITS AND MOPP GEAR

DONNING

1. Receive medical check.
2. Remove your duty belt, notepad, pens, uniform shirt and tie, and jewelry. Place in a plastic bag in a secure location.
3. Open the sealed pouches, and inspect the suit pants and suit coat (seams and zippers) and air-purifying respirator (APR). Check the filter canister to ensure that it is appropriate for use with the contaminant in the hot zone. Check the end of service life indicator on the canister(s) (or determine the appropriate service life time) and make sure the canister(s) is not damaged.
4. Place a shoe/boot cover over each shoe or combat boot.
5. Don the LANX/MOPP trousers. Adjust the suspenders by sliding the adjustment clip up or down. Fasten the pants by snapping the front inner and outer snaps and zipping up the trousers. Adjust the waistband for a snug fit (do not fasten the bottom of the trousers at this time).
6. Don the boots. Secure the trouser legs over the boots by folding excess suit material over, making a snug fit, and attaching the Velcro straps.
7. Don the LANX/MOPP jacket and hood: Open the sealed bag and put on the jacket. Zip up the jacket. Reach between your legs and grab the elastic loop attached to the back of the jacket. Pull the loop under your legs and secure it with the front snap closure. Adjust the drawstrings at the bottom of the jacket by pulling snug and tying in a bow knot. Secure the jacket by attaching the Velcro in the front and by pulling the drawstrings snug using the adjustment clips to secure them. Lifting your chin up will help you in securing the Velcro and creating a good APR to jacket seal.
8. Don the air-purifying respirator (APR) and check for a good seal.

9. Pull the hood up and over the head, making sure the hood overlaps the APR.
10. Reattach duty belt.
11. Don nitrile innermost gloves.
12. Don chemical protective outer gloves. Secure the jacket sleeve over the gloves by folding excess suit material over, making a snug fit, and attaching the Velcro straps.
13. Check all zippers, ties, and Velcro, particularly around the APR and under the chin, and make sure they are all secured.

DOFFING

1. If you have a firearm, secure the weapon and place in the designated secure weapon-drop area or hand directly to the weapons officer.
2. Drop your duty belt and any tools into the designated tool drop area.
3. Loosen all Velcro closures, ties, zippers, and the boot straps.
4. Pull the hood off your head.
5. Remove the outer gloves and place them into an overpack drum (keep the inner gloves on).
6. Drop the coat off your back and place it into the overpack drum.
7. Unbuckle the suspenders, and unsnap and unzip the trousers.
8. Roll the trousers down, touching only the inside of the trousers, and pull your feet out of the boots and place them into the overpack drum.
9. Take a large step away from the potentially contaminated PPE toward the cold zone.
10. In one motion, remove the APR.
11. Remove the inner gloves and discard them.
12. Verify that decontamination was successful using monitoring equipment.
13. Perform post-entry medical monitoring and rehabilitation.

EMERGENCY PROCEDURES

Emergency procedures should be practiced for in-suit emergencies such as an air emergency or failure of an exhalation valve on level A suits. Believe it or not, there is still breathable air inside a level A suit, even if your SCBA is completely empty or not functioning properly. Typically there will be 5 to 10 minutes of breathable air. The key is to stay calm and move to the decontamination line as quickly as possible and let the decon personnel know of your emergency. It is a good idea to train on low-air emergencies under controlled situations so you know what to do. When a one-way valve becomes inoperable on a level A suit, it will continue to inflate until you look like the Michelin man. In this case, you must attempt to reset the valve by burping the suit. This can be accomplished by slowly and methodically bending over with your arms crossed to attempt to force the air out of the valve without puncturing the suit elsewhere. After burping the suit, you should exit the hot zone to determine why it malfunctioned.

PPE Program

HAZWOPER requires that a written PPE plan be in place prior to using personal protective equipment. This plan should cover PPE selection, maintenance, testing, inspection, and storage of the PPE. The donning and doffing procedures just discussed should be part of this written PPE plan as well.

MEDICAL SURVEILLANCE PROGRAM

Hazardous materials incidents place first responders in uncontrolled environments with the potential for chemical exposure. HAZWOPER therefore requires that all employers, including public safety agencies, must establish a medical surveillance program that expects their employees to use personal protective equipment (PPE). The medical surveillance program must be designed to determine whether personnel are healthy and fit enough to safely and effectively use PPE. Therefore, the medical exam should include a medical and work history with an emphasis on hazardous chemical exposure and fitness to wear PPE. The medical surveillance program should be administered by, or under the supervision of, a licensed physician that is informed of the expected job duties and the types of PPE personnel will be wearing. Medical exams must be performed prior to assignment to an organized hazardous materials response team, annually after that (or biannually with approval of the physician), upon chemical exposure, and after personnel leave the hazardous materials response team (exit physical). The medical surveillance program is extremely important not only to determine fitness for duty but also to detect any chemical exposures early so that treatment can be started as soon as possible.

MAINTAINING PPE

Personal protective equipment must be maintained properly to ensure its effectiveness when used. Proper maintenance includes visually inspecting the PPE periodically; refolding the chemical protective clothing to limit damage from deep creases; and inspecting valves, zippers, and seams.

In addition, vapor- and gas-tight suits, or level A suits, need to be pressure tested per manufacturer's recommendations, which is usually once a year. Pressurized cylinders that are used with supplied air respirators must be hydrostatically tested at intervals determined by the manufacturer based upon cylinder construction.

PPE should be stored per the manufacturer's recommendations. This usually means storing the PPE in a cool, dry place away from sunlight, dust, and thermal stresses. PPE should also be kept away from solvents, abrasive materials, and sharp materials that may cause damage. Periodical inspections should be documented in written form, as should be the use of reusable PPE, any training use of the PPE, and any use at hazardous materials or WMD incidents. Most PPE has an expiration date after which it should not be used, except for use in training that does not involve live chemical agents.

Summary

Personal protective equipment is the last defense between you and the hazardous material. Therefore, it is essential to understand how to use it properly. This includes selecting chemical protective clothing based upon the chemical hazards that are present and the chemical compatibility with the suit material (breakthrough time). You must also be able to quickly and effectively don and doff the personal protective equipment. First responders at the operational level need to receive hands-on training from their agency in the reliable use of the personal protective equipment that is provided. Between incidents, PPE must be maintained, stored, inspected, and tested per the manufacturer's recommendations. Remember, your PPE is your last defense against hazardous materials after engineering controls and workplace practice changes have failed. Treat your PPE like your best friend!

Review Questions

1. What are the four levels of protection according to the EPA, and what are their distinguishing features?
2. What are the three ways PPE may be compromised?
3. What types of physiological stress can PPE use cause?
4. When must a decontamination plan be in place?
5. When should medical surveillance be performed on entry team members?

Problem-Solving Activities

1. Determine which of the PPE that your agency uses is compatible with the following materials: acetone, chlorine, ammonia, gasoline, and sulfuric acid. What are the breakthrough times for each?
2. What type of PPE would you select to investigate a chlorine leak deep inside a building? What are your considerations?
3. What type of PPE would you select to clean up a large benzene spill inside a warehouse? What are your considerations?
4. What type of PPE would you use to investigate a tanker carrying sulfuric acid on a 97°F (36°C) day with 100% humidity? What are your considerations?
5. How does your agency avoid heat stress? How would your agency deal with a heat stress emergency?
6. Find the donning and doffing procedures for your chemical protective clothing. Can you follow it? If you have training suits available, find a partner and don and doff the PPE.

References and Further Reading

Centers for Disease Control and Prevention. (2007). *NIOSH Pocket Guide to Chemical Hazards.* Washington, DC: U.S. Government Printing Office.

Forsberg, Krister, and S. Z. Mansdorf. (2007). *Quick Selection Guide to Chemical Protective Clothing* (5th ed.). New York: John Wiley.

National Fire Protection Association. (2008). NFPA 472, *Standard for Competence of Responders to Hazardous Materials/Weapons of Mass Destruction Incidents.* Quincy, MA: Author.

National Fire Protection Association. (2008). NFPA 473, *Standard for Competencies for EMS Personnel Responding to Hazardous Materials/Weapons of Mass Destruction Incidents.* Quincy, MA: Author.

National Fire Protection Association. (2008). NFPA 1999, *Standard on Protective Clothing for Emergency Medical Operations.* Quincy, MA: Author.

National Fire Protection Association. (2007). NFPA 1994, *Standard on Protective Ensembles for First Responders to CBRN Terrorism Incidents.* Quincy, MA: Author.

National Fire Protection Association. (2005). NFPA 1991, *Standard on Vapor-Protective Ensembles for Hazardous Materials Emergencies.* Quincy, MA: Author.

National Fire Protection Association. (2005). NFPA 1992, *Standard on Liquid Splash-Protective Ensembles and Clothing for Hazardous Materials Emergencies.* Quincy, MA: Author.

Occupational Safety and Health Administration. (1990). 29 CFR 1910.120, *Hazardous Waste Site Operations and Emergency Response (HAZWOPER).* Washington, DC: U.S. Department of Labor.

Occupational Safety and Health Administration. (1984). 29 CFR 1910.134, *Respiratory Protection Standard.* Washington, DC: U.S. Department of Labor.

Weber, Chris. (2007). *Pocket Reference for Hazardous Materials Response.* Upper Saddle River, NJ: Pearson/Brady.

7

Operations Level Responders Mission-Specific Competencies: Mass Decontamination

KEY TERMS

ambulatory, *p. 218*

dilution, *p. 222*

evidence preservation, *p. 226*

fire engine alley, *p. 220*

mass casualty incident, *p. 218*

non-ambulatory, *p. 218*

triage, *p. 218*

OBJECTIVES

- Define a mass casualty incident.
- Plan for a mass casualty incident involving hazardous materials or WMD.
- Describe the equipment and methods needed to set up a mass decontamination line.
- Set up a mass decon line using available equipment.
- Describe crowd control procedures at a mass casualty incident.
- Perform scene control activities at a mass casualty incident.
- Describe why proper documentation is crucial during mass decon operations.

Resource Central

For additional review and practice tests, visit **www.bradybooks.com** and click on Resource Central to access book-specific resources for this text! To access Resource Central, follow directions on the Student Access Card provided with this text. If there is no card, go to **www.bradybooks.com** and follow the Resource Central link to Buy Access from there.

You Are on Duty! Large Vapor Cloud from Production Plant

Dalton, GA

A 911 call is received at 9:34 P.M. on April 12, 2004, from the local chemical manufacturing plant that there has been a spill of allyl alcohol during a production run. Dispatch information indicates the spill is inside the building. However, a lack of accurate information leads to a series of first responder exposures: The first arriving fire unit drives through the vapor cloud but with good presence of mind redirects follow-on responding units. Police and EMS units attempt to evacuate residents in the vapor cloud without any personal protective equipment (PPE). Ultimately 154 people required decontamination and medical treatment, including 15 first responders.

- How would your agency handle the decontamination of over 150 individuals?
- What is the appropriate decontamination solution and procedure for allyl alcohol?
- How would the incident area and evacuation zone be controlled to ensure all contaminated individuals receive decontamination?
- How would you handle the weapons that law enforcement officers carry? Would your local law enforcement officers hand over their weapons to decon personnel? If so, how would they be secured?
- How would your hospital ensure that patients are properly decontaminated? Does your hospital have the appropriate facilities, equipment, and trained personnel?
- How would the personal property of the victims be documented and secured?

Let's try to answer these questions in this chapter.

mass casualty incident ■ An emergency incident that presents more patients than the responding agency can handle at the time.

ambulatory ■ Able to walk on one's own.

non-ambulatory ■ Not able to walk on one's own.

triage ■ The process of prioritizing patient care based upon available resources.

Mass casualty incidents are a challenge for any agency to handle, especially if they involve the release of hazardous materials or the dissemination of weapons of mass destruction (WMD). Decontaminating a large number of victims, both **ambulatory** and **non-ambulatory**, along with first responders would be difficult for even the most well-equipped and trained agency. Such a challenge requires using different techniques from the much more common technical decontamination, which will be discussed in Chapter 8. Mass decontamination combines crowd control, **triage**, and rapid decontamination. Because resources by definition will be limited, decontamination will almost always be imperfect and require further judgment as far as exposure triage and decontamination verification are concerned. In this chapter, we discuss how to maximize the effectiveness of mass decontamination with limited resources.

Mass Decontamination Operations

Mass decontamination operations must be carried out quickly, often with incomplete information, using an agency's available equipment. The mass decontamination line should be located upwind, uphill, and upstream of the hazardous materials or WMD release. Several different techniques can be used for mass decontamination operations (Figure 7-1). Fire departments, for example, typically carry water on their fire engines and trucks. Responders mainly use water during a mass decontamination incident because water is considered the universal solvent; however, if soap is available in sufficient quantities, it should also be used. EMS, fire department, and law enforcement organizations may have specialized decontamination tents or trailers at their disposal. This equipment has been designed to decontaminate large numbers of people. However, unless they are readily available and personnel are well trained in their deployment, they may take longer to deploy.

The easiest population to decontaminate will be ambulatory victims that can follow verbal commands. Using a bullhorn, or another means of public address, ambulatory victims can be told to exit the hot zone, remove their contaminated clothing, move to the

FIGURE 7-1 Mass decontamination operations. *Courtesy of Mike Becker, Longmont (CO) Fire Department.*

FIGURE 7-2 Diagram of a fire engine alley. *Art by David Heskett.*

triage area, and walk through a mass decontamination line (Figure 7-2). Non-ambulatory victims and those that cannot follow verbal commands will have to be removed from the hot zone. Hot zone rescue will be covered in detail in Chapter 12, Operations Level Responders Mission-Specific Competencies: Victim Rescue and Recovery.

PPE SELECTION

The personal protective equipment (PPE) that is available will dictate the type of mass decontamination and follow-up medical care that can be performed. For example, a fire department that has no other PPE besides structural firefighter protective gear may still perform mass decontamination, but must remain upwind and uphill, and a sufficient distance away from the contaminated victims. In this case, the firefighters will not be able to provide medical care, and crowd control is of utmost importance. In contrast, if chemical protective clothing and respiratory protection are available, the department may perform a greater level of decontamination and medical care. For mass decontamination, chemical protective clothing (CPC) and an air-purifying respirator (APR) with an NIOSH CBRN-rated cartridge or SCBA would generally be acceptable personal protective equipment.

DECONTAMINATION SUPPLIES AND EQUIPMENT

A vast array of decontamination equipment and supplies are available thanks to the grant money available recently from the U.S. Department of Homeland Security (DHS). Although preconfigured tents and trailers, showers and pools, and specialized decontamination solutions are helpful and make the mass decontamination process faster and more efficient, all that is truly needed is a good water supply.

Preconfigured Tents and Trailers

Preconfigured tents and trailers boost the efficiency and effectiveness of the mass decontamination process, which increases the safety of the first responders as well as the victims that are being decontaminated. Typical features of such equipment include prearranged shower lanes, non-ambulatory decontamination corridors, pre-plumbed hot and cold water, pre-plumbed soap, and air heating and cooling capabilities. This type of equipment is typically easy and fast to set up with proper training. One major advantage of using this equipment is the increase in both throughput and effectiveness of the decontamination procedure itself. Jurisdictions that have a higher risk for terrorist incidents or hazardous materials incidents that could involve mass casualty incidents should seriously consider using preconfigured tents and trailers as part of their response capability.

Fire Engine Alley

With a little practice, every fire department can set up mass decontamination relatively quickly and effectively. The simplest way to perform mass decontamination is to set up a **fire engine alley**. Figure 7-3 shows the operation of a fire engine alley. In its simplest form, two fire engines are parked side by side approximately 10 to 15 feet (3 to 4.5 m) apart, and the fog nozzle is applied to each side discharge. Ambulatory victims are directed to disrobe at one end, walk through the shower produced by the fog nozzles, and re-dress in clean clothing at the other end where triage can occur. This system can be enhanced by providing a soap solution using a foam eductor, increasing privacy by adding tarps above and beside the shower corridor, and adding a rinse station by providing another fire engine alley. This technique is especially useful in urban areas where reliable water supplies are common (Box 7-1).

Pre-Plumbed Decontamination Corridors at Fixed Site Facilities

Hospital and clinic emergency rooms increasingly have pre-plumbed decontamination corridors. Such fixed site facilities are very helpful at locations where victims acquiring decontamination are expected to congregate. Although health care professionals are typically trained in the use of the pre-plumbed decontamination corridors they work with, they may require the assistance of local public safety first responders such as the fire

fire engine alley ▪ Emergency decontamination using two fire engines that are positioned side by side with their side discharges operating, that are used to wash victims as they walk through the resulting corridor.

FIGURE 7-3 A common fire engine alley configuration that is used for mass decontamination activities, especially for ambulatory victims.

BOX 7.1 OPERATION OF A FIRE ENGINE ALLEY

1. From a distance, instruct the victim to move away from the hazardous material or WMD.
2. From a distance, instruct the victim to remove his or her contaminated clothing completely. This includes underwear and jewelry. Attempt to ensure victim privacy using tarps or tents as long as scene resources permit.
3. Have the victim enter the mass decontamination corridor (tent, or fire engine alley). Having pre-printed materials helps to speed up the process. Have him or her briefly wash using lukewarm or cool water (1 minute max). Cooler water closes pores, but be careful not to induce hypothermia. Some mass decon tents have the ability to deliver soap and water, but most mass decontamination systems do not.
4. Have the victim move to the next station and rinse with cool or lukewarm water.
5. Have the victim re-dress in a disposable Tyvek suit or other suitable clothing.
6. Have the victim move to the triage area where a primary assessment can be conducted.

department or hazardous materials response teams to effectively decontaminate a large number of victims.

Improvised Equipment and Methods

Mass decontamination can also be carried out using improvised methods with the resources already on hand in the community. Swimming pools, both municipal and private, provide excellent sources of water. Even if pumping equipment is not available, ambulatory victims can be asked to walk through the pool. The chlorination or other disinfectants in the water are often efficient at breaking down chemicals, especially some chemical agents and most biological agents. Other excellent sources of decontamination are the emergency showers located at industrial facilities, research facilities, and laboratory teaching facilities. These facilities have reliable sources of water, and the runoff is channeled away from the victims and responders through the showers' drainage system. The owners and managers of these facilities will usually be very open to helping first responders in times of emergency. However, preplanning for the use of these types of facilities should be performed. In addition, coordination and training should be conducted with facility staff to ensure a smooth operation in the event of an emergency.

DECONTAMINATION METHOD SELECTION

As important as selecting the decontamination equipment and supplies are, even more critical is choosing the appropriate decontamination method. The decontamination method is the process by which harmful and dangerous contaminants are removed from the victims and their personal effects. For example, using a simple water spray from a hose is a very effective decontamination method for water-soluble contaminants. However, this method would not work for a viscous oil-like substance that does not dissolve in water. For this

SOLVED EXERCISE 7-1

In the incident described in the opening scenario, how would you set up the mass decon line near the production plant? At the hospital?

Solution: At the production plant or the hospital, an outdoor fire engine alley or tent system would likely work well in Georgia in April. The hospital may have a dedicated decon room or lane set up and/or have the training and equipment to set up a decon tent. Determine your local hospital's decontamination equipment and capabilities now, before an incident!

oil-like material, a decontamination method consisting of brushing with soap and water would be necessary. Let's take a look at some common decontamination methods.

Dilution

Dilution is the process of reducing the concentration of harmful contaminants below a dangerous level. Dilution is most commonly accomplished using water for water-soluble contaminants, or soap and water for water insoluble contaminants.

Dilution is very effective for removing contaminants from victims. For example, a patient has been contaminated with muriatic acid, a water-soluble corrosive, can be effectively decontaminated by diluting the acid with copious amounts of water. We typically attempt to flush corrosive exposures for at least 15 minutes. At a mass casualty incident, this method may not be practical for large numbers of contaminated victims. However, if a manageable number of victims have been contaminated, it may be feasible.

Water is the universal solvent and fortunately is typically readily available across the United States. Unfortunately, mass decontamination efforts in rural areas, arid climates, or very cold climates may be hindered due to the lack of readily available water. Under such circumstances, dilution may not be the most effective decontamination method.

One alternative to supplying large amounts of water is to use swimming pools for decontamination. A very fast and efficient process for mass decontamination is having victims disrobe, walk into the pool, briefly wash, and then re-dress on the other side. Pools have the advantage of being chlorinated, which is a disinfectant that breaks down many chemicals through oxidation. Indoor pools can also be used in cold climates where outdoor, wet decontamination is hazardous to the patient and otherwise not practical. Keep in mind that it is essential to have mutual aid agreements with facilities that have large indoor pools to facilitate the decontamination process.

Washing

When the contaminant is viscous and difficult to remove, brushing or another mechanical means of contaminant removal may be necessary. Washing is more labor and time intensive than dilution. However, washing is typically necessary in order to thoroughly remove most contaminants (Figure 7-4). During mass decontamination operations, it is

dilution ▪ The process of reducing the concentration of a hazardous material by mixing it with a nonhazardous material (usually water) in the hopes of rendering the entire mixture nonhazardous. Typically large quantities of water are needed to dilute a hazardous material, especially corrosives, to safe levels.

FIGURE 7-4 An example of mass decontamination of a non-ambulatory patient using washing. *Courtesy of Mike Becker, Longmont (CO) Fire Department.*

important to instruct ambulatory patients to wash themselves because this allows the use of scarce personnel to decontaminate the non-ambulatory patients or others requiring assistance. It is helpful to have printed materials or signs directing the ambulatory victims to the appropriate area and instructing them in the intended washing procedure. Preplanning and training are essential components to a successful mass decontamination event.

Isolation

When dilution and washing have failed to remove the harmful contaminants, or a biological agent is involved, isolation may be the only option at the first responder operations level. Contaminated or infected victims should be isolated in an area that is away from other emergency operations, is in a safe location, and allows for easy crowd control with minimal personnel. The area should be upwind of the hot zone, but downwind of the cold zone.

SOLVED EXERCISE 7-2

In the incident described in the opening scenario, what type of decontamination solution would you use and why?

Allyl alcohol	Formula: $CH_2=CHCH_2OH$	CAS#: 107-18-6	RTECS#: BA5075000	IDLH: 20 ppm
Conversion: 1 ppm = 2.38 mg/m^3	DOT: 1098 131			

Synonyms/Trade Names: AA, Allylic alcohol, Propenol, 1-Propen-3-ol, 2-Propenol, Vinyl carbinol

Exposure Limits: NIOSH REL: TWA 2 ppm (5 mg/m^3) ST 4 ppm (10 mg/m^3) [skin] OSHA PEL †: TWA 2 ppm (5 mg/m^3) [skin]	Measurement Methods (see Table 1): NIOSH 1402, 1405

Physical Description: Colorless liquid with a pungent, mustard-like odor.

Chemical and Physical Properties: MW: 58.1 BP: 205°F Sol: Miscible FI.P: 70°F IP: 9.63 eV Sp.Gr: 0.85 VP: 17 mmHg FRZ: −200°F UEL: 18.0% LEL: 2.5% Class IB Flammable Liquid	Personal Protection/Sanitation (see Table 2): Skin: Prevent skin contact Eyes: Prevent eye contact Wash skin: When contam Remove: When wet (flamm) Change: N.R. Provide: Quick drench	Respirator Recommendations (see Tables 3 and 4): NIOSH/OSHA 20 ppm: Sa:Cf*/PaprOv*/CcrFOv/ GmFOv/ScbaF/SaF §: ScbaF: Pd,Pp/SaF:Pd,Pp;AScba Escape: GmFOv/ScbaE

Incompatibilities and Reactivities: Strong oxidizers, acids, carbon tetrachloride
[**Note:** Polymerization may be caused by elevated temperatures, oxidizers, or peroxides.]

Exposure Routes, Symptoms, Target Organs (see Table 5): ER: Inh, Abs, Ing, Con SY: Eye imit, tissue damage; imit upper resp sys, skin; pulm edema TO: Eyes, skin, resp sys	First Aid (see Table 6): Eye: Irr immed Skin: Water flush immed Breath: Resp support Swallow: Medical attention immed

Courtesy of National Institute for Occupational Safety and Health.

Solution: Water or soap and water would be an excellent decontamination solution. According to the NIOSH guide, allyl alcohol is miscible with water, which means it is completely soluble in all proportions. Water alone should be adequate to remove the allyl alcohol. Soap will enhance the effectiveness of the water, but it is not absolutely necessary.

PATIENT TREATMENT IN THE HOT AND WARM ZONES

EMS personnel should be prepared to treat victims in the warm zone after they emerge from the mass decontamination line in a separate triage and treatment area. EMS personnel must be in the appropriate PPE and have the suitable training to safely perform medical treatment in the warm zone. Remember, patients who have just emerged from a mass decontamination line may still be contaminated. The degree of product removal will depend on the type of decontamination solution, first responder oversight during the process, number of patients who pass through, the speed with which they pass through, and whether decon verification could be completed before they enter the secondary triage site (unlikely).

Ideally, rudimentary patient treatment has already begun in the hot zone. If the ABCs—airway, breathing, and circulation—were not attended to quickly there, the patient will not be viable by the time he or she arrives in the warm zone. The victim may not even be contaminated and have injuries that are completely unrelated to any chemical exposure. If basic lifesaving measures were started in the hot zone, continue them uninterrupted even before decontamination is complete. Although ascertaining vital signs in full chemical protective clothing may initially appear impossible, breathing may be assessed by watching for chest movement, and perfusion may be assessed by observing capillary refill. Potentially contaminating the patient further is the lesser worry compared to whether he or she is not breathing or does not have a pulse. After the patient has been decontaminated, more advanced treatment is possible.

CROWD MANAGEMENT AND COMMUNICATIONS

Crowd control is an essential skill to master when responding to mass decontamination incidents (Figure 7-5). Imagine facing the prospect of all of the occupants of a large sports stadium or convention center exiting the facility at one time after a hazardous materials incident or terrorist attack. It is impossible to know which people have been exposed, which people have been contaminated, or which people are suffering from agent exposure. The first priority must be to contain the crowd and prevent them from dispersing

FIGURE 7-5 Crowd control at a hazardous materials incident. Cones and barrier tape are combined with signage to lead victims into the mass decontamination tent. Large crowds may require more substantial barriers and a law enforcement presence. *Courtesy of Mike Becker, Longmont (CO) Fire Department.*

across the area to their homes and hospitals. The second priority must be to determine which people have been contaminated and need field decontamination. The third priority must be to determine which people need medical treatment and transportation to a hospital for definitive care. Accomplishing these tasks relies heavily on crowd control.

Crowd control must be initiated immediately upon arrival at a mass casualty incident. Law enforcement has extensive experience in crowd control and will be helpful during the planning process, training, and the actual emergency. Larger events that have high attendances already have their own routine crowd control methods in place. Take a look at those methods, and integrate your emergency plans with those of the facility. Typically, law enforcement and the public works department work closely together for routine crowd control procedures. These systems will be critical to the crowd control process during an emergency at such an event. With proper planning and training, the existing crowd control infrastructure can be quickly converted to direct people to the mass decontamination and triage areas.

The mass decontamination corridor needs to be set up quickly in the path of the moving crowd. First responders must clearly and concisely communicate with the victims, or they will disperse and create a larger problem. Key information that must be quickly disseminated includes:

1. Location of the triage area
2. What will happen in the triage area
3. How to perform decontamination
4. What will happen to their personal effects
5. When they will get medical treatment
6. How they can find loved ones they were separated from on-site during the emergency
7. How they can contact loved ones off-site

This information can be distributed using preprinted fliers, loudspeakers, bullhorns, or electronic signs (Figure 7-6). Crowd control resources, equipment, and procedures should be determined well in advance of large events. These same resources and procedures can then be used for mass casualty incidents even when those incidents are not part of a larger organized event.

EVIDENCE PRESERVATION AND PROCESSING

Mass decontamination incidents involving criminal or terrorist events pose additional dangers and problems. One danger is that the terrorists or criminals may be among the suspected victims. They may be contaminated, or uncontaminated posing as victims. The terrorists or criminals may still be extremely dangerous at this point, and, of course, it would be optimal to apprehend them. How can this be accomplished? If it is feasible, recording the mass decontamination operation on video may be helpful

FIGURE 7-6 An example of the use of information signs for crowd control.

In the incident described in the opening scenario, how would you manage crowd control near the production plant? At the hospital?

Solution: It is important to ensure contaminated victims receive decontamination to avoid victim injury, and to avoid secondary contamination and subsequent injuries. Near the production plant, the evacuating residents should be channeled toward the mass decon area using roadblocks, fencing, and information. At the hospital, arriving patients should be screened for contamination before they enter the facility. The hospital may need to be locked down and extra security personnel will be necessary. Ideally, screening would be performed in the parking lot at a designated triage area that has decontamination capabilities. Allyl alcohol can be detected at low concentrations using a photoionization detector (PID) because its ionization potential (IP) is 9.63 eV.

to law enforcement for suspect identification and emergency medical services for patient accountability.

The personal property of the victims that is removed during the mass decontamination operation must be carefully tracked for several reasons. First, the property eventually should be returned to the victims. There may be very valuable property such as wallets, diamond rings, and other jewelry. In addition, the clothing and other property may be valuable evidence during the investigation. It is extremely important that everyone on the scene of a WMD incident or crime scene practices **evidence preservation**. The victims' clothing may have residue that will aid in identification of the agent and prosecution of the case. When feasible, law enforcement should be present at the decontamination line to maintain the documented chain of custody of the personal effects of the victims.

evidence preservation ■ The process of minimizing damage to evidence while performing other essential duties. At crime scenes, evidence preservation includes minimizing the number of entry personnel, disturbing as little as possible while performing work, good documentation through photography and videography, and maintaining a chain of custody.

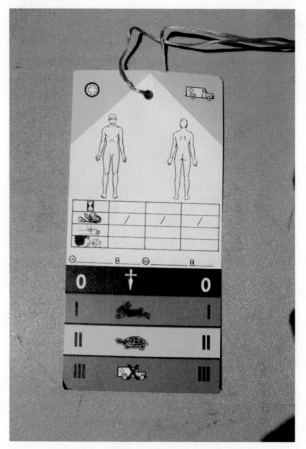

FIGURE 7-7 A triage tag with multiple detachable numbered bar codes that make tracking victim property easier.

All clothing and other property should be removed prior to showering, placed in a plastic bag, and then double bagged for safety. The contents must eventually be matched up with the victim in the hours, days, weeks, or even months after the incident. The easiest way to track patient belongings is to use bar-coded triage tags that have removable stickers or tags that can be placed directly on valuable property and into the bag containing the artifacts (Figure 7-7). The triage tag number should remain part of the permanent patient record. In addition, for stable, ambulatory patients, decontamination personnel should record a name and triage tag number immediately after patients exit the mass decontamination corridor.

Law enforcement officers may also be contaminated and need to move through the mass decon line. Their weapons will need to be decontaminated and handled in a safe manner according to their agencies' standard operating procedures (SOPs).

In the incident described in the opening scenario, how would you manage the weapons of the law enforcement officers?

Solution: Standard operating procedures (SOPs) that address firearm security and decontamination methods need to be developed in conjunction with law enforcement agencies prior to the incident. Options for police officer decontamination include having a sworn officer as part of the decon crew that would take custody of the weapon and unload it. The second option is to have a secure area where the weapons are placed that is under law enforcement oversight or has locking capabilities. Most law enforcement agencies prohibit their officers from handing over their weapons to anyone other than another sworn officer. This policy must be addressed through training.

This can be a bigger challenge than you may think. Many law enforcement SOPs prohibit officers from turning over their weapons to others, mandating them to maintain control of their weapons at all times. You should address this situation with the law enforcement agencies in your area during the planning stage to avoid any surprises at the scene of a hazardous materials or weapons of mass destruction incident.

EVALUATING DECONTAMINATION EFFECTIVENESS

Before patients are transported off-site, the effectiveness of the patient decontamination must be evaluated. There are several ways of doing so:

1. Visually inspect the patient to determine whether the product has been removed. Although this is the simplest technique, it is rarely completely effective. Most chemicals, and certainly CBRNE agents, are toxic well below the level they can be visualized, so this method is not recommended for evaluating decontamination effectiveness. Relying on visual inspection alone will almost certainly lead to secondary contamination and possible injuries to first responders, the contaminated victim, and the public.

2. Use air-monitoring or solid and liquid sample identification equipment to analyze the rinse solution. This is a better method and, depending on the sensitivity of the monitoring equipment used, can be very effective. The drawback of this technique is that because each patient is not monitored individually, spot contamination may go undetected. The advantage of this technique is that it is not resource intensive: Large numbers of decon personnel and monitoring equipment are not necessary.

3. Use air-monitoring or solid and liquid sample identification equipment to analyze the patient directly (Figure 7-8). This is the safest method to use both for the patient and for the first

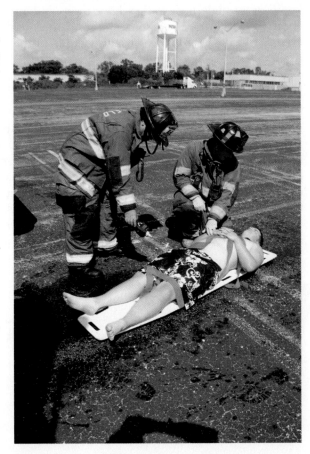

FIGURE 7-8 Decontamination verification during mass decontamination using air-monitoring equipment.

responders on the scene and the receiving facility. If the resources on scene permit, each victim should be individually monitored for thorough decontamination.

PATIENT TREATMENT AREA

EMS personnel should be prepared to treat victims after they have passed through the mass decontamination line in a designated patient treatment area. The mass decontamination line should be arranged so that the people pass directly into a triage area where decontamination effectiveness and injuries can be rapidly assessed. A triage system must be employed that can quickly categorize victims into delayed or immediate care categories, as resources permit. The START triage system is one such example (see Box 12-1). EMS personnel must be in the appropriate PPE and have the suitable training to safely perform triage and medical treatment in the warm zone. Ideally, immediate lifesaving patient treatment has already begun in the hot zone, although this is most likely unfeasible in mass casualty incidents.

Mass Decontamination Management

Due to the large number of victims and the large amount of resources needed for mass decontamination, effective management is critical. First, it is very important to place the mass decontamination facility in a practical and safe location. As mentioned before, it will be critical to funnel as many of the victims as possible into the mass decontamination lanes. If the location of the mass decontamination line is not chosen properly, the contaminated crowd will quickly scatter and endanger themselves and others. Also, a sufficient amount of resources should be made available for mass decontamination efforts to minimize wait time. If wait times are extensive, particularly if the victims know or strongly believe they have been contaminated, they will scatter and find help elsewhere. If sufficient resources cannot be obtained, as is typical for mass casualty events, it will be crucial to triage victims properly to prioritize the decontamination order.

INCIDENT MANAGEMENT

The mass decontamination operation will typically be managed by a mass decontamination branch director or mass decontamination group supervisor under a NIMS-compliant incident management system. As has been discussed, because mass decontamination operations are very equipment and personnel intensive, resource management is critically important to successfully manage mass decontamination events. Resources must be ordered early and used effectively. It is also very important to rapidly notify area hospitals that they should expect a large influx of people self-transporting to the emergency department, and that you may need to transport a large number of patients to their location. Always ask what their capacity is to avoid overloading their emergency department.

SOLVED EXERCISE 7-5

In the incident described in the opening scenario, how did weather considerations play a role in your mass decon setup decision?

Solution: Winter weather would require dry decontamination or a tent system with heated water and air to safely perform mass decontamination. The victims will need warm clothing and a sheltered triage area as well. Hypothermia is a primary concern because cold water will rapidly conduct heat away from the body, and evaporative cooling will exacerbate the problem.

In the incident described in the opening scenario, how would the runoff from decontamination operations affect your decisions?

Solution: The EPA has stated that runoff does not need to be contained when life safety is at stake and there are insufficient resources to manage the runoff. However, the mass decon line could be set up in an area where runoff can be easily contained. Such locales include a grassy area where the soil can be removed later or, better yet, a surfaced area that is curbed and where storm drains can be diked, or a location that can easily be diked to contain the runoff. When possible, attempt to prevent runoff into sensitive areas such as sewers, storm drains, and bodies of water.

REPORTS AND SUPPORTING DOCUMENTATION

When properly managed, mass decontamination incidents are documentation intensive. Above and beyond the typical incident report information, it must be carefully documented which responders were on scene and what their roles were in order to properly track any chemical exposure that might have occurred. In addition, depending on the size of the incident, a large number of victims need to be decontaminated. It is vital to track all of the patients that come through the mass decontamination line for patient and property accountability and optimal medical treatment, and to facilitate subsequent criminal prosecution by law enforcement agencies.

Personnel Exposure Records

Personnel exposure records must be kept for all first responders that are involved in mass decontamination efforts. These data include the name of the first responder, the employing agency, the employee identification number, the location, the length of time, and the job function the first responder was performing at the incident. In addition, routine air-monitoring readings at the incident should be recorded as well as any documented chemical exposures the first responder may have received. Any exposure records must be kept by the employer for a minimum of 30 years after termination of employment per HAZWOPER (29 CFR 1910.120).

Activity Log

An activity log is a list of all of the actions performed during the mass decontamination operation. It should include a description of the activity, the times it was started and completed, and the personnel that performed the activity. Examples of activities that are logged include setting up and performing triage, setting up and performing decontamination, donning and doffing personal protective equipment (including on-air times), patient treatment and transport, and chain of custody transfers of personal effects and clothing. This list is certainly not all-inclusive, and any significant activities should be logged.

Filing Documentation and Maintaining Records

Any documentation generated during a mass decontamination incident should be filed with the rest of the incident documentation. Federal regulatory agencies require that certain records, such as personnel exposure reports, be kept for a lengthy time. Other documentation, such as costs and equipment used, will be required by federal agencies and state agencies for reimbursement. The type of contamination and how it was removed should become part of the patient record. Good documentation is especially critical in mass casualty incidents due to the volume of patients that pass through the decontamination line and the large amount of personal effects collected and separated from their owners. Without accurate documentation, it will be impossible to return these items to their rightful owners. Also, at criminal or terrorist mass casualty events, personal effects become evidence that may make or break a criminal prosecution.

In the incident described in the opening scenario, how would you manage and track personal property?

Solution: Bags and numbered triage tags or specially designed mass decon kits can be effectively used. The kits contain all of the necessary components, including numbered bags and identification as well as disposable garments for re-dressing. Large bags and triage tags with serrated serial numbers work as well. Disposable Tyvek® suits can be used to re-dress the decontaminated victims.

SPECIAL CONSIDERATIONS FOR HOSPITALS

Hospitals will be the second "incident" in mass casualty incidents. It is almost inevitable that many people will self-transport to the hospital emergency department after a significant hazardous materials or WMD incident. The sarin attack in the Tokyo subway system is a stark example of that fact. Five thousand people reported to local hospitals within several hours of the attack. Some were contaminated, but most were not. How would your hospital handle that many potentially contaminated victims? It is imperative that contaminated people do not enter the hospital.

Summary

Mass casualty incidents requiring mass decontamination are very stressful and chaotic events. First responders will be pushed to the limit. Initial actions upon arrival and preplanning will determine whether mass decontamination can be successfully accomplished. How well the crowds are managed and the perimeter is controlled will dictate the scope of off-site problems. If the appropriate resources are available and used effectively, mass decontamination efforts on the scene will prevent further victim injuries, contamination of first responders on scene, and cross contamination of hospitals and other first receiving facilities. Although mass decontamination must be performed quickly due to resource and time constraints, it must also be executed effectively for victim and first responder safety. Training and preplanning will make or break any mass decontamination operation.

Review Questions

1. What are the advantages and disadvantages of mass decontamination?
2. How could you set up an improvised decontamination shower using fire department equipment?
3. What types of decontamination methods are available for mass decontamination?
4. What is the importance of evidence preservation at mass decontamination incidents?
5. What type of documentation is important at mass decontamination operations? Why?

Problem-Solving Activities

1. With the mass decontamination equipment available to your agency (both commercial and improvised), plan a mass decontamination operation for a high school football game.
2. With the mass decontamination equipment available to your agency (both commercial and improvised), plan a mass decontamination operation for an elementary school.
3. With the mass decontamination equipment available to your agency (both commercial and improvised), plan a mass decontamination operation for a school bus full of children.
4. If your agency is a fire department, construct a functional fire engine alley and flow water.

References and Further Reading

National Fire Protection Association. (2008). NFPA 472, *Standard for Competence of Responders to Hazardous Materials/ Weapons of Mass Destruction Incidents*. Quincy, MA: Author.

Occupational Safety and Health Administration. (1990). 29 CFR 1910.120, *Hazardous Waste Site Operations and Emergency Response (HAZWOPER)*. Washington, DC: U.S. Department of Labor.

Weber, Chris. (2007). *Pocket Reference for Hazardous Materials Response*. Upper Saddle River, NJ: Pearson/Brady.

8

Operations Level Responders Mission-Specific Competencies: Technical Decontamination

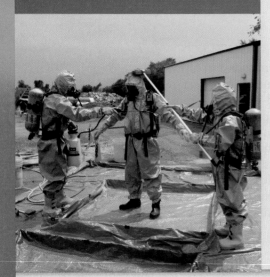

OBJECTIVES

- Explain why technical decontamination is necessary at hazardous materials incidents.
- Explain the differences between dry decontamination and wet decontamination.
- Describe the equipment and tools necessary to perform technical decontamination for both ambulatory and non-ambulatory personnel.
- Set up a technical decontamination line.
- Perform technical decontamination on ambulatory personnel.
- Perform technical decontamination on non-ambulatory personnel.
- Explain how to properly dispose of contaminated PPE, tools, equipment, and decontamination runoff.

You Are on Duty! An Explosion at the University Laboratory

Lubbock, TX

On a January morning, a 911 call is received stating that a graduate student has been severely burned in a laboratory accident. As you respond to the incident you wonder what kind of chemical was involved in the release. Is the call referring to a chemical burn or a thermal burn? How many chemicals have been released? Is there a fire involved? Are other patients involved? You transmit your questions to dispatch, and the report back is that an explosion occurred in a lab occupied by two researchers. Dispatch is unaware of the number or type of chemicals involved. Upon arrival, you notice the building is not being evacuated, and you are notified that the patient is in the hallway outside the lab. Reports state that the patient and a second person are covered in a liquid and a white powder.

- As a law enforcement officer, how would you handle this emergency?
- As an EMS professional, how would you approach this patient? Are you concerned about cross contamination?
- As a firefighter, how would you handle this emergency? Should the building be evacuated? How would the patient be rescued? How would cross contamination be minimized?
- As a member of the university hazmat response team, how would you handle this incident?

Let's try to answer these questions in this chapter.

Technical decontamination is the backbone of hazardous materials operations (Figure 8-1). Decontamination is the chemical and/or physical process of removing contamination from

FIGURE 8-1 Technical decontamination is required at all hazmat incidents in which personnel enter the hot zone.

people and equipment. Most important, technical decontamination is the decontamination of first responders and victims. The primary reasons to perform technical decontamination are to protect first responders and victims from exposure to the hazardous material, to minimize the cross contamination of clean areas, and to prevent the contaminants from permeating into personal protective equipment (PPE) and other mitigation equipment.

Setting up the decontamination line is often time consuming and can cause delays in hot zone entry by technician level hazardous materials response teams. When operations level first responders are able to set up the decontamination line prior to the arrival of the hazardous materials response team, entry operations are greatly accelerated. When first responders at the operations level are able to set up technical decontamination lines, they provide invaluable support to technician level hazardous materials response teams.

Decontamination Plan

HAZWOPER (29 CFR 1910.120) requires that a decontamination plan be in place before offensive, entry team operations begin. Technician level hazardous materials response teams have interpreted this to mean that a decontamination line must be set up prior to hot zone entry. The decontamination plan should be written and cover the following key components of the technical decontamination process:

1. The number and layout of decontamination stations
2. The decontamination equipment
3. The decontamination methods
4. Procedures that minimize **cross contamination** of clean areas
5. Procedures that minimize cross contamination during doffing of PPE
6. Disposal procedures for hazardous waste, including disposable PPE and equipment
7. Patient treatment location and personnel

cross contamination ■
The transfer of contaminants from one person or location to another.

The decontamination plan should be revised whenever there is a change in scene conditions, incident hazards, or the type or level of PPE.

A warm zone containing the decontamination line must be located upwind, uphill, and upstream of the hazardous materials release. In addition, the decontamination stations should be arranged in a straight line going from the hot zone side to the cold zone side whenever possible. To avoid cross contaminating the warm zone, decontamination personnel need to remain at their assigned station and should not wander back and forth between stations. Entry and exit points should be conspicuously marked.

Your standard operating guidelines or procedures (SOGs/SOPs) should cover technical decontamination methods. This helps to ensure that all first responders understand the decontamination methods and equipment to be used in response to a hazardous materials incident or terrorist attack. Guidelines or procedures allow training to be carried out

SOLVED EXERCISE 8-1

In the incident described in the opening scenario, how do you formulate your decontamination plan? What other information would you like to know?

Solution: Given the fact that the building is not being evacuated, the explosion was likely relatively small and contained to the initial lab or fume hood. Laboratories are designed to have many air exchanges per hour and evacuate the air through a laboratory fume hood. That having been said, approach the area cautiously. A decontamination plan needs to take into account preexisting facilities such as emergency eyewash stations and emergency showers; laboratory buildings will have many of these. Initial emergency decon could be performed using an emergency shower of a neighboring lab or one on a different floor (if the general area is determined to be safe). Preexisting decon facilities will greatly speed up the decontamination process and minimize chemical injury to the victim. As a backup plan, a technical decon line may also be set up outside.

It is beneficial to know what the chemicals and their properties are, especially when it comes to water reactivity and water solubility. Knowledge of these properties aids in determining the most effective decon method. Knowing the identity of the materials also allows us to determine a detection method to verify decontamination. The hospital will want to know the chemical identity to determine likely injuries, signs and symptoms, and treatment and antidote.

consistently prior to the call, and they may be referred to in the decontamination plan for a specific hazmat incident.

Technical Decontamination Operations

Technical decontamination involves decontaminating hot zone entry personnel and their equipment. It may also involve decontamination of a small number of victims, including non-ambulatory victims. The last chapter covered mass decontamination operations, which involve the expedient removal of hazardous materials from a large number of victims. The mass decontamination process is typically less thorough and takes less time per person than does the technical decontamination process.

Occasionally, technical decontamination of animals must be performed. This may be the result of a hazardous materials spill in a rural area and the contamination of large livestock such as horses, cows, and pigs; or the result of a hazardous materials spill in a residential area where one or more pets need to be decontaminated. Not too long ago, we responded to a mercury spill where we needed to decontaminate a dog. We used dandruff shampoo (containing selenium sulfide), which binds mercury quite well. This was a relatively easy process because the dog was docile, but other animals likely will not be. It is always a good idea to consult with a veterinarian or animal control officer when faced with such a situation.

The decontamination area is located in the warm zone, which is carved out of the cold zone before any personnel enter the hot zone. The decontamination area should be located upwind, uphill, and upstream of the hazardous materials release to minimize the chance of exposure to hazardous materials due to the wind or gravity allowing a liquid to flow toward the support zone and decontamination line.

Decontamination is performed in a systematic fashion in which more heavily contaminated items, such as outer gloves and boots, are removed first closest to the hot zone. Decontamination stations are located in a linear array stretching from the hot zone to the cold zone. The heaviest contamination is removed first, at the stations closest to the hot zone. By the time the last station near the cold zone is reached, the contamination should be completely removed (Figure 8-2). In order to prevent cross contamination between the decontamination stations, workers should remain at their assigned station and not wander between stations. There needs to be sufficient physical separation between the stations to avoid cross contamination from over spray.

Once the entry team operations have ended and technical decontamination is no longer needed, the decontamination line must be broken down and the decon workers need to be decontaminated themselves in a systematic fashion. The decontamination workers at the dirtiest station, closest to the hot zone, are decontaminated first. They will step into the next closest station in the direction of the cold zone and proceed with decontamination. Then the workers at this station (the second station in from the hot

FIGURE 8-2 A more complex technical decontamination layout that includes a heated tent for cold weather operations.

zone) proceed to the next closest station toward the cold zone and are decontaminated there. This process continues until the last remaining decon workers are reached; they should be uncontaminated because they were using instrumentation to verify that decontamination operations were effective, and should not have had any appreciable contact with the hazardous materials. They can therefore doff their own PPE and exit the warm zone.

PPE SELECTION

Personal protective equipment must be carefully selected for decontamination operations. The PPE must be compatible with the spilled hazardous material as well as the decontamination solution. Some equipment may be decontaminated with a strong decontamination solution; for example, toluene diisocyanate (TDI), a component of foam manufacturing, is often decontaminated using an ammonia solution. If not handled properly, the corrosive ammonia decontamination solution poses its own hazards. Therefore, we must wear PPE that is compatible with both toluene diisocyanate and ammonia in this case. The ammonia solution should never be used to decontaminate victims directly.

Typically, the same level of protection, or at a minimum one level of protection lower than what the entry team is wearing, is chosen for decon workers nearest the hot zone. However, if the entry team is in level A protection, the primary wash and rinse station personnel may be in level B protection, and the secondary wash and rinse station personnel may be in level C protection. Thus, the level of protection may be downgraded as the decontamination personnel are stationed closer to the cold zone (farther away from the hot zone).

DECONTAMINATION SUPPLIES AND EQUIPMENT

The supplies and equipment needed to set up the decontamination line will vary based on the type of decontamination to be performed and nature of the hazardous materials incident (Figure 8-3). Decontamination supplies and equipment may be off the shelf, homemade, or specifically designed for hazardous materials decontamination operations. Make sure that you train with your supplies and equipment to ensure that your layout and decontamination method are practical. When the hazardous materials call comes in is not the time to find out that equipment is missing or that the job cannot be done the way you planned!

When life safety is not at stake, the Environmental Protection Agency (EPA) requires that the runoff from the decontamination line be contained. In practical terms, this means placing a tarp underneath all of the decontamination stations. The tarp should have a berm (constructed using an uncharged hose line tucked under the tarp) around the edges that can contain any over spray from the wash and rinse stations. In addition, each wash and rinse station should have a pool of sufficient size to contain the runoff. On long and complex hazardous materials incidents, provisions should be made for pumping off the pools into suitable secondary containment vessels, such as bladders, totes, or 55-gallon drums.

Dry Decontamination

Dry decontamination is becoming the method of choice for technical decontamination (Figure 8-4). Dry decon may have a number of advantages over wet decon, but it is not appropriate for all hazardous materials. A hazardous material that has a moderate

SOLVED EXERCISE 8-2

You determine that the chemical involved in the explosion in the opening scenario is nickel hydrazine perchlorate. How does this information affect your decon plan?

Solution: Try to determine the chemical and physical properties of nickel hydrazine perchlorate using available reference materials and the Internet. It is very difficult, if not impossible, to find this information because the scientist was experimenting and creating new compounds the day he was injured. Technical experts can be very helpful in situations such as this. The lab director or other researchers are a first source to turn to. Another source would be the university or company health and safety director who should have a copy of the MSDS sheet. With little or no information, flooding amounts of water, followed by soap and water, are the best decontamination option.

FIGURE 8-4 Dry decontamination. The entry team member is standing in a bag and will remove the outer layer of PPE and then step toward the cold zone.

vapor pressure and is skin absorptive needs to be removed using wet decontamination. Many hazardous materials have either a very high or a reasonably low vapor pressure and are not skin absorptive. For example, chlorine is a gas with a high vapor pressure that will normally completely evaporate before the entry team personnel reach the decontamination line.

On the other hand, sulfuric acid has an extremely low vapor pressure and will not generate significant vapors during the decontamination process. Therefore, chlorine and sulfuric acid are excellent candidates for dry decontamination. Dry decontamination may also be chosen when monitoring equipment indicates the entry team has not been contaminated with the hazardous material. Box 8-1 describes the dry decontamination process in a step-by-step format. The doffing procedures for different types of PPE are described in detail in Chapter 6, Operations Level Responders Mission-Specific Competencies: Personal Protective Equipment.

Resource
Central

View an illustrated step-by-step guide on the Dry Decontamination Process

BOX 8.1 DRY DECONTAMINATION PROCESS

1. Remove outermost gloves (third layer) and place in bag.
2. Step into bag.
3. Nonencapsulating PPE: Remove SCBA from back while staying on air.
4. Have partner unzip suit.
5. Carefully peel suit back from zipper. Do NOT allow outside of suit to contact clothing or skin. Do NOT allow the outer gloves to contact the inside of the suit or body.
6. Peel suit down to the boots.
7. Step out of the bag toward the cold zone.
8. Go off air and remove face piece (mask) with inner gloves (typically latex or nitrile).
9. Remove inner gloves.

Wet Decontamination

Wet decontamination supplies include water at the minimum. An emergency decontamination line set up for emergency rescue may be a nearby emergency shower, a garden hose, or a fire department hose line operated at low pressure. Typically, technical decontamination supplies also include soap, brushes, buckets, tarps, wash and rinse pools, sprayers, and trash cans. Figure 5-25 illustrates a common technical decontamination line configuration, and Figure 5-23 shows examples of technical decontamination lines at hazardous materials incidents.

A standard decontamination line includes a tool drop area in the hot zone. This area allows multiple entry teams to reuse potentially contaminated tools to mitigate a more complex hazardous materials or WMD incident. This area may also include a boot rinse or overboot/booty removal area.

The first station commonly located just inside the warm zone is the primary wash and rinse station. Decon personnel will first wash the entry team personnel, using brushes if necessary, and rinse them using sprayers or hoses. Most of the contaminants are removed at this location. It is a good idea to use monitoring equipment to determine the level of contamination prior to decontamination. This serves two purposes: the documentation of potential exposures and the determination of the effectiveness of the decontamination method. Assess the contamination level before and after performing decontamination, and if a significant improvement is not seen, modify the decon method or choose an entirely different one.

The next station is the secondary wash and rinse station that may or may not be necessary based upon the results of the monitoring equipment. After the secondary

SOLVED EXERCISE 8-3

Draw the decon line you would set up for the incident in the opening scenario, and explain your choices.

Solution: The first choice for patient decontamination would be the laboratory shower and eyewash in an adjoining lab or on the floor below (on the way out). This is the fastest, most climate-controlled and privacy-controlled way to decontaminate the victim. A technical decon line should be set up for first responders and the patient if he requires additional decontamination (as determined by screening tools such as oxidizer paper). Wet decon is the best choice because there is limited information and flooding amounts of water can be used. Because the temperature can get rather cold in Lubbock, Texas, in January, the wet technical decon line should be set up in a heated tent, and the water used to decontaminate the victim should also be heated to a lukewarm temperature to avoid inducing hypothermia.

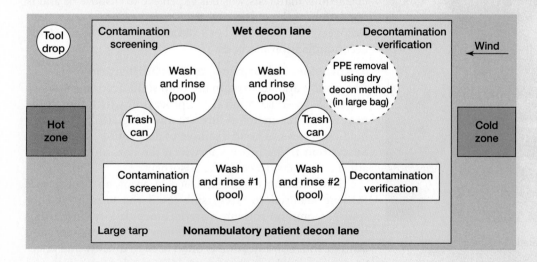

View an illustrated step-by-step guide on the Wet Decontamination Process

BOX 8.2 WET DECONTAMINATION PROCESS

1. Have entry team drop their tools in bucket at edge of hot zone.
2. Decon team members visually inspect the entry team member being decontaminated by having the entry team member rotate 360° and lift his or her feet.
3. If possible, evaluate the extent of contamination using monitoring instruments. If the personnel are not contaminated, they do not need wet decontamination and they can go straight to dry decontamination to doff their PPE.
4. Remove outermost gloves (third layer) and boot covers, and place in first recovery drum.
5. Wash/scrub any contaminated areas in the first pool and then rinse.
6. Have the entry member step into the second pool, and wash and rinse.
7. Remove PPE per the appropriate suit doffing procedure (dry decontamination).

adsorption ■ A physical phenomenon characterized by the adherence of a substance to the surface of another substance.

solidification ■ The process of trapping a liquid in a solid form by absorption or concretion.

wash and rinse station, the PPE is doffed appropriately based upon the level of protection and the type of chemical protective clothing (CPC) used. Respiratory protection is the final piece of PPE to be removed. Respiratory protection removal happens at the station closest to the cold zone. An optional field wash station may be included in the cold zone. It is a great idea to have the entry team members' footwear handy after the final PPE is removed, especially during inclement weather. Box 8-2 illustrates the wet decontamination process in a step-by-step format.

DECONTAMINATION METHOD SELECTION

The decontamination method selected will depend on the resources available, the urgency with which decontamination is needed, the weather conditions, and most important the chemical and physical properties of the hazardous material involved. In practical terms, the work assignments of entry team members, the levels of contamination that can be expected, the toxicity of the contaminants, and the propensity for cross contamination should all be considered before choosing a decontamination method. It is important to carefully consider the decontamination method to maximize safety and efficiency, and to minimize waste generation. Hazardous materials waste is expensive to dispose of and is regulated cradle to grave by the Resource Conservation and Recovery Act (RCRA). Let's discuss some of the available decontamination methods.

Absorption, Adsorption, and Solidification

Absorption and adsorption are both solidification processes (Figure 8-5). Absorption is the process in which the absorbent retains the material through a wetting effect. A sponge holding water is a good example of absorption. In contrast, **adsorption** is the process in which the hazardous material adheres to the surface of the adsorbent.

Solidification is accomplished by using an appropriate sorbent material to irreversibly bind the hazardous substance. Common sorbent materials include kitty litter, diatomaceous earth, clay, powdered lime, shredded paper, and sand or soil. Solidification is often used to quickly and effectively remove the bulk of contaminants that have adhered

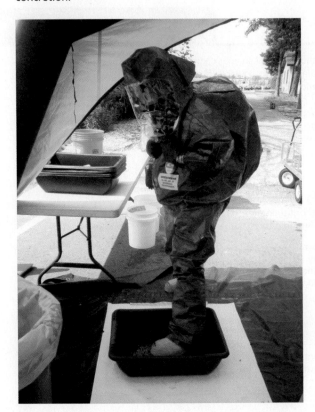

FIGURE 8-5 Solidification materials can be used to absorb contaminants from boots.

to footwear. Do not forget to check for chemical compatibility of the sorbent material! Also be aware that many sorbents contain moisture and should be carefully considered before being used to solidify water-reactive materials. Solidification can typically be used on PPE, equipment, and patients per the manufacturer's recommendation.

Dilution and Washing

Dilution is the process of reducing the concentration of harmful contaminants below a dangerous level (Figure 8-6). It is most commonly accomplished using water for water-soluble contaminants, or soap and water for water-insoluble contaminants.

Washing adds the physical or mechanical removal of contamination to the process of dilution. Brushes or sponges are most commonly used in the washing process with or without soap. Pressure washing may be used on equipment durable enough to withstand the process, such as heavy equipment or metal tools. Not all materials can be decontaminated by washing due to the phenomenon of permeation, which is the penetration of a chemical into a material at the molecular level. Clearly, washing is not going to remove a hazardous material that has permeated into a material. Factors that affect the permeation rate are:

- Contact time
- Concentration of contaminant
- Temperature
- Contaminant size
- Molecular pore size of the material
- Physical state of the contaminant (solid, liquid, or gas)

Evaporation

Evaporation is the removal of hazardous materials by volatilization. Evaporation is a successful decontamination method for hazardous materials that have high vapor pressures. Gases and cryogenic liquids evaporate extremely rapidly, whereas high vapor pressure solvents such as acetone evaporate quickly. Dry decon is often used when evaporation is a viable option.

Isolation and Disposal

Isolation and disposal are the primary decontamination methods used in dry decon (Figure 8-7). Isolation refers to placing the contaminated PPE or equipment in an impermeable container and disposing of it as hazardous waste. The contamination is not removed or treated in any other way. Isolation and disposal are a very effective decontamination method for disposable PPE and equipment, but they are not practical for expensive equipment such as air monitors and reusable PPE.

Chemical Degradation

Chemical degradation is a very effective decontamination method. It removes contamination by reacting with a hazardous material and rendering it inert or less hazardous. Because these chemicals are usually hazardous materials themselves, great caution must be used. Examples of chemical degradation include the inactivation of toluene diisocyanate using ammonia, the binding of mercury using zinc amalgams, and the inactivation of chemical warfare agents using bleach. Because the decontamination solutions, such as bleach and ammonia, are highly reactive and may generate heat, they should never be used for patient decontamination and should be carefully considered for use on PPE. Chemical degradation is often used to decontaminate equipment.

Disinfection and Sterilization **Disinfection** is a form of chemical degradation. Disinfecting agents, such as chemical disinfectants, high heat, and radiation, are used to kill biological organisms. Chemical disinfectants include bleach, ammonia solutions, and quaternary ammonium-based commercial disinfectants. Heat sterilization includes autoclaves, boiling water, and steaming. Sterilization using radiation is accomplished using ultraviolet (UV) radiation or gamma radiation such as cobalt-60 (Co^{60}). Needless to say, disin-

FIGURE 8-7 Isolation and disposal are often the final steps during the technical decontamination process because most PPE is disposable, and rinse water and absorbents must be disposed of properly according to EPA regulations.

chemical degradation ■ The breakdown of a material through the action of a hazardous material.

disinfection ■ The process of killing a disease-causing microorganism. Methods of disinfection include autoclaving, irradiation, boiling, and chemical inactivation (such as with bleach).

neutralization ■ An exothermic reaction between an acid and a base, resulting in the production of heat, water, and a salt. The reaction between a concentrated acid and a concentrated base may be extremely violent.

fection and sterilization are not appropriate for use on patients. Some forms of chemical disinfection may be used with PPE if compatibility is not an issue. As a rule of thumb, disinfection and sterilization are used only with equipment.

Neutralization **Neutralization** is a form of chemical degradation in which an acid is reacted with a base to yield a noncorrosive product or, vice versa, a base is neutralized by an acid. Neutralization reactions yield water and a salt:

$$\text{Acid} + \text{Base} \rightarrow \text{Water} + \text{Salt}$$
$$\text{HCl} + \text{NaOH} \rightarrow \text{H}_2\text{O} + \text{NaCl}$$

Neutralization reactions generate a great amount of heat and should *never* be used on patients or PPE to prevent injury or damage. However, neutralization is a very effective decontamination method for equipment that is not sensitive to or damaged by corrosives or heat.

Vacuuming

Vacuuming may be used to remove liquids and solids from contaminated surfaces. Wet vacuum cleaners remove liquid contamination from PPE and equipment. HEPA vacuum cleaners remove fine particulates and dusts from PPE and equipment. Specialized vacuum cleaners, such as those designed to remove mercury contamination, may have applications in certain situations.

Given the lack of information about nickel hydrazine perchlorate in the opening scenario, what type of decon method would you choose?

Solution: Washing with flooding amounts of water followed by soap and water. Some information about nickel hydrazine perchlorate can be obtained from the name of the material itself: By looking up perchlorates, we find they are strong oxidizers. Hydrazines are used in rocket fuel. Together this points to an explosive material that must be treated with care. Although this is not detailed information about the material, it offers an idea of its general behavior.

EVIDENCE PRESERVATION AND PROCESSING

In criminal or terrorist events, special care must be taken during the decontamination process to preserve evidence. In a terrorist attack, traces of chemical agent on victim clothing or equipment may be important for laboratory identification and may be seized as evidence at a later time. Decontamination of the deceased should only be carried out after consultation with the appropriate law enforcement agency having jurisdiction. Because harsh decontamination methods such as chemical degradation, disinfection, sterilization, and neutralization will destroy the evidence, they should be avoided when a criminal or terrorist incident is suspected.

Care should be taken when decontaminating chemical samples. Double bagging the chemical samples minimizes contact with any decontamination solutions. However, be careful to handle the secondary containment carefully to avoid puncturing or otherwise damaging it. In addition, the chain of custody of these samples should remain intact, which may require the presence of a law enforcement officer in or near the warm zone.

SUPPORTING HOT ZONE OPERATIONS

Technical decontamination is primarily a hot zone support operation, so decontamination activities should be closely coordinated with entry team operations. This teamwork will allow optimal decontamination line layout, ensure that the appropriate decontamination methods are selected, and make certain that entry team members are familiar with the decontamination line layout.

Entry Team Members

Entry team members often come into contact with hazardous materials in the hot zone while performing product control operations. Your job as an operations level responder trained to this mission-specific competency will be to decontaminate, or assist in the decontamination of, these entry team members. This involves decontaminating personnel in level C, level B, or level A personal protective equipment (PPE). Each level of PPE has its unique challenges during the decontamination process. For example, with level C PPE, avoid getting the air-purifying respirator (APR) cartridges wet because most of them tend to clog with excessive moisture. With level B PPE, the self-contained breathing apparatus (SCBA) can be challenging to decontaminate. Remember always to ensure that decontamination is complete using detection equipment.

Some entry team members will require special consideration. For example, law enforcement officers may have weapons that need to be decontaminated and handled in a safe manner according to their agencies' standard operating procedures (SOPs). This can be a bigger challenge than you may think. Many law enforcement SOPs prohibit law enforcement officers from turning over their weapons to others, mandating them to maintain control of their weapons at all times. You should address this situation with the law enforcement agencies in your area during the planning stage to avoid any surprises at

In the incident described in the opening scenario, how would you manage the weapons of the law enforcement officers?

Solution: SOPs that address firearm security and decontamination methods need to be developed in conjunction with law enforcement agencies prior to the incident. Options for police officer decontamination include having a sworn officer as part of the decon crew that would take custody of the weapon and unload it. The second option is to have a secure area where the weapons are placed that is under law enforcement oversight or has locking capabilities. Most law enforcement agencies prohibit their officers from handing over their weapons to anyone other than another sworn officer. This policy must be addressed through training.

the scene of a hazardous materials or weapons of mass destruction incident. Other entry team members that require special consideration are canines and bomb squad members.

Entry team members also carry many tools into the hot zone, including air-monitoring instrumentation, sampling tools, product control tools and equipment, and sample identification equipment. Some of this equipment is disposable, such as shovels; but most of this equipment is expensive and certainly not disposable, such as air-monitoring instrumentation and sample identification equipment. The decontamination plan should address how to decontaminate not only PPE but also the equipment used in the hot zone.

If the hazardous materials incident is a crime scene, such as weapons of mass destruction incidents and environmental crimes, evidence will have to be decontaminated. Evidence such as biological samples poses a challenge because it may be easily destroyed. Typically, most evidence is triple bagged and should be relatively easy to decontaminate. However, because the chain of custody must be maintained, the presence of a law enforcement officer may be required in the warm zone to accompany the evidence through the decontamination process.

Ambulatory and Non-ambulatory Victims

Victims are among the most challenging people to decontaminate. Because they are not emergency responders, they will not be familiar with the decontamination line layout or the decontamination methods and procedures. You will likely have to explain what you are about to do and possibly calm them down if they feel they have been exposed or are contaminated and are showing signs and symptoms of exposure. This adds time to the decontamination process! The good news is that ambulatory patients (the "walking wounded") may be able to assist with their own decontamination. Box 8-3 illustrates the ambulatory patient decontamination process.

Non-ambulatory patients, on the other hand, require specialized equipment and a significantly increased amount of time to perform thorough decontamination. Specialized equipment such as sawhorses, backboards, and rollers will be required for

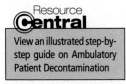

Resource
Central

View an illustrated step-by-step guide on Ambulatory Patient Decontamination

BOX 8.3 AMBULATORY PATIENT DECONTAMINATION

1. From a distance, instruct the patient to move away from the hazardous material.
2. From a distance, instruct the patient to remove his or her contaminated clothing.
3. Wash the patient using lukewarm or cool water. Cooler water closes pores, but be careful not to induce hypothermia.
4. Rinse the patient with cool or lukewarm water.
5. Repeat until the patient is contaminant free.
6. Re-dress the patient in a disposable Tyvek suit or other suitable clothing.

BOX 8.4 NON-AMBULATORY PATIENT DECONTAMINATION

Resource Central

View an illustrated step-by-step guide on Non-Ambulatory Patient Decontamination

1. Evaluate ABCs. If possible, open the airway, assist ventilations, and secure c-spine (in the case of trauma). Depending on the hazard of the product and route of exposure, some lifesaving measures may need to be delayed briefly while rapid decontamination is performed on targeted areas (such as the face).
2. Remove clothing.
3. Wash patient.
4. Rinse patient.
5. Repeat until patient is contamination free (determined using appropriate monitoring instruments).
6. Re-dress the patient in a disposable Tyvek suit or other suitable clothing.

non-ambulatory patient decontamination. Decontamination personnel will need to decontaminate around the face first to ensure that hazardous materials are not continued to be inhaled or ingested. Next, the top portion of the victim, which in the supine position will be the chest, groin, and top of the legs, is decontaminated. Then the non-ambulatory patient must be rolled onto his or her side; and the back, buttocks, and back of the legs must be decontaminated (Figure 7-4). As you can see, this procedure is time and resource intensive. Each non-ambulatory patient decontamination station requires a minimum of two decon personnel. Box 8-4 illustrates the non-ambulatory patient decontamination process.

Several parts of the body, such as the hair, tend to collect and hold contaminants. Skin and hair should be flushed 3 to 5 minutes with water or soap and water. Because the eyes are especially sensitive and absorb chemicals very effectively, they should be flushed for at least 5 minutes with water or sterile saline solution. Contact lenses should be removed if it is practical to do so because they tend to trap contaminants underneath them, causing more damage and absorption of the hazardous material. Once again, ensure that you use detection equipment to verify that decontamination has been thorough.

PATIENT TREATMENT IN THE WARM ZONE

EMS personnel should be prepared to treat victims in the warm zone immediately after decontamination or partial decontamination as their medical condition warrants. Ideally, transfer of care will occur from the hot zone. The decontamination line should be arranged to allow EMS access while decontamination is taking place. EMS personnel must be in the appropriate PPE and have suitable training to safely perform medical treatment.

Ideally, rudimentary patient treatment has already begun in the hot zone. If the ABCs—airway, breathing, and circulation—were not attended to quickly there, the patient will not be viable by the time he or she arrives in the warm zone. The victim may not even be contaminated and/or have injuries that are completely unrelated to any chemical exposure. If basic lifesaving measures were started in the hot zone, continue them uninterrupted even before decontamination is complete. Although ascertaining vital signs in full chemical protective clothing may initially appear impossible, breathing may be assessed by watching for chest movement, and perfusion may be assessed by observing capillary refill. Potentially contaminating the patient further is the lesser worry compared to whether he or she is not breathing or does not have a pulse. After the patient has been decontaminated, more advanced treatment is possible. Box 8-5 summarizes some of the warm zone patient treatment guidelines.

BOX 8.5 WARM ZONE PATIENT TREATMENT GUIDELINES

1. Protect yourself, using proper personal protective equipment (PPE).
2. Maintain airway, breathing, and circulation (ABCs). Even if the patient is contaminated, ensure adequate perfusion.
   a. Quickly remove as much contamination from the face (nose and mouth area) as possible.
   b. If trauma is suspected, protect the cervical spine (c-spine).
   c. Assess for respirations. If the patient is not breathing, open the airway.
   d. Assess for respirations. If the patient is still not breathing, start artificial ventilation using an approved device (such as a bag valve mask). Do not perform mouth-to-mouth ventilation on a potentially contaminated patient!
   e. Perform cardiopulmonary resuscitation (CPR) if the patient is pulseless.
3. Decontaminate the patient as soon as possible. Continue ventilations and CPR during the decontamination process. Field decontamination is necessary to prevent cross contamination of medical personnel, the ambulance, and the hospital. Rapid field decontamination limits further injury to the patient from the hazardous material.
4. After wet decontamination is complete, ensure that the patient is warm and has not become hypothermic.

EVALUATING DECONTAMINATION EFFECTIVENESS

Before the decontaminated entry team members can leave the warm zone, the effectiveness of the decontamination process must be evaluated (Figure 8-8). There are several ways of accomplishing this:

1. Visually inspect the PPE to determine whether the product has been removed. Although this is the simplest technique, it is rarely completely effective. Most chemicals, and certainly CBRNE agents, are toxic well below the level they can be visualized, so this method is not recommended for evaluating decontamination effectiveness. Relying on visual inspection will almost certainly lead to secondary contamination and possible injuries to the entry team member and other first responders.
2. Use detection equipment to analyze the rinse solution. This is a better method and, depending on the sensitivity of the monitoring equipment that is used, can be very effective. The drawback of this technique is that because each patient is not monitored individually, spot contamination may go undetected. The advantage of this technique is that it is not resource intensive: Large numbers of decon personnel and monitoring equipment are not necessary.
3. Use air-monitoring or solid and liquid sample identification equipment to analyze each entry team member directly. This is the safest method to use both for the entry team member and for the first responders on the scene. If the resources on scene permit, the PPE of each entry team member should be individually monitored for thorough decontamination.

SOLVED EXERCISE 8-6

Given the information in the preceding Solved Exercises in this chapter about nickel hydrazine perchlorate, how would you evaluate decon effectiveness?

Solution: Oxidizer paper. Perchlorates are strong oxidizers used in the explosives industry.

Technical Decontamination Management

Sound management of technical decontamination operations is crucial. Technical decontamination requires a lot of equipment and may be difficult to set up. Its methods and procedures are also complicated. Training and a good entry team briefing will be required to convey the intricacies of the decontamination plan to both the entry team members and the decon personnel. Therefore, the technical decontamination group supervisor should be an experienced member of your agency.

INCIDENT MANAGEMENT

The technical decontamination operation will usually be managed by a technical decontamination group supervisor under a NIMS-compliant incident management system. As has been discussed, because technical decontamination operations can be equipment and personnel intensive, resource management is very important to successfully carry out technical decontamination. Resources must be ordered early and used effectively.

FIGURE 8-8 It is essential that decontamination of personnel and equipment is verified using test equipment, such as the air-monitoring equipment pictured here.

Technical decontamination efforts should be closely coordinated with entry team operations to ensure everyone's safety. The entry team members must understand the decontamination line layout and be able to confidently pass through the individual stations. The decontamination group supervisor and the decon personnel should attend the entry team briefing to ensure that both the decon personnel and the entry team members are on the same page.

REPORTS AND SUPPORTING DOCUMENTATION

The technical decontamination group supervisor should include a description of the decontamination line layout and the decontamination methods used in the incident report. In addition, all personnel involved in decontamination efforts should be listed, as well as their job functions. Special care should be taken to note when and how the hazardous waste generated during the decontamination process was disposed of. The hazardous waste manifest should become part of the incident documentation. Remember, hazardous waste is tracked from cradle to grave per RCRA!

Personnel Exposure Records

Personnel exposure records must be kept for all first responders involved in technical decontamination efforts. These include the name of the first responder, the employing agency, the employee identification number, the location, the length of time, and the job function the first responder was performing at the incident. In addition, routine air-monitoring readings at the scene should be recorded as well as any documented chemical exposures the first responder may have received. Exposure records must be kept by the employer for a minimum of 30 years after termination of employment per HAZWOPER (29 CFR 1910.120).

Activity Log

An activity log is a list of all of the actions performed during the technical decontamination operation. It should include a description of the activity, the time it was started and completed, and the personnel that performed the activity. Examples of logged activities include setting up and performing decontamination, donning and doffing personal protective equipment (including on-air times), and chain of custody transfers of samples. This list is certainly not all-inclusive, and any significant activities should be logged.

Filing Documentation and Maintaining Records

Any documentation generated during the technical decontamination process should be filed with the rest of the incident documentation. Federal regulatory agencies require that certain records, such as personnel exposure reports, be kept for a lengthy time. Other documentation, such as costs and equipment used, will be required by federal agencies and state agencies for reimbursement. Also, at criminal or terrorist events, any chemical samples and personal effects of victims become evidence and may make or break a criminal prosecution.

SOLVED EXERCISE 8-7

What type of documentation is required for the incident described in the opening scenario?

Solution: Personnel exposure records should be kept in case of an undetected exposure or latent effect such as cancer. An activity log of the decon procedures should be kept that includes the use of the emergency shower and what was done with the victims' personal property as well as decon verification methods, instrument readings, and results. Any additional routine documentation such as run reports and patient forms must also be generated and should contain pertinent information from the activity log.

Summary

Technical decontamination is required when down-range or hot zone operations are performed at hazardous materials or terrorist incidents. First responders at the operations level can perform a vital support function to technician level hazardous materials response teams by setting up and performing technical decontamination. This will not only speed up the hot zone entry process but also make vital resources available to the hazardous materials response team that would otherwise be used in the technical decontamination operation.

Review Questions

1. What components should be covered in the decontamination plan?
2. What are the advantages and disadvantages of dry decontamination?
3. What are the special considerations when decontaminating law enforcement officers?
4. How does the decontamination of ambulatory versus non-ambulatory victims differ?
5. How can decontamination effectiveness be evaluated?

Problem-Solving Activities

1. With the technical decontamination equipment available to your agency (both commercial and improvised), set up a dry decon line according to your standard operating guidelines.
2. With the technical decontamination equipment available to your agency (both commercial and improvised), set up a wet decon line according to your standard operating guidelines.
3. With the technical decontamination equipment available to your agency (both commercial and improvised), plan a technical decontamination operation for an industrial accident involving an acid.
4. With the technical decontamination equipment available to your agency (both commercial and improvised), plan a technical decontamination operation for an overturned gasoline tanker.
5. With the technical decontamination equipment available to your agency (both commercial and improvised), plan a technical decontamination operation for a hazmat incident involving several abandoned drums.

References and Further Reading

National Fire Protection Association. (2008). NFPA 472, *Standard for Competence of Responders to Hazardous Materials/Weapons of Mass Destruction Incidents.* Quincy, MA: Author.

Occupational Safety and Health Administration. (1990). 29 CFR 1910.120, *Hazardous Waste Site Operations and Emergency Response (HAZWOPER).* Washington, DC: U.S. Department of Labor.

Weber, Chris. (2007). *Pocket Reference for Hazardous Materials Response.* Upper Saddle River, NJ: Pearson/Brady.

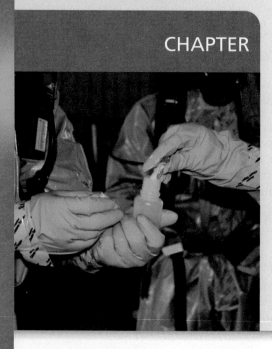

9

Operations Level Responders Mission-Specific Competencies: Evidence Preservation and Sampling

OBJECTIVES

- List four types of hazardous materials and WMD crime scenes.
- Describe the components of an effective site safety plan.
- Describe how to avoid destroying or damaging evidence at hazardous materials and WMD incidents.
- Describe methods to collect evidence that avoid cross contamination.
- Explain field screening techniques for evidentiary samples.
- Explain how the two-person sampling technique helps avoid cross contamination of evidentiary samples.
- Describe the tools and equipment used in evidence collection.
- Explain the significance of chain of custody.
- Describe the packaging and shipping requirements for transporting potentially contaminated evidence.

You Are on Duty! Ricin Found in the Local Motel

LAS VEGAS, NEVADA

On February 29, 2008, a maid cleaning a hotel room discovered suspicious chemical and laboratory materials. Local law enforcement officials were quickly notified by the hotel manager. The room, in an extended stay hotel in Las Vegas, belonged to a man that had neither been seen in many days nor paid for his room. Therefore, hotel management decided to evict him. Arriving police officers found extremist literature and manuals detailing the production of toxins. The officers backed out of the room and called local hazardous materials response teams, which subsequently identified a vial containing ricin using handheld assays. It turns out that the man had been hospitalized due to respiratory distress for 2 weeks prior to the incident. The man had admitted himself into the hospital on February 14, 2008, due to difficulty breathing, and he subsequently lapsed into a coma. (From interviews of members of the Las Vegas Metro Police Department ARMOR and the Nevada WMD-CST)

- How would your law enforcement agency handle this investigation?
- How would the evidence be collected?
- How and where would you package, decontaminate, and store this evidence?
- How would your fire department and EMS agency handle this situation if the man had called 911 from his motel room?

Let's try to answer these questions in this chapter.

When operations level personnel respond to hazardous materials or weapons of mass destruction (WMD) incidents that require **evidence** collection and/or sampling, they should operate using standard operating guidelines (SOGs), or under the guidance of a hazardous materials technician or an allied professional. Those operations level personnel that are expected to perform evidence collection and sampling should also be trained in the mission-specific competency of personal protective equipment (Chapter 6) and the mission-specific competency of air monitoring and sampling (Chapter 11).

> **evidence** ■ Anything that may be linked to a suspect and used to convict the suspect in a court of law.

Types of Hazardous Materials and WMD Crime Scenes

Hazardous materials incidents can become crime scenes for many different reasons (Figure 9-1). For example, if hazardous materials were not transported or stored properly, laws may have been violated and a criminal investigation will ensue. If hazardous materials were disposed of illegally, in other words, dumped illegally, environmental crimes have been committed. Or hazardous materials may be used in the process of committing a crime, such as in an **illicit laboratory** synthesizing methamphetamine drugs. Of course, weapons of mass destruction incidents are by definition a crime scene and fall under the FBI's jurisdiction. The key question is what makes a hazardous materials incident criminal in nature and how do we recognize that fact. Let's explore some of the different types of hazardous materials and WMD crime scenes you may encounter.

> **illicit laboratory** ■ A location where a final product is made using precursor chemicals, reagents, and solvents. Illicit labs can be configured to make illegal drugs, chemical warfare agents (CWAs), biological warfare agents (BWAs), or explosives. Also known as clandestine laboratory, clan lab, or illicit lab.

SUSPICIOUS LETTERS AND PACKAGES

Suspicious letters and packages may take many forms. You are probably already familiar with the infamous anthrax letters that were mailed to the offices of the *National Enquirer* in Florida, the New York offices of NBC News, and Senator Daschle's office on Capitol Hill in the fall of 2001 (the Amerithrax investigation). These letters were extremely dangerous due to the presence of processed anthrax spores. They caused the death of five

FIGURE 9-1 Crime scenes involving hazardous materials or WMDs can come in many forms, including illegal hazardous waste dump sites (left) and illicit laboratories (right).

people, millions of dollars of direct property damage, and hundreds of millions of dollars of emergency response costs over the next several months due to hoaxes and an anxious public. Only as recently as 2008 was the suspected perpetrator identified. These types of responses pose a real danger to responders at the operations level and must be carefully managed. The most significant hazard of suspicious letters that may contain biological agents is the potential of a grave inhalation hazard. Suspicious letters may contain other dangerous hazardous materials or weapons of mass destruction, such as the readily available toxin ricin.

Other suspicious letters or packages may be bombs. Letter bombs may contain explosives such as Primasheet, a commercial plastic explosive, and are designed to kill or injure individuals. Packages may contain larger quantities of explosives and create significant structural damage. If the letter has traveled in the mail, the explosive it contains is typically fairly stable to withstand the rigors of transport. Letter bombs may be triggered by several different means: upon opening, remotely, on a time delay, or at a precise time. These types of suspicious letters are extremely dangerous due to the energy released by the explosives. An explosive ordnance disposal (EOD) team, or bomb squad, should always be called when dealing with a suspicious letter that could be a bomb.

Suspicious letter calls are still fairly common for most communities. The vast majority of these suspicious letters are harmless. It must be determined which ones constitute credible threats and which are almost certainly hoaxes or misperceptions. Some indications of a credible threat are:

1. The suspicious envelope or package contains a threat (written threat, phoned-in threat, e-mail threat, media-disseminated threat, threatening photographs, etc.).
2. The intended recipient or the location at which the item was received is a potential target (judge, politician, federal building, courthouse, post office, monument, abortion clinic, celebrities, etc.).
3. An unusual material is detected in or around the suspicious envelope or package (white powder, oily residue, leaking solid or liquid, etc.).

Remember, always assume the threat is real until it can be ruled out using identification equipment such as biological agent identification equipment (antibody-based handheld assays or DNA-based polymerase chain reaction [PCR] tests) or explosives detection equipment.

ILLICIT LABORATORY

Illicit laboratories, or clandestine laboratories, are a very common crime scene in many communities, with the most prevalent ones being methamphetamine laboratories (refer to Figure 9-1). These dangerous laboratories are becoming more and more common throughout North America and contain a witch's brew of hazardous materials: corrosive materials, flammable materials, water- and/or air-reactive materials, as well as many **toxic** materials. In addition, the individuals making the drugs often deploy booby traps to dissuade would-be thieves, do not possess proper laboratory or chemistry training, have poor housekeeping skills, and dispose of hazardous waste improperly. The Drug Enforcement Administration (DEA), the state police, or local law enforcement has jurisdiction over clandestine labs producing illegal drugs.

toxic ■ Having the ability to cause disease, injury, or death in exposed individuals at comparatively low concentrations.

Illicit explosive laboratories are becoming more common in the United States and have been prevalent through parts of the Middle East, Asia, and Europe. Over the years, these laboratories have become very sophisticated and can produce large quantities of powerful and sensitive explosives. In the United States, terrorists have traditionally used commercially acquired and homemade ammonium nitrate–fuel oil (ANFO) mixtures in their attacks because ANFO is a very simple explosive to produce. Recently, foreign terrorists have begun to use organic peroxide–based explosives such as triacetone triperoxide (TATP) and hexamethylene triperoxide diamine (HMTD). These explosives are a little bit more complicated to produce and are a lot more dangerous due to their sensitivity to heat, shock, and friction. An explosive ordnance disposal (EOD) team should always be called when dealing with a suspected illicit explosives laboratory. When dealing with illicit laboratories that are producing weapons of mass destruction such as these, the Federal Bureau of Investigation (FBI) and the Bureau of Alcohol, Tobacco, Firearms and Explosives (ATF) must be notified per the National Response Framework and presidential directives.

Other illicit laboratories may produce even more sinister materials such as chemical agents or biological agents. These extremely toxic materials should be handled with the utmost of care. Due to their unusual nature, specialists should be called when dealing with this type of laboratory.

It is highly recommended that the personnel expected to deal with illicit laboratories be trained to the Mission-Specific Competency: Illicit Laboratory Incidents (Chapter 13).

WMD RELEASE OR ATTACK

Weapons of mass destruction (WMD) may be released or used in an attack in an almost endless variety of ways. WMD releases and attacks are therefore more common than might be thought. In 1983 cyanide was placed in packages of Tylenol, resulting in the death of several people. This incident spurred the creation of the tamperproof closures for pharmaceutical and food containers with which everyone is familiar.

Ricin, a toxin that is isolated from the castor bean, has been used in numerous attacks throughout the United States. In 1991, ricin was used in a thwarted attempt to assassinate a U.S. marshal in Minnesota. In 2004, a man identifying himself as the "Fallen Angel" sent a vial of ricin to the federal government. In 2008, the occupant of a Las Vegas motel room, who was also the suspected perpetrator of producing ricin, became ill, likely due to ricin toxicity. He subsequently plead guilty to possessing a biological toxin.

WMD attacks have also been carried out on a larger scale. In 1995, the Aum Shinrikyo doomsday cult in Japan carried out an attack on the Tokyo subway system. Cult members synthesized sarin, placed the approximately 25% pure sarin into small plastic bags, and punctured them on crowded subway cars. This attack led to the death of 12 people, sickened dozens more, and caused thousands of people to crowd the emergency rooms of Tokyo hospitals.

By law, the FBI is the agency in charge of investigating any incident involving weapons of mass destruction (WMD). The FBI must be notified as soon as possible for crimes involving WMDs. Each FBI field office has a WMD coordinator that receives and acts on the notification of any WMD incidents. You may contact a WMD coordinator by calling the FBI field office in your area. The WMD coordinator can assist through having the reach-back capability to national assets and subject matter experts who are available 24 hours a day, 7 days a week. The WMD coordinator can convene a conference call in as little as 15 minutes, during which these national assets can be called upon to provide guidance on WMD matters.

The National Guard's WMD civil support teams (WMD-CSTs) specialize in responding to weapons of mass destruction incidents. CSTs have the capability of initiating secure communications and of performing mass decontamination, technical decontamination, hot zone entries, **air monitoring**, sampling and evidence collection, as well as advanced sample identification in a self-contained mobile laboratory. The mobile lab has advanced immunology, microscopy, PCR, and GC/MS capabilities among many other capabilities. Each state has at least one CST, which is a great resource to have at the scene of WMD incidents. Contact your state WMD-CST team; get to know the team members and train with them!

air monitoring ■ A systematic analysis of atmospheric hazards using detection equipment. The atmosphere should be screened for radiation, corrosives, oxygen, combustible gases, and toxic materials.

ENVIRONMENTAL CRIMES

Given the large amount of chemicals used in North America annually, it is not surprising that environmental crimes are very common. These crimes include illegal dumping of chemical waste and discharging untreated chemical waste into waterways.

Examples of environmental crimes involving illegal discharge into waterways are introducing chemicals into sanitary sewers or storm drains. For example, a plant manager at a plating facility directs employees to bypass the wastewater treatment unit to save money. This sends untreated wastewater directly into the sewer system in violation of the permit issued by the municipal sewer authority. The plant manager is guilty of a criminal violation of the Clean Water Act.

Examples of environmental crimes involving illegal dumping include abandoning 55-gallon drums full of waste chemicals by the side of the road or disposing of chemicals in an improper waste stream (refer to Figure 9-1). For example, to save money, a cleaning products manufacturer decides to dump several buckets of corrosive waste into its municipal waste dumpster. The responsible party within the company is guilty of a criminal violation of the Resource Conservation and Recovery Act (RCRA).

Environmental crime investigations often involve tracing pollutants to the originator (responsible party). This may involve the analysis of soil, water, or air samples that contain low pollutant concentrations using sensitive laboratory equipment. Proper sample collection and rapid transport are essential for successful prosecution of many environmental crimes. Because the discharge site, such as a storm sewer outflow into a lake, may be distant from the source of the pollutant, good detective work is necessary to trace the path of the pollution back to its source for identification of the responsible party.

Environmental crimes are investigated by local, state, and federal authorities. At the local level, the public health department or the local department of environment (DOE) is usually in charge. At the state level, the state Department of Environmental Quality (DEQ) or the state Environmental Protection Agency is responsible for investigating and prosecuting environmental crimes. At the federal level, the U.S. Environmental Protection Agency (EPA) has the responsibility for environmental crimes. Any time an environmental crime is suspected, local law enforcement and the local agency responsible for the environment must be notified. If a quantity greater than the reportable quantity (RQ) has been released, the state and federal governments must be notified as well. The federal government is notified by calling the National Response Center (NRC) at 1-800-424-8802 or reporting it online at www.nrc.uscg.mil.

What type of crime scene is described in the opening scenario?
Solution: A WMD release that falls under the jurisdiction of the FBI.

Evidence Collection Versus Public Safety Sampling

What do all of these criminal incidents involving hazardous materials have in common? There is evidence that needs to be collected. But what is evidence? There are two parts to this question: How do we recognize evidence in the first place, and at what incidents do we even have evidence? Imagine you are called to an apartment complex by a landlord because he has found a funny-looking contraption in the basement storage area. He leads you down there, and you immediately recognize it as an incubator, the type commonly used to brew beer. But why is it there? There are two possibilities: either the tenant was brewing beer and forgot to pack his equipment when he moved out or it is a clandestine lab. In either case, it is vital to sample the contents in order to err on the side of safety. So what type of sample is being collected: a public safety sample or an evidentiary sample?

Evidence may be collected only by sworn law enforcement officers and fire investigators, or by other first responders under their direct supervision. Evidence may also be collected only after probable cause has been established or a warrant has been issued. Therefore, if a fire department–based hazmat team collects a sample as "evidence" at an emergency incident, that sample may be thrown out of court due to improper search and seizure. Because the hazmat team was called to an emergency, probable cause may not exist for a law enforcement operation, which evidence collection is by definition. Non-law-enforcement agencies should therefore always call their sampling activities "public safety sampling" instead of "evidentiary sampling." Nevertheless, agencies' public safety sampling protocols should adhere to high standards of documentation, certified tools and containers, and good sampling protocols that minimize cross contamination. Such guidelines will not only give more reliable field test results and lab analysis but also allow law enforcement to seize these public safety samples into evidence at a later time, after a warrant has been obtained. So in practice, the performance of public safety sampling and **evidentiary sampling** looks very similar; the difference is intent. The intent of a public safety sample is to identify a substance to mitigate an emergency, whereas the intent of an evidentiary sample is to solve and prosecute a crime. Chapter 11 covers public safety sampling in greater detail.

To finish answering the question about what is evidence, a law enforcement agency called to the scene by the landlord is collecting evidence. On the other hand, the fire department is collecting a public safety sample irrespective of the final outcome of field identification. If it does turn out to be a clandestine laboratory, and public safety sampling has been performed in a rigorous fashion, it will be relatively easy for law enforcement to seize those samples as evidence at a later time. If the fire department collected "evidence" without proper law enforcement authority, that sample may eventually be thrown out of court on a technicality.

evidentiary sampling ■ The collection of chemical samples that will be entered into evidence. Evidentiary sampling must be performed using very precise and controlled methods in order to be admissible in court and withstand cross-examination.

Preparation and Planning

After life safety and incident stabilization considerations have been met, the next priority is to preserve evidence. That activity starts with isolation of the area and denying entry to everyone, including first responders, who is not absolutely necessary to perform mission-critical functions on scene.

Proper documentation of evidence is necessary to ensure admissibility and defensibility in a court of law. This means ensuring probable cause exists and obtaining any and all necessary search warrants. If time permits, obtain a search warrant even if sufficient probable cause exists to remove any judgment questions from law enforcement's shoulders and ensure court approval during later legal proceedings.

It is important to identify the agency with investigative authority to guarantee that the appropriate procedures are implemented at the crime scene. Many procedures—such as sampling, type and quantity of sample containers, evidence collection, documentation, and laboratory analysis—may be agency specific. Improper evidentiary laboratory analysis procedures may hamper criminal prosecution.

To ensure the viability of potentially unstable samples, the evidence collection process should be started as quickly as possible after emergency operations have ceased. It is ideal to collect within 2 hours all transient evidence that has the potential of disappearing or degrading. All other crime scene operations should be completed very methodically and not be hurried. This is a major difference between public safety sampling and crime scene processing. With crime scene processing, we can and should take our time to ensure it is done correctly, methodically, and in such a manner as to not overtax the crime scene processors that are in personal protective equipment. There is only one chance to process a crime scene; once the crime scene is relinquished, we cannot go back and do it again.

SITE SAFETY PLAN

Evidence collection and sampling activities should begin with a review of the relevant site safety plan (SSP). This is a critical component of the planning process. The site safety plan contains vital safety information that is important for the first responder at the hazmat operations level who will be performing evidence collection and sampling to know. The site safety plan describes the following components of a hazardous materials or weapons of mass destruction incident:

- Incident command system (ICS)
- Safety procedures and notifications
- Scene safety precautions
- Continuous air-monitoring procedures
- Hazardous materials and chemical site inventory
- Decontamination procedures
- Personal protective equipment (PPE) selection
- Sampling procedures
- **Backup team** procedures

backup team ■ A group of trained responders assigned to perform rescue of the entry team in the event of an emergency; also called a rapid intervention team (RIT).

SOLVED EXERCISE 9-2

What are the key components of the site safety plan for the incident described in the opening scenario?
Solution: Key SSP components include:

Signs and symptoms of ricin exposure by route of entry
Emergency contact information (hospital, etc.)
Access control procedures
Personal protective equipment needed by zone
Decon line location and decontamination methods and solution
Ricin identification and detection equipment and procedures
Downrange stay times
How the scene will be documented pre-entry and post-entry (photography, videography, sketching, etc.)
Evidence collection methodology

Incident Command System

HAZWOPER (29 CFR 1910.120) mandates the use of the incident command system (ICS) at all hazardous materials incidents, including environmental crimes, illicit labs, and incidents involving weapons of mass destruction (WMD). Furthermore, a NIMS-compliant ICS must be used per the National Response Framework (NRF) and presidential directive. Determine who the incident commander is, and what you and your agency's role within the incident command system will be. It is very important to maintain the established chain of command.

Decontamination

A decontamination plan must be in place before anyone enters the hot zone per HAZWOPER (29 CFR 1910.120). At a minimum, this means there must be the capability to perform dry decontamination, with a source of water available to remove contaminants from entry personnel if the need arises (Figure 9-2). Typically this requires that a technical decontamination line be in place that includes a source of water; the ability to use soap; tools and equipment to perform decontamination; a secondary containment system for the used decontamination solution; and a disposal method for disposable PPE, used decontamination solution, and other decontamination waste. Personnel must be trained in proper technical decontamination of entry personnel. Typically, a hazardous materials response team performs decontamination functions. Be familiar with this decontamination plan before entering the hot zone. Technical decontamination is discussed in greater detail in Chapter 8, Operations Level Responders Mission-Specific Competencies: Technical Decontamination.

Documentation

Proper documentation is critical at crime scenes. It is crucial that every piece of evidence recovered be completely and accurately documented. Become familiar with the agency's evidence documentation procedures. An example of an evidence collection form is shown in Figure 9-3.

SOLVED EXERCISE 9-3

What type of decon line should be set up for the entry team collecting and processing evidence in the incident described in the opening scenario?

Solution: Dry decon would be effective for a toxin such as ricin. Wet decon using soap and water could be considered in case unexpected hazards are found.

FIGURE 9-3 An evidence collection form.

EVIDENCE [ ]	SAMPLED [ ]	DESTROYED [ ]	
DESCRIPTION		NIK	
		Negative Positive N/A	
LOCATION FOUND		**PH**	
FOUND BY	**SAMPLED BY**	**ITEM #**	

BOX 9.1 EVIDENCE COLLECTION

1. Photograph area before entry.
2. Photograph evidence to be collected.
3. Collect evidence. Chemical samples may be collected as evidence using the two-person sample collection technique.
4. Secure container with evidence tape.
5. Photograph area after collection.
6. Document chain of custody.

The evidence collection form should, at a minimum, contain the following information:

- Photographs of the scene before manipulation
- Photographs of the scene after manipulation
- Date and time of evidence collection
- Names and ranks of personnel performing sampling and evidence collection
- A reference to the relevant standard operating guidelines (SOGs) governing the evidence collection method in use at the incident
- Chain of custody
- Tracking number

The step-by-step process of evidence collection is outlined in Box 9.1.

Response

The crime scene should be accessed and processed in a systematic fashion. Always work from one side of the scene to another, or have a systematic work method in place. Following a systematic method will make the crime scene processing safer, make it easier to document your actions, and make it easier to testify in court.

During a response to any crime scene, the following must be accomplished:

1. Coordinate crime scene operations with the law enforcement agency having jurisdiction.
2. Ensure that all agencies secure and preserve the evidence.
3. Carefully select the appropriate level and type of personal protective equipment (PPE).
4. Perform the first entry and reconnaissance jointly with special weapons and tactics (SWAT), EOD, and hazardous materials teams to avoid booby traps, secondary devices, and hazardous atmospheres.

5. Perform continuous air monitoring during the operation.
6. Have a decontamination plan in place that can effectively and safely deal with all of the deployed assets, including SWAT, EOD, and K-9 units.
7. Coordinate post–crime scene remediation operations.
8. Thoroughly document all completed operations.

Although most of these actions will be carried out by other teams, it is the responsibility of the first responder at the hazardous materials operations level, acting in the evidence preservation and sampling capacity, to ensure they are accomplished for everyone's safety.

ENSURING SCENE SAFETY

Before entering the crime scene, a thorough hazard assessment should be conducted based upon available information such as dispatch information, interviews of witnesses and neighbors, and a visual overview of the site. Always be aware of the potential for booby traps inside the building itself as well as in the exterior yard and grounds.

Booby Traps and Secondary Devices

Booby traps and secondary devices are a very real risk at crime scene operations. These devices come in many forms, including explosives, chemicals, firearms, and mechanical devices (Figure 3–37 and Figure 9–4). Common booby traps include containers full of hazardous chemicals propped above doorways that are designed to fall on anyone entering, fishhooks strung at eye level both indoors and outdoors, shotguns rigged with a pull string tied to doorknobs, lightbulbs partially filled with gasoline inserted into light sockets, and secondary devices constructed from explosives. It is therefore very important to watch where you walk, be careful opening doors, and not to turn on light switches.

Many booby traps and secondary devices are explosive in nature. It is therefore very important to have an explosive ordnance disposal (EOD) team, or bomb squad, clear the area before evidence collection and sampling activities begin.

Continuous Air Monitoring

Atmospheric monitoring should be performed for the five types of hazards that may be encountered (Figure 9-5):

1. Radiation
2. Corrosive atmospheres
3. Oxygen-deficient or -enriched atmospheres
4. Flammable atmospheres
5. Toxic atmospheres

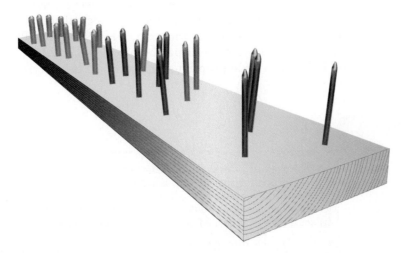

FIGURE 9-4 This booby trap consists of a piece of wood, such as plywood with nails driven through it. The board is then camouflaged and placed on the ground in a high traffic area by the criminal.

FIGURE 9-5 Air-monitoring activities at a crime scene. It is important to monitor the atmosphere for hazards before entering.

The atmosphere should be monitored carefully upon initial approach and entry into the crime scene, and then continuously air monitored during the evidence collection process. It is important to continuously monitor the atmosphere because conditions can change as containers are manipulated and personnel move around the crime scene.

Radioactive materials may be encountered in laboratories, suspicious packages, environmental crimes, and most certainly in radiological dispersal device (RDD) incidents. Geiger counters should be used to monitor for dangerous levels of radiation.

Corrosive atmospheres may be encountered at just about any crime scene that has been discussed. Wetted pH paper should be used to monitor for corrosive atmospheres.

Flammable atmospheres may also be encountered at any crime scene that has been discussed. Monitor for oxygen deficiency to ensure that a combustible gas indicator (CGI) functions correctly. Level C PPE is commonly used in collecting evidence because hazards can usually be mitigated sufficiently using engineering controls. Remember, a minimum of 19.5% oxygen must be present in order to use level C PPE. Most instrumentation, such as multi-gas detectors, that monitors for flammable atmospheres also contains an oxygen sensor.

Toxic atmospheres are also frequently encountered at the crime scenes discussed. Many different types of detectors are appropriate for low-level detection of toxic materials. Some of these are broad-range detectors such as the photo-ionization detectors (PID); others are very specific such as hydrogen sulfide sensors and colorimetric tubes. Toxic contaminants must be identified and remain below IDLH levels in order to use level C PPE; otherwise, a higher level of PPE must be used (such as level A or B). Air-monitoring instrumentation and strategies are discussed in greater detail in Chapter 11, Operations Level Responders Mission-Specific Competencies: Air Monitoring and Sampling.

Personal Protective Equipment (PPE)

Respiratory protection and personal protective equipment must be carefully selected. Initial characterization of an unknown crime scene should be conducted in level A personal protective equipment (Figure 5-19). However, having a completely uncharacterized site is very uncommon once the evidence collection stage has been reached. If there is some initial hazard information, a lower level of PPE can usually be used. For example, if you have been dispatched to a suspected methamphetamine laboratory, level B PPE may be appropriate for an initial characterization. However, keep in

SOLVED EXERCISE 9-4

What type of air-monitoring activities should be performed in the opening scenario?

Solution: Air-monitoring equipment is not available for large toxin molecules such as ricin because it would not be effective. However, air monitoring should be conducted for unexpected hazards. Ideally all five hazards should be screened in the appropriate order to ensure there are not any additional hazards. At minimum, a multi-gas meter and a PID should be used to clear the area upon initial entry.

What type of PPE should be worn by entry personnel in the opening scenario?

Solution: Assuming no other hazards are found, level C protection consisting of chemical protective clothing and an N95 respirator would suffice. The respirator should be used in case the ricin has been processed into a fine powder.

mind that if unexpected hazards are encountered, you may have to exit the laboratory and upgrade your PPE.

Evidence collection is often carried out in level B or level C PPE, depending on the hazard assessment. Once the hazards and chemicals have been characterized during the initial reconnaissance using air monitoring, the level of PPE is often downgraded. During the evidence collection stage, unknown materials will be visually examined, handled, and sampled. After the containers have been tentatively identified and we are ready to begin sampling and evidence collection, we may downgrade to level C protection if the hazard assessment warrants such actions.

Always ensure that the PPE you are using is compatible with the hazardous materials found in the laboratory. Remember, we always want to be more cautious in unknown situations! Selection and use of personal protective equipment is discussed in greater detail in Chapter 6, Operations Level Responders Mission-Specific Competencies: Personal Protective Equipment.

PRESERVING THE CRIME SCENE

Always use the least number of people to perform evidence collection. The least number of people will be determined by safety considerations as well as the work that needs to be done in the crime scene area. Ensure other agencies do not decontaminate vital evidence unless absolutely necessary, such as for life safety reasons. Do not touch or remove items until the crime scene has been photographed, and then only if it is absolutely necessary. If it is necessary, remove any easily accessible, nonessential materials that may become trip hazards first. Then remove the highest hazards first to increase the safety of the evidence collection operation.

IDENTIFICATION OF SAMPLES FOR EVIDENCE COLLECTION

Finding evidence is the ultimate goal for law enforcement purposes. Identifying, characterizing, and sampling the evidence are your primary roles. Before sampling a container, ensure that it is safe to do so! Keep in mind that some containers may be booby-trapped, other containers may be overpressurized or weak, and yet other containers should not be opened because you are in the inappropriate level of PPE or ventilation is not sufficient.

Evidence

What is evidence? As discussed earlier, there are many different types of evidence, but in the hot zone we will be collecting physical evidence, some of which may be trace evidence. The physical evidence at a hazardous materials or WMD crime scene will include typical crime scene evidence such as fingerprints, footprints, tool marks, and the like. Whenever processing evidence, care should be taken not to disturb traditional forensic evidence such as fingerprints, hairs and fibers, DNA, and the like. As little manipulation as possible should be used when sampling and packaging evidence. This is why it is crucial to consider areas where the perpetrator may have held a container, leaving his or her fingerprints. Although it is important to identify the analyte of interest, it is extremely important to get the traditional forensic evidence from the processing equipment, notebooks, computers, and other guidance documents. The traditional forensic evidence will be exploited at a forensic laboratory, but the collection team must be diligent in not ruining this delicate evidence in its collection and processing of the crime scene.

In addition, chemical samples will need to be collected. This physical evidence may be liquids or solids found in an abandoned drum or containers in the lab, residue found on surfaces or on equipment, or gas and vapor samples obtained from the scene that may contain illegal emissions or biological agents. The primary difference between operating at a routine crime scene versus a hazardous materials or WMD crime scene is the level of care that must be taken to protect yourself.

Container Handling

The first step before handling the container is a damage assessment. This assessment includes looking for distended or bulging containers, indicating a possible pressure buildup inside; leaking or weeping container seams; or other visible damage to containers such as cracks, dents, or corrosion. Also make sure that the container closures are present and not damaged. For example, caps on corrosive containers are often cracked and leak, whereas gaskets on solvent containers are often corroded or brittle and leak. For your safety, all of these issues should be addressed before handling these containers in any way.

Distended or bulging containers are often the most dangerous due to the possibility of pressure buildup. If possible, these containers should be moved out of the lab using remote handling before personnel continue with laboratory characterization or remediation. If this is not possible, a decision needs to be made whether the bulging container is stable and other containers can be characterized and removed while it is still in place, or whether this container needs to be dealt with first. Consult an expert before making these extremely important decisions!

Some common dangerous containers found at hazardous materials and WMD incidents include:

- Acid gas generators (Figure 9-6)
- Anhydrous ammonia containers (Figure 9-6)
- Hot glassware
- Corroded gas cylinders

Containers that require the use of self-contained breathing apparatus (SCBA) should be addressed and disposed of first. Always use the buddy system. When handling any container, use dry gloves; hold the container with both hands; and do not handle the

FIGURE 9-6 Examples of dangerous containers found at crime scenes include propane tanks that may contain anhydrous ammonia (left) and acid gas generators (right). Ammonia turns brass a bluish-green color and corrodes the gas cylinders from the inside out, structurally weakening them. Acid gas generators, which consist of salt and acid inside a makeshift container, can be reactivated by movement.

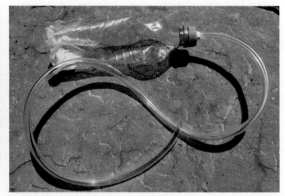

container by its cap, neck, or any other delicate area. Move the containers out of the immediate work area, and make sure to segregate the chemicals by compatibility. For example, do not place strong acids and strong bases together, and do not place oxidizers and fuels (solvents) together in the same immediate area. The chemical staging area should have good ventilation; plenty of room; and easy, unhindered access to avoid accidents. Before placing chemicals in the staging area, lay down a sheet of plastic as a secondary containment area in case any of the containers leak.

Sample Identification

Once it is determined safe to handle and open a container to perform sampling, choose the appropriate tools and receiving containers. Generally, chemicals should be placed in certified clean glass containers, and materials of biological origin should be put in sterile plastic containers. Liquids are typically sampled using pipettes, whereas solids are usually sampled using spatulas. Keep in mind that other techniques might be appropriate, such as swabbing. When dealing with liquids, carefully examine the container to determine whether multiple layers are present. If they are, be sure to sample each layer. For example, in a meth lab, the final product such as meth oil may be a thin layer atop an aqueous solution. This layer may be very easy to miss, and it is, of course, one of the most important samples to recover. Depending on your jurisdiction's sampling protocols, you may be able to place multiple layers in a single container (as you found them); or you may have to place each individual layer in a separate container. Consult the analytical laboratory for the appropriate method in your area. Also sample any sediment found at the bottom of containers.

Next, consider how to identify the material, choosing among the different techniques available. Air-monitoring equipment may be used to assess the vapors that the material is emitting by carefully monitoring above the container and/or its headspace. The liquid itself may be assessed using pH paper, oxidizer paper, and M8 or M9 papers. Field identification techniques include Raman spectroscopy, Fourier transform infrared (FTIR) spectroscopy, gas chromatography/mass spectrometry, and X-ray fluorescence. Not all agencies have field access to this type of advanced equipment. Therefore, we will ultimately collect a sample to send to an analytical laboratory, especially if we suspect we are dealing with the final product—whether that is an illegal drug such as methamphetamine, an explosive, a chemical agent, or a toxin. Sample identification is discussed in greater detail in Chapter 11, Operations Level Responders Mission-Specific Competencies: Air Monitoring and Sampling.

FIELD SCREENING OF SAMPLES

All evidence that has the possibility to contain hazardous materials must be field screened for the following hazards before being removed from the site:

1. Corrosiveness
2. Radioactivity
3. Explosiveness
4. Flammability
5. Oxidizing potential

There are several reasons samples and evidence must be field screened. First, the U.S. Department of Transportation (DOT) requires that hazardous materials be properly packaged, labeled, and placarded for transport. Second, most analytical laboratories will not accept samples unless they have been field screened for potential hazards. Third, is it wise to have an uncharacterized piece of evidence sitting in the evidence locker for days, months, or years at a time? The use of **field screening** tools is described in more detail in Chapter 11, Operations Level Responders Mission-Specific Competencies: Air Monitoring and Sampling.

Each sample should be screened in the indicated order. It is important to determine the pH of the material before exposing sensitive electronic equipment to potentially

field screening ■ The process of investigating a sample that will be sent to an analytical lab for hazardous conditions such as corrosiveness, radioactivity, explosiveness, flammability, and oxidizing potential.

corrosive atmospheres and/or materials. The pH paper should be wetted, and a drop of the sample placed on the pH paper. Never dip the pH paper into the sample itself! Otherwise, dyes in the pH paper will leach into the sample, contaminating it.

Radioactive hazards should be identified next. At this point, we are primarily looking for alpha and beta radiation and weak gamma radiation that may have been missed earlier. Remember, screening for strong gamma radiation has already taken place during the initial air-monitoring activities.

Finally, the evidence must be screened for explosiveness, flammability and oxidizing potential. Explosives can be identified using handheld Raman and FTIR identifiers as well as with chemical test kits, such as the DropEx kit, which has the capability of detecting nitro-based explosives, organic peroxides, nitrate, chlorates, and bromides. Flammability may be determined by using a combustible gas indicator (CGI) or placing a small amount of the material on a watch glass and attempting to ignite it. For safety reasons, this should be done with only a small amount of material (a pea-sized amount or less) and well away from the original sample or its container. Destructive testing is the less desirable method than nondestructive testing (such as using air-monitoring techniques), and should be avoided whenever possible. If destructive testing is absolutely necessary, use the smallest amount practical. Oxidizer paper is used to screen for oxidizing potential. The oxidizer paper is used in a similar fashion to pH paper, although it should not be wetted. Place a drop of the unknown material on the oxidizer paper. Once again, never dip the oxidizer paper into the sample or container itself to prevent contamination of the evidence.

Solid and liquid field sample identification equipment has become much more common in recent years and is beginning to augment, if not outright replace, many chemistry-based field screening techniques. The two most common types of sample identification equipment are Fourier transform infrared (FTIR) spectroscopy and Raman spectroscopy (Figure 11-13). These complementary technologies are very effective tools for presumptive identification of unknown liquid and solid materials.

EVIDENTIARY SAMPLE COLLECTION

The evidence should be sampled, identified, quantified, and collected. After the essential components of the crime scene have been processed, the rest of the materials can be identified and disposed of properly. Always keep in mind that you may find the evidence in unexpected places later during the processing operation.

Evidence collection should be prioritized in the order of final product followed by essential precursors. The other chemicals—reagents and wastes—will likely not be sampled as evidence, although their presence will be documented. However, consult with the local crime lab to determine the appropriate procedures. Remember, there is limited space in the evidence locker! Do not forget to collect an equipment blank and a sample blank, collect photographic documentation, and seal the sample container with evidence tape.

Evidence collection should generally be prioritized in the following order:

1. Agent, dissemination devices, and precursors
2. Environmental samples
3. Biomedical specimens from individuals

SOLVED EXERCISE 9-6

To what type of field screening should the evidentiary samples from the opening scenario be subjected?

Solution: Whatever field screening the sentinel laboratory that the samples will be sent to requires. This should have been determined prior to the incident. Typically, samples should be screened at a minimum for radiation, corrosiveness, and flammability.

FIGURE 9-7 The two-person evidence collection technique in use.

BOX 9.2 TWO-PERSON EVIDENCE COLLECTION TECHNIQUE

1. Scribe photographs the area before anyone makes entry.
2. Handler prepares tools.
3. Sampler retrieves sterile tool.
4. Sampler collects sample.
5. Sampler places sample into container and seals it.
6. Handler over packs container.
7. Scribe documents sample.
8. Sampler removes outermost gloves and replaces before moving to next sample.

Two-Person Evidence Collection Technique

The two-person sampling technique, colloquially referred to as the clean person–dirty person technique, minimizes cross contamination of samples and was specifically designed for evidentiary sample collection (Figure 9-7). This technique ensures that samples are as pristine as possible to withstand court scrutiny, which will also lead to better field sample identification results and analytical laboratory test results. The two-person evidence collection technique is outlined in Box 9.2.

The two-person sampling technique uses a minimum of two people per team. The "handler" handles only the clean sampling tools and equipment as well as the double-bagged sample. With double bagging, each bag should contain the identifying information about the evidence. The "sampler" collects the sample of interest using the opened tools handed to him or her by the handler. All sampling tools are disposable and used for one sample only. The sampler wears a minimum of two layers of gloves. The outer

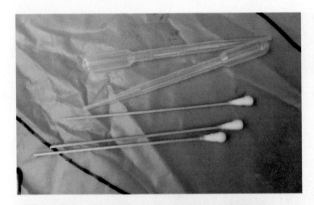

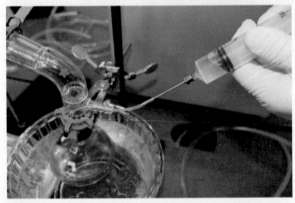

FIGURE 9-8 Liquid sample collection. The top photo illustrates tools that can be used to collect liquids, such as plastic pipettes and swabs. The middle photo shows a syringe and tubing being used to access a vacuum adapter of a distillation apparatus to collect a liquid sample. The bottom photo shows a swab being used to collect a liquid sample from the surface of a table.

layer will always be changed between sampling points to eliminate any chance of cross contaminating the sample. Often a third member will be added to the team, the scribe, who documents the scene by jotting down notes, taking photographs, and sketching the scene.

Documentation is very important during sampling activities. The names of the sampler and handler, the time of sampling, the sample location, and the sample tracking number should all be noted for each individual sample. When the sampling techniques and sampling personnel are well documented, the evidentiary samples will be much more likely to stand up in court. A control blank consisting of the sampling tools and containers should be maintained. The certifying letter (that the equipment is clean or sterile) must also be maintained with the lot number recorded for all collection and sampling materials.

Sampling Gases or Vapors

Gas samples may include the contents of abandoned cylinders, air samples within buildings or open areas, headspace samples of containers, atmospheric samples, gaseous contents of bags, and so on. Sampling gases can be a difficult process. Hazardous materials technicians or environmental contractors may use pumps or vacuum canisters (such as SUMMA canisters), but these are typically not readily available to personnel operating at the hazardous materials operations level. A neat trick is to use a photoionization detector (PID) as a pump: PIDs typically have a threaded gas outlet, and most manufacturers supply a nipple that can be screwed into the outlet and connected to a gas collection bag. Remember, PIDs use a nondestructive form of ionization in the detection process, and the sample train is usually made of stainless steel and glass, which minimally absorb chemicals.

Sampling Liquids

Liquid samples may include contents of abandoned containers, freestanding liquids, surface waters, liquid runoff, sewage, liquids found in clandestine laboratories, pastes, and so forth. Liquids are typically sampled using plastic or glass pipettes (Figure 9-8). A cotton or Dacron swab can be used to sample small amounts of liquid, or liquids in hard-to-reach places (Figure 9-8). Pastes and gels can be scraped up using a spatula or other device.

Sampling Solids

Solid samples may include pellets, powders, dusts, granules, gels, sediments, soils, plants, residues, clothing, and many other solid materials. Solids are typically sampled using a spatula (Figure 9-9). Some solids may be fine powders, which aerosolize easily and may be a potential inhalation hazard. Typically, solids do not have a high vapor pressure and do not pose a significant respiratory hazard. However, under certain circumstances, such as white powder incidents or when fine particulates are known to be involved, the highest level of respiratory protection may be warranted.

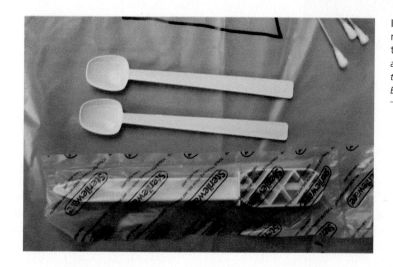

FIGURE 9-9 Equipment that can be used to sample solids. *Photo by author. Sterilewear® sampling tools courtesy of Bel-Art Products*

DECONTAMINATION OF EVIDENCE

At criminal or terrorist incidents, special care must be taken during the decontamination of evidence. Harsh decontamination methods such as chemical degradation, disinfection, sterilization, and neutralization may destroy the evidence and should be avoided when decontaminating evidence. Care should be taken when decontaminating biological and chemical samples. Although the evidence should have been double bagged, be careful to handle the packages carefully to avoid puncturing or otherwise damaging them.

It is very important that the chain of custody of the evidence remains intact. This may require the presence of a law enforcement officer from the authority having jurisdiction (AHJ) in or near the decontamination line. Samples of the wash and rinse solutions from the decontamination line may be kept for later analysis.

PACKAGING EVIDENCE FOR TRANSPORT

Evidence may be transported one of two ways: by official government vehicle or using commercial shippers. In either case, the evidence must be transported safely. The Department of Transportation (DOT) regulates the shipment of hazardous materials. Biological or chemical evidence may be considered a hazardous material depending on its properties.

If the evidence is classified as a hazardous material through field screening, and it will be transported using a commercial shipper, it must be properly packaged, labeled, and documented. Proper packaging may include packaging that has been certified to DOT standards. In addition, DOT requires specialized training for anyone who presents hazardous materials for commercial shipping. If samples are shipped to laboratories using the U.S. Postal Service, or any other commercial carrier, pay special attention to the hazardous materials shipping regulations.

If hazardous materials are transported in government-owned vehicles, these DOT regulations do not apply. Government vehicles are exempt from DOT placarding requirements. However, remember that the DOT regulations were put in place to ensure the safety of drivers and the public, and make sure that potentially hazardous evidence is appropriately contained!

In either case, samples may need to be transported under environmentally controlled conditions, especially when dealing with biological samples. This may require packaging the samples with dry ice, ice, cold packs, or heaters to maintain the appropriate temperature.

Summary

Evidence collection is vital for successful prosecution of crimes involving hazardous materials or weapons of mass destruction (WMD). Crimes involving dangerous materials include suspicious letters and packages, illicit laboratories, the release of WMD agents, and environmental crimes. When conducting evidence collection and sampling at these locations, you must perform a hazard risk analysis. A site safety plan must be developed that addresses the hazards possible in the hot zone (crime scene). Continuous air monitoring must be performed inside the hot zone during evidence collection activities. Sampling activities must be carried out in ways that do not cross contaminate the evidence; the two-person technique is an effective way to sample potential evidence. The tools and equipment used during sampling operations should be sterile and disposable, and must be changed between sampling points. The evidence must be double bagged and decontaminated before leaving the hot zone. Special regulations apply when shipping hazardous evidence to analytical laboratories or other agencies.

Review Questions

1. Name six components of a well-written site safety plan.
2. Before evidence collection technicians enter the hot zone, what safety procedures should be in place?
3. Why should continuous air monitoring be performed during evidence collection at crime scenes involving hazardous materials or weapons of mass destruction?
4. Explain how the two-person sample collection technique works.
5. What is the difference between public safety sampling and evidentiary sampling?

Problem-Solving Activities

1. Find your agency's evidence collection form, and determine what information is to be recorded.
2. Determine what type of air-monitoring equipment your agency uses, and review how to operate it.
3. Determine what type of personal protective equipment your agency uses, and practice donning and doffing it.
4. Review your agency's standard operating guidelines (SOGs) covering evidentiary sampling activities at crime scenes involving hazardous materials incidents or incidents involving weapons of mass destruction (WMD).
   a. Determine when sampling will be performed.
   b. List the sampling equipment your agency uses.

References and Further Reading

E2458 Practices for Bulk Sample Collection and Swab Sample Collection of Visible Powders Suspected of Being Biological Agents from. Nonporous Surfaces. (2010). ASTM International.

E2770 Guide for Operational Guidelines for Initial Response to a Suspected Biothreat Agent. (2010). ASTM International.

National Fire Protection Association. (2008). NFPA 472, *Standard for Competence of Responders to Hazardous Materials/Weapons of Mass Destruction Incidents.* Quincy, MA: Author.

Occupational Safety and Health Administration. (1990). 29 CFR 1910.120, *Hazardous Waste Site Operations and Emergency Response (HAZWOPER).* Washington, DC: U.S. Department of Labor.

U.S. Department of Transportation. (2008). *2008 Emergency Response Guidebook.* Washington, DC: Pipeline & Hazardous Materials Safety Administration.

Weber, Chris. (2007). *Pocket Reference for Hazardous Materials Response.* Upper Saddle River, NJ: Pearson/Brady.

10

Operations Level Responders Mission-Specific Competencies: Product Control

KEY TERMS

alcohol resistant foam (ARF), *p. 281*

aqueous film-forming foam (AFFF), *p. 279*

booming, *p. 277*

damming, *p. 277*

diking, *p. 276*

diversion, *p. 277*

fixed site facility, *p. 271*

fluoroprotein foam, *p. 281*

foam eductor, *p. 279*

high expansion foam, *p. 281*

remote valve shutoff, *p. 273*

vapor dispersion, *p. 282*

vapor suppression, *p. 278*

OBJECTIVES

- Describe the planning requirements to successfully apply product control techniques at hazardous materials incidents.
- List eight different product control options available to the hazardous materials operations level responder.
- Explain the importance of remote valve shutoff in product control.
- Explain three techniques to control moving liquids.
- Explain three techniques to control hazardous materials in bodies of water.
- Explain two techniques to control vapors.
- Describe the types of foam and their uses.

Resource Central

For additional review and practice tests, visit **www.bradybooks.com** and click on Resource Central to access book-specific resources for this text! To access Resource Central, follow directions on the Student Access Card provided with this text. If there is no card, go to **www.bradybooks.com** and follow the Resource Central link to Buy Access from there.

You Are on Duty! Acid Spill at the Local Production Plant

LYNDEN, WA

On a cold January afternoon, a 911 call is received stating about 15 to 20 gallons (57 to 75 L) of sulfuric acid have been spilled out of one of the warehouses and into the road. Because the spill has migrated off site, it is considered an emergency, and local, state, and federal authorities must be notified. The fire department and police department are dispatched. Upon arrival, you are alerted that there are no injuries and the warehouse has been evacuated. A cracked pipe and/or valve is apparently the cause of the slow leak that is continuing. There is a stream of acid coming out of the warehouse, into the road, and it is slowly migrating toward the sanitary and storm sewer systems.

- How will your crew handle this incident?

Let's try to answer this question in this chapter.

When hazardous materials or weapons of mass destruction are released, the incident is ultimately resolved by controlling the product and getting it back into its container (Figure 10-1). This is typically considered an offensive, technician level response. However, before the arrival of specialized teams, many defensive product control techniques can be applied to minimize the ultimate impact of the release to first responders, citizens, property, and the environment. In this chapter we cover actions the operations level first responder may take to minimize the spread of the release.

Let's first explore the difference between defensive and offensive product control techniques. Defensive product control techniques can typically be performed without direct contact with the product. Examples of defensive product control techniques are

FIGURE 10-1 A hazardous materials release from an overturned waste hauler on the highway. Inside were 2 dozen containers ranging from several hundred gallon totes to 55-gallon drums, several of which were damaged during the accident and were leaking.

placing absorbent material away from the liquid release to intercept the product flow, diking a storm drain or sewer opening, and placing a foam blanket over a flammable liquid release. Each of these techniques can be accomplished from a safe distance without contacting the hazardous material directly. On the other hand, offensive product control techniques, such as closing valves, over packing containers, or neutralizing products, typically require working in close proximity to the hazardous material. Personnel trained only to the operations level should generally not perform offensive product control techniques.

Preparation and Planning

Preparation and planning is crucial before initiating defensive product control techniques. The first step in planning a product control operation is to review the **fixed site facility** plans or shipping papers to gather as much product information as is available (Figure 10-2). Fixed site facility plans may include the SARA Title III paperwork the facility provides to the LEPC and the local fire chief annually, facility maps, and any specific facility preplans formulated by local agencies. These facility plans are important to identify any safety hazards posed by other chemicals, industrial equipment, or building layout. Product information can be obtained from material safety data sheets (MSDS); shipping papers; facility inventories, which can be found in the SARA Title III paperwork or Tier 2 forms; and research materials. Once we understand the hazards of the released product, and the dangers the facility may pose, a safe and effective defensive control operation can be planned.

Although the goal of a defensive product control operation is to have no contact with the released hazardous material, sometimes the wind shifts or the product moves faster than anticipated. Therefore, we must use the appropriate personal protective equipment (PPE), set up a decontamination line, and have a backup team in place in case rescue is needed. Planning for a worst-case scenario increases the safety level of the operation.

fixed site facility ■ An installation that stores and/or uses hazardous materials.

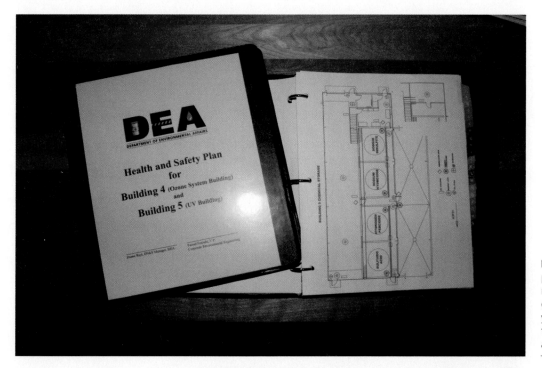

FIGURE 10-2 Emergency plan for a hazardous materials facility. It is a good idea to familiarize yourself with these facilities before an accident happens.

SOLVED EXERCISE 10-1

How would you determine how much product the facility from the opening scenario has on site?

Solution: There are several options. You could consult the SARA Title III plans that list the hazardous materials on site and their maximum quantities. However, there are two problems with this: (1) The maximal quantity may not be on site, and (2) these plans are based on the previous year's production process and statistics. In fact, some of the materials listed in the plan may no longer be on site, and new materials may be on site that are not listed in the plan. The better option is to consult directly with the plant manager or another knowledgeable employee who has a reasonably good idea of how much sulfuric acid is currently in the tank.

SOLVED EXERCISE 10-2

How can contact with the acid that was spilled in the opening scenario be minimized through good work practices?

Solution: The HAZWOPER standard prescribes using engineering controls, safe work practices, and personal protective equipment (in that order) to minimize exposure. An engineering control that may be used is remote valve shutoff, if one exists. Safe work practices include working upwind and uphill, not walking in visible product, and using tools to maximize the distance between you and the product. As a last line of defense, proper PPE must be chosen and worn in the hot zone.

SOLVED EXERCISE 10-3

What other safety considerations are there when responding to the incident described in the opening scenario?

Solution: What other materials are in the path of the flowing acid? Acids often form flammable gases such as hydrogen when they come into contact with metals. Also, acids react violently with many materials. These reactions are often exothermic and may produce toxic by-products.

WORK PRACTICES

Work practices should be chosen to minimize contact with the spilled hazardous materials or the deployed weapons of mass destruction. First and foremost, do not walk through the contaminants, which include such obvious materials as spilled liquids or powders, or less obvious materials such as gases and vapor clouds. Although the spilled liquids and solids will typically be visible, as long as we are looking for them, gases and vapor clouds often are not. This is where air-monitoring equipment is essential for any personnel entering the hot zone.

Whenever possible, do not come into contact with the hazardous material. For technician level personnel, often that is not possible because they are performing offensive product control techniques. However, personnel at the operations level should rarely if ever be in a position to contact spilled hazardous material directly. If your job duties at a hazardous materials incident bring you in such close proximity to the hazardous material, you are likely performing technician level duties. However, if your agency has the appropriate equipment and you have been trained to use it properly, you may be asked to perform job functions in closer proximity to hazardous materials. Good work practices dictate that you remain as far as is practical from the hazardous material while accomplishing your objectives. One way to avoid direct contact with the hazardous material is to use remote sampling and handling tools such as drum grapplers whenever possible. For example, when spreading solidification materials, use a long-handled shovel to put distance between you and the spilled material (see the section on Absorption, Adsorption and Solidification below).

How would you determine what resources the plant from the opening scenario has to help stop the leak?
   Solution: Ask the plant manager or another knowledgeable employee, or consult your pre-incident response plan or the SARA Title III plan.

Control Options

Control options should be carefully chosen based primarily upon safety considerations, available equipment and training, the chemical and physical properties of the product, and the nature of the release. Operations level responders should keep in mind that they may not always be able to safely and effectively control hazardous materials or weapons of mass destruction releases. Remember, when the hazardous materials release is beyond the scope of your training and available equipment, quickly notify the appropriately trained hazardous materials response team and isolate the immediate area. If, however, after a thorough incident analysis, it is determined that you can safely and effectively slow or stop the progress of the hazardous material, several defensive product control options are at your disposal.

When selecting a product control option, identify the purpose for implementing it. It is vital to have defined operating procedures that have been trained on, access to the appropriate product control equipment, and knowledge of the appropriate safety precautions. Product control options should be carefully selected based upon scene considerations to maximize the safety of the operations level personnel that will be engaging in defensive product control procedures in or near the hot zone.

REMOTE VALVE SHUTOFF

Remote valve shutoff devices are one of the most effective ways to quickly stop the release of a hazardous material from its container. Remote valve shutoff switches can be found at fixed site facilities and in the transportation industry (Figure 10-3). Often facility workers or drivers will have used the remote shutoff before responders even arrive on the scene of the hazardous materials incident. This action is often the critical step to prevent a small release from spiraling into a community-wide disaster. Sometimes, however, the workers or drivers do not have the time or are incapacitated and cannot work the remote shutoff valves. In these cases, the operations level responder is the next line of defense.

remote valve shutoff ■ A method to safely stop a leak issuing from a pipe by closing an outlet at a distance from the leak.

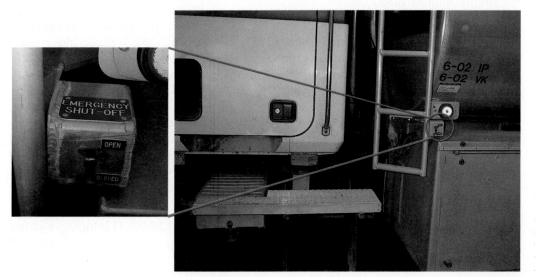

FIGURE 10-3 The pneumatically operated remote shutoff switch on an MC 306 highway cargo tanker is located on the driver's side near the front of the tank.

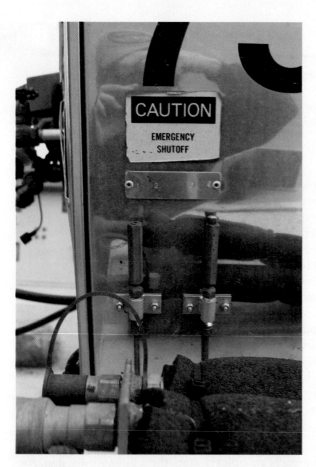

FIGURE 10-4 A hydraulically operated remote shutoff valve on a DOT 407 highway cargo tanker. The tube is manually broken off, which releases the hydraulic pressure and closes the valve. Note that it will be very difficult to reopen the valve for subsequent off-loading of the product.

Transportation

Highway cargo tankers are equipped with remote valve shutoff switches. These switches are typically located on the left front side of the tank and on the right rear side of the tank, kitty-corner to each other. They operate an internal emergency shutoff valve completely contained within the tank, so that even if external piping is sheared away, the internal valve should stay intact and operate. The remote shutoff devices are typically mechanically, pneumatically, or hydraulically operated and contain a fusible link that fails at a predetermined temperature in case of fire.

DOT 406/MC 306 Highway Cargo Tanker The DOT 406/MC 306 highway cargo tankers typically carry flammable liquids such as gasoline, diesel fuel, and E85 (Figure 3-16). These tankers are typically constructed of aluminum and the remote shutoff valves are usually operated mechanically or pneumatically (Figure 10-3).

DOT 407/MC 307 Highway Cargo Tanker The DOT 407/MC 307 highway cargo tankers are known as the workhorse of the chemical industry (Figure 3-17). These tankers are generally constructed of stainless steel, and the remote shutoff valves are typically hydraulically operated (Figure 10-4).

MC 331 Highway Cargo Tanker The MC 331 highway cargo tankers carry compressed liquefied gases such as propane and anhydrous ammonia (Figure 3-19). These tankers are constructed of steel. MC 331 tankers with a capacity of over 3500 gallons (13,250 L) must have two remotely operated emergency shutoff switches, whereas those with a capacity of less than 3500 gallons must have at least one emergency shutoff switch (Figure 10-5). In addition, while offloading MC 331 tankers, the driver must have a remote shutoff switch that can be operated 100 feet away from the vehicle and must be able to shut off the engine and close the internal shutoff valve.

MC 338 Highway Cargo Tanker The MC 338 highway cargo tanker is designed to carry relatively nontoxic, liquefied compressed gases (Figure 3-20). Typically these gases are cryogenic materials, which means they have a boiling point below −130°F (−90°C). These tankers are designed to vent small amounts of their contents under normal operating conditions, which is known as autorefrigeration. The MC 338 tanker can be considered a thermos bottle on wheels. The extremely cold product is designed to cool the tank by evaporation if the internal pressure rises too high. Thus, it is not uncommon to see a white cloud coming off these tankers. This white cloud is actually water vapor that the cold gaseous product condenses out of the air. Not surprisingly, more humid atmospheres tend to produce larger white clouds. Unfortunately, some Good Samaritans may call 911 because they think the tanker is leaking. The good news is that the tanker is likely operating normally, and all is well. So do not overreact!

DOT 412/MC 312 Highway Cargo Tanker The DOT 412/MC 312 highway cargo tanker is designed to carry corrosive liquids (Figure 3-18). These tankers are not required to have emergency remote shutoff valves.

FIGURE 10-5 A remote shutoff valve on an MC 331 highway cargo tanker.

Fixed Site Facilities

Remote shutoff valves are often difficult to locate at fixed site facilities (Figure 10-6). Facility preplans and routine training tours are essential to gain familiarity with fixed site facilities in your jurisdiction. Also discuss the use of these remote shutoff valves with the facility technical experts in case of emergency. The use of any fixed site facility remote shutoff valves should be covered in the appropriate facility emergency response procedures for your agency. Remote shutoff valves should be indicated on all facility maps.

It is important not to confuse remote shutoff valves with other valves that the facility may have. It can be extremely dangerous to turn on or off valves that are not designed to be emergency remote shutoff valves. Doing so may make chemical reactions unstable or cause over pressurization of pipes or vessels and lead to further and more severe hazardous materials releases.

DILUTION

When large amounts of water are available, dilution is a low-tech method to attempt to bring down the concentration of a hazardous material below a dangerous level. Dilution typically is not a practical solution due to the large amounts of water required to make a hazardous material nonhazardous. In addition, many hazardous materials cannot be rendered nonhazardous by dilution with water. Nevertheless, sometimes dilution is the answer. For example, alcohols, which are water soluble, may be diluted to the point at which they are no longer flammable.

FIGURE 10-6 A remote shutoff valve in a fixed site facility. Handwritten signs or minimal marking of the shutoff switches are commonplace.

FIGURE 10-7 Using absorbents at a hazardous materials incident involving flammable liquids.

Acids and bases may be diluted to the point that the pH is no longer considered corrosive. For example, as a rule of thumb, each change of 1 pH unit requires 10 times the amount of water. Therefore, to dilute 100 gallons of sulfuric acid with a pH of 2 to pH 7 (a change of 5 pH units), would require 1 million gallons of water. This volume was arrived at by multiplying 100 gallons by 10,000, which is $10 \times 10 \times 10 \times 10 \times 10$ (for each of the 5 pH unit changes). In both of these examples, large amounts of water would be needed, which is typically not practical.

ABSORPTION, ADSORPTION, AND SOLIDIFICATION

Absorption, adsorption, and solidification are three processes by which a liquid can be more easily handled and disposed of (Figure 10-7). Absorption refers to trapping a liquid in the pores of another material such as water in a sponge or a paper towel. Adsorption is the process of a liquid adhering to the surface of the adsorbent material such as when oil coats the surface of solid plastic beads. In the chemical process of solidification, the liquid being disposed of reacts with the solidifying agent to form a nonhazardous, solid product. For example, the hazardous liquid hydrofluoric acid may be treated with the sorbent calcium chloride to form a solid calcium fluoride product, which is nonhazardous. Each of these processes has its advantages and disadvantages.

Absorption and adsorption merely contain the hazardous material within the fibers or on the surface of the sorbent material. The disadvantage is that the hazardous material, if released, is still dangerous. The used sorbent material must be disposed of as hazardous waste.

Solidification, on the other hand, may render the hazardous material nonhazardous, possibly allowing for normal solid waste disposal. Always check with the manufacturer for proper disposal guidelines before disposing of any spent solidification matrix. If spent sorbent is not properly disposed of, such action may be a violation of RCRA and incur large civil and possibly criminal penalties. Specialists should always be consulted before disposing of hazardous waste or potentially hazardous waste. The procedure for applying absorbent, adsorbent, or solidification materials onto a chemical spill is outlined in Box 10-1.

DIKING

diking ■ The procedure of keeping a spilled hazardous material in a defined location using barriers such as sand, berms, or absorbent material.

Diking is the process of protecting sewer or other drainage openings from hazardous runoff. Diking materials can be purchased commercially and are very effective for

BOX 10.1 ABSORPTION, ADSORPTION, AND SOLIDIFICATION

1. Choose the appropriate material.
2. Test a small amount to ensure compatibility.
3. Apply material from the outside in.
4. Continue to apply absorbent toward the middle or source of the leak.
5. Rake absorbent back and forth to maximize absorption.
6. Shovel absorbent into appropriate waste container.

a wide range of hazardous materials. Some diking materials are designed for water-soluble materials, whereas others are designed for oily or water-insoluble materials. It is important to use the proper diking material on the appropriate hazardous material; otherwise, the diking material will be ineffective at stopping the hazardous material or may lead to an unexpected chemical reaction. Diking materials often have embedded sorbent materials.

RETENTION, DAMMING, AND DIVERSION

If a hazardous material is leaking out of its container, the simplest way to retain it is to use catch basins, tubs, buckets, pools, or other readily available containers (Figure 10-8). However, before choosing a container, ensure that the container material is compatible with the hazardous material you are attempting to retain. Otherwise, the hazardous material may eat through the container or, worse yet, may react violently with the container. If a flammable material is involved, static electricity is generated when the droplets of flammable liquid move through the air. The drier the air, the more static electricity is generated. Flammable materials should be retained in metal containers that are grounded and bonded whenever possible. When approaching the area of the leak, always be very careful not to contact a hazardous material directly.

Dams are used to retain liquid in a specific area on land, or to retain a heavier than water liquid at the bottom of a stream or river as shown in Figure 10-9. Materials used in dam construction are usually obtained locally, often at the hazardous materials incident itself. Very common **damming** materials are soil and sand. When the soil or sand is very porous, the dam may be lined with an impermeable material such as plastic sheeting or clay to increase its effectiveness. Dams work best on slow-moving waterways. Because the hydraulic energy that is carried by moving bodies of water can be enormous, always have a backup plan in place downstream.

Diversion involves digging temporary ditches and using pipes or other channels to redirect the flow of liquids. Diversion may be as simple as a few shovel loads of sand placed at a strategic location to redirect the flow of a hazardous material into a low-lying area that can be used as a retention pond. Or it may be as complex as digging a deep channel with a backhoe to redirect the flow of a large amount of hazardous materials into a retention pond.

BOOMING

On a body of water, floating booms are used to retain a lighter than water liquid in a specific area, as shown in Figure 10-10. **Booming** materials can be purchased commercially and are very effective for a wide range of hazardous materials. Because booms will be deployed in water, they are typically designed for water insoluble hazardous materials with a specific gravity of less than 1 (which means they float on water). By far the most commonly boomed hazardous materials are petroleum products such as crude oil, diesel fuel, and gasoline.

Depending on the affected body of water, booming operations can be logistically very difficult. A small pond may be effectively boomed using a few personnel from shore, whereas a large lake or river may require watercraft and a

damming ■ The process of restricting the flow of a spilled hazardous material by building a barrier to redirect or contain it.

diversion ■ The procedure of redirecting the flow of a spilled hazardous material away from a sensitive area.

booming ■ The process of using floating barriers to contain nonpolar, lighter-than-water hydrocarbons on bodies of water.

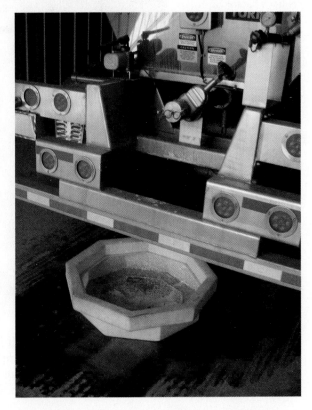

FIGURE 10-8 Catch basins are effective at minimizing the effect of slow leaks at hazmat incidents. Be aware that static electricity can build up as the liquid drips through the air and may eventually ignite flammable liquids if proper grounding and bonding procedures are not followed.

FIGURE 10-9 Schematic drawing of an underflow and an overflow dam that can be used to trap substances that are not water soluble in flowing bodies of water. An underflow dam is used to trap materials that are lighter than water, and an overflow dam is used to trap materials that are heavier than water.

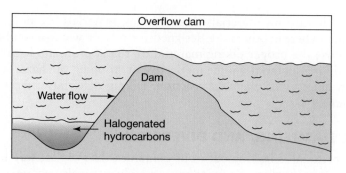

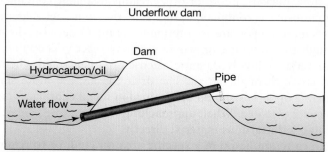

FIGURE 10-10 Floating booms can be used to control a hazardous materials release in a river.

large number of personnel. It is extremely important to avoid becoming exposed to the hazardous material while accomplishing booming operations. This means erecting booms well downstream of the hazardous materials release, especially on faster moving bodies of water. Be aware of the fact that booms do not work well in rough water or fast-moving rivers.

VAPOR SUPPRESSION

vapor suppression ■
The process of limiting the evaporation of a pooled liquid.

Vapor suppression is the technique of choice to limit the amount of vapors that enter the atmosphere after a hazardous materials release. Vapor suppression may take the form of covering the released material with either plastic sheeting or vapor-suppressing foam.

FIGURE 10-11 Plastic sheeting can be used to suppress vapors. In this case, plastic sheeting is being used to cover a mercury spill until cleanup contractors arrive to mitigate the spill.

Plastic sheeting, such as Visqueen, can be placed over a released hazardous material to reduce its evaporation rate (Figure 10-11). Keep in mind always to make sure the hazardous material is compatible with the plastic sheeting being used.

Firefighters often use foam to suppress vapors when volatile materials such as gasoline are spilled. Depending on the type of chemical that has been released, several different types of foam may be required. The most important chemical and physical property to consider when choosing foam is the spilled product's water solubility. Water insoluble materials, such as the common hydrocarbons gasoline and diesel fuel, will typically require the use of an aqueous film-forming foam (AFFF). Water-soluble materials, such as alcohols, will require alcohol resistant foam (ARF) and are usually used in higher concentrations because water-soluble materials tend to attack and degrade the foam much more rapidly than do water-insoluble materials.

Foam is applied using standard fire hoses with a **foam eductor** applying a consistent and measured amount of foam concentrate into the water stream using the Venturi effect (Figure 10-12). Firefighters are typically well versed in foam applications.

The foam layer will tend to degrade over time no matter which foam product is used or which hazardous material has been released. Before any foam application is started, be certain that you have enough foam on hand to maintain the foam blanket; otherwise, dangerous vapors will quickly start to build up and may find an ignition source and ignite unexpectedly, leading to an explosion. Many hydrocarbon fires and releases have turned tragic because the protective foam blanket could not be maintained.

Aqueous Film-Forming Foam (AFFF)

Aqueous film-forming foam (AFFF) is the most common foam used by the fire service today (Figure 10-13). AFFF is used to suppress the vapors generated by hydrocarbon liquids. Hydrocarbons are the most common types of flammable liquids transported in the United States and Canada. Fuels such as gasoline, diesel fuel, kerosene, and Jet A are all liquid hydrocarbons. Hydrocarbons are water insoluble and are the easiest types of liquid hazardous material to keep a foam blanket intact on. AFFF is typically used in either a 3% or a 6% formulation, and is designed for optimal flow properties. AFFF foam forms

foam eductor ■ A device used to mix foam concentrate and water in the correct proportion.

aqueous film-forming foam (AFFF) ■ A frothy liquid-air mixture prepared by mixing a concentrate with water; intended to extinguish fires involving water insoluble, flammable, nonpolar hydrocarbon liquids. Examples of nonpolar hydrocarbon liquids are gasoline, crude oil, kerosene, diesel fuel, and heating oil.

FIGURE 10-12 Foam is being used to suppress flammable liquid vapors.

BOX 10.2 RESPONSE TO HYDROCARBON TANKER FIRES

Gasoline tanker fires are an impressive sight. When 9000 gallons of volatile fuel burn, large amounts of smoke, fire, and heat are released. The first urge that most firefighters have is to rapidly extinguish the fire. This can be a more difficult proposition than it appears at first, because gasoline is not soluble in water and is lighter than water. Therefore, the burning gasoline tends to float on top of the water and flows along the path of least resistance, usually into the sewer system, storm drains, and basements. This tends to spread the fire, and cause property and environmental damage, if not worse.

Sometimes, it is better to let a burning gasoline tanker burn itself out. Although this creates air pollution, the resulting environmental damage is significantly less than when hydrocarbons get into the groundwater or other water sources. DOT 406 and MC 306 type fuel tankers usually are constructed of aluminum and burn down like candle wicks as the aluminum melts or burns away. Because these tankers are not pressurized vessels (like LPG tankers), the risk of a boiling liquid expanding vapor explosion (BLEVE) is minimal.

There are instances when the fire must be extinguished. When life safety or significant property damage are at stake, extinguishment has to be attempted. Examples of life safety include trapped occupants in vehicles after a crash, or occupants of nearby dwellings that cannot be safely evacuated, such as a hospital. An example of significant property damage includes infrastructure such as bridge overpasses that cost months of time and millions of dollars to replace. In this case, the fire can be extinguished using the appropriate foam. It is important to have a sufficient amount of foam on hand before extinguishment operations begin to completely extinguish the fire and maintain a foam blanket for an extended period of time over the volatile gasoline. Difficulties include accessing the burning gasoline when the fire has entered the tank after it has been torn open, extinguishing a moving fire, and maintaining a foam blanket.

It is also vitally important to select the appropriate foam. Gasoline is a nonpolar hydrocarbon and can be extinguished with standard aqueous film-forming foam (AFFF). Fuel mixtures containing ethanol, which is a polar alcohol, will require alcohol resistant foam (ARF). When AFFF foam is applied to fuels such as E85 (85% ethanol, 15% gasoline), the foam blanket is quickly degraded and the problem becomes worse. After the fire is extinguished, the remaining fuel must be transferred to another container as quickly as possible to prevent reignition. Transferring heated, volatile fuels also carries risks, so hazmat technicians or cleanup contractors will likely perform this function.

Carefully consider the cost–benefit analysis when attempting to extinguish gasoline tanker fires. It is sometimes a better idea to let it burn itself out.

a thin film layer on top of the hydrocarbon surface that spreads quickly and suppresses vapors. AFFF foam is most commonly used for hydrocarbon fuels, can be used with freshwater or seawater, and can be used without an aspirator (although the results will be suboptimal). Because a solid film is formed by the foam on the hydrocarbon surface, it is important to avoid disturbing it; for example, by walking through the foam blanket.

Alcohol Resistant Foam (ARF)

Alcohol resistant foam (ARF) is designed to be used on polar, water-soluble solvents such as alcohols. ARF is becoming much more important with the advent of E85 fuels—a mixture of 85% ethyl alcohol (ethanol) and 15% gasoline. Standard AFFF foam would be virtually useless fighting an E85 fuel fire. Alcohol resistant foams are made of a combination of synthetic detergents, fluorochemical surfactants, and polysaccharide polymers. Alcohol resistant foams can be used on both hydrocarbon fuels and polar fuels, making this foam a good choice due to its flexibility of use. ARF is typically used in either a 3% or 6% formulation, depending on the type of fuel. When ARF is used on polar solvents, the polysaccharide polymer forms a durable membrane that keeps the fuel and the foam separate, extending the lifetime of the foam blanket. Because a solid film is formed by the foam on the surface of the polar liquid, it is important to avoid disturbing it; for example, by walking through the foam blanket.

alcohol resistant foam (ARF) ▪ A frothy liquid-air mixture prepared by mixing a concentrate with water, intended to extinguish fires involving water-soluble, flammable polar hydrocarbon liquids. Examples of polar hydrocarbon liquids are E85, ethanol, methanol, and acetone.

Fluoroprotein Foam

Fluoroprotein foam has a fluorochemical surfactant additive mixed in with protein foam. This makes the foam fairly heat resistant, improves resistance to fuel pickup, and makes the foam dry chemical compatible. Fluoroprotein foam is most commonly used for hydrocarbon fuels, can be used with freshwater or seawater, and requires using an aspirator.

fluoroprotein foam ▪ A frothy liquid-air mixture consisting of a fluorinated protein polymer, intended to extinguish fires involving flammable liquids.

High Expansion Foam

High expansion foams have an expansion ratio above approximately 200:1. These foams are designed to be used primarily on class A fuels (common combustible materials such as paper, wood, and plastics) in small, well-defined spaces. High expansion foam is made from synthetic foaming agents; is designed to be used with a high expansion foam generator; and is often used in confined spaces, such as warehouses, mines, and ships.

high expansion foam ▪ A type of foam that expands at least 200:1 when mixed in a foam eductor.

Since high expansion foam is designed to be used on class A fuels, it is not particularly effective on flammable liquid fires (class B fuels).

VAPOR DISPERSION

vapor dispersion ■ The process of moving and diluting a gas or vapor.

When vapor suppression operations have failed or cannot be safely and effectively performed, **vapor dispersion** is used to reduce the airborne concentration of gases or vapors. There are several different methods of vapor dispersion: The simplest method is to use a fan or other mechanical device to create airflow that pushes the vapors out of the immediate area and dilutes them in the process. This process is also called ventilation.

When larger volumes of air need to be moved, fire department hose lines can be effective at dispersing vapors. This technique relies on the Venturi effect, which will pull air along with the moving water. It has long been used successfully during ventilation operations at structure fires and is called hydraulic ventilation in that application. Hydraulic vapor dispersion can be used for both water-soluble and water-insoluble gases or vapors. When this technique is used with water-insoluble gases and vapors, it removes them from the immediate area. When this technique is used with water-soluble gases and vapors, the chemical will dissolve in the water and the resulting runoff may be hazardous and should be contained (corrosive vapors are one example).

Any of these techniques may be used indoors or outdoors, but keep in mind that the smaller the volume of vapors and the smaller the space that needs to be ventilated, the more effective the operation will be. Generally these techniques, when used outdoors, are designed to protect exposed downwind populations that cannot be effectively evacuated or sheltered in place. When ventilating indoor gas or vapor releases to the outside, keep in mind what populations are downwind. It is a good idea to have downwind air monitoring in place before starting vapor dispersion or ventilation activities to avoid creating a bigger problem downwind.

SOLVED EXERCISE 10-5

What is the best option for stopping the acid leak described in the opening scenario?
Solution: Remote valve shutoff.

SOLVED EXERCISE 10-6

If a remote shutoff valve is not present, what other techniques can be used to stop the leak described in the opening scenario?
Solution: Pipe patch kit and a catch basin (pool or bucket depending on the rate of flow).

SOLVED EXERCISE 10-7

What are several options to deal with the spilled acid in the incident described in the opening scenario?
Solution: The acid could be absorbed (kitty litter, for example), neutralized (soda ash or sodium bicarbonate), or diluted with water. Ensure that any absorbent used is compatible with the acid released. Dilution is the least attractive option due to the large amounts of water that would be needed to raise the pH to an acceptable level. The local water treatment plant should be called to determine what an acceptable level is. The neutralized runoff should not be allowed to enter the environment without specific permission from environmental authorities such as the state Department of Environmental Quality or the U.S. EPA.

What is the best option to prevent the acid from reaching the drains in the opening scenario?
Solution: Drain covers and/or diking and damming.

How could acid vapors be suppressed in the incident described in the opening scenario?
Solution: Vapors could be suppressed using absorbents and/or covering the spill with plastic sheeting. Ensure that materials that come into contact with the acid are compatible.

Summary

Product control is a very important aspect of hazardous materials response. Although product control typically is considered an offensive, technician level response, there are many defensive techniques that the operations level responder can use. One of the most effective product control techniques is remote valve shutoff. Highway cargo tankers and fixed site facilities typically have remote emergency valve shutoff switches located at a distance from likely hazardous materials release points. You should become familiar with the location and operation of these switches. Other effective techniques for product control are absorption and solidification; these can be as simple as putting kitty litter on a spill or using absorbent pads. Diking, damming, diversion, and retention are more complicated product control techniques that can effectively protect exposures from a moving release. These techniques typically require larger amounts of materials and possibly heavy equipment, making planning an essential consideration. Gases and vapors must be dealt with in populated areas. Vapor suppression is the technique of choice to limit the evaporation, whereas vapor dispersion can be effectively used to reduce the hazard to the public after vapors have been generated or the gas has been released. Whenever you perform product control techniques, keep in mind that your safety must come first! Remember, at the operations level, you should avoid contact with the hazardous material you are trying to control.

Review Questions

1. What are the primary differences between offensive and defensive product control techniques?
2. Where are the remote shutoff valves located on DOT 406/MC 306, DOT 407/MC 307, and MC 331 highway cargo tankers?
3. How can you identify the location of emergency remote shutoff valves at fixed site facilities?
4. Why should diking be performed well in advance of a flowing hazardous material?
5. What are two common vapor suppression techniques?

Problem-Solving Activities

1. What product control techniques would you use to contain a small leak from a gasoline tanker (MC 306)?
2. What product control techniques would you use to contain a small leak from a cold storage warehouse containing anhydrous ammonia?
3. What product control techniques would you use to contain a small leak from a propane tanker (MC 331)?
4. What product control techniques would you use to confine 12 leaking 5-gallon (19-L) pails of muriatic acid in the middle of a busy road?
5. What product control techniques would you use to confine a chlorinated hydrocarbon spill leaking into a stream?

References and Further Reading

National Fire Protection Association. (2008). *NFPA 472, Standard for Competence of Responders to Hazardous Materials/Weapons of Mass Destruction Incidents*. Quincy, MA: Author.

Noll, Gregory G., Michael S. Hildebrand, and James Yvorra. (2005). *Hazardous Materials: Managing the Incident*. Chester, MD: Red Hat Publishing.

Occupational Safety and Health Administration. (1990). 29 CFR 1910.120, *Hazardous Waste Site Operations and Emergency Response (HAZWOPER)*. Washington, DC: U.S. Department of Labor.

Weber, Chris. (2007). *Pocket Reference for Hazardous Materials Response*. Upper Saddle River, NJ: Pearson/Brady.

11

Operations Level Responders Mission-Specific Competencies: Air Monitoring and Sampling

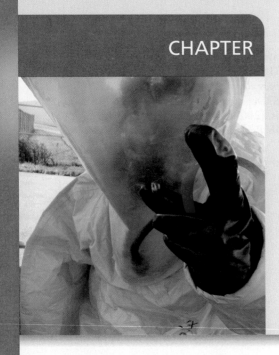

KEY TERMS

OBJECTIVES

- Describe five reasons air monitoring is important.
- State the appropriate order of air monitoring and the reasons for this order.
- Describe the advantages and disadvantages of Geiger counters.
- Describe the advantages and disadvantages of pH paper.
- Describe the advantages and disadvantages of combustible gas indicators.
- Describe the advantages and disadvantages of electrochemical sensors.
- Describe the advantages and disadvantages of photoionization detectors.
- Describe the advantages and disadvantages of colorimetric tubes.
- Perform effective air-monitoring operations.
- Describe the advantages and disadvantages of Raman spectroscopy.

- Describe the advantages and disadvantages of FTIR spectroscopy.
- Perform effective sample identification operations.

sampling ■ The process of gathering a small amount of material for analysis or as evidence.

sample identification ■ The process of determining the identity of an unknown material.

You Are on Duty! Train Derailment in the Center of Town

Graniteville, SC

Early on the morning of January 6, 2005, two trains collided in Graniteville, South Carolina, causing the rupture of a chlorine tank car that killed 9 people and injured 250. The chlorine cloud immediately killed several people near the derailment, including a number of people in a nearby textile plant, a residence, and a vehicle. Over 5000 residents were evacuated for 2 weeks while hazmat crews mitigated the incident. The accident occurred because a switch had been left in the wrong position.

- What concerns do you have before entering this area?
- What type of air monitoring must be conducted prior to entering this area for search and rescue?

Let's try to answer these questions in this chapter.

Air monitoring, **sampling**, and **sample identification** are among the most important activities at any hazardous materials incident (Figure 11-1). Any emergency response should start with a hazard and risk assessment. How can we assess the hazards and the consequent risks if we do not know what we are dealing with? We must use air-monitoring instrumentation to determine whether the hot zone boundaries have been chosen appropriately. Many hazardous materials are odorless above harmful concentrations. Then how can we be certain that the hot zone is large enough without using air-monitoring procedures? On top of that, before anybody enters the hot zone, a thorough scene assessment should be conducted using monitoring instrumentation. Hazardous materials releases are dynamic incidents affected by changing release rates and weather conditions.

Likewise, sampling and sample identification are critical for a thorough hazard and risk assessment of any unknown release. In this chapter, we cover various solid, liquid, and gas sampling techniques as well as the equipment that can

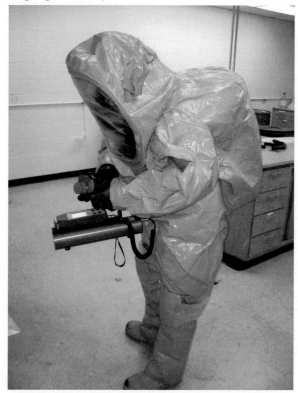

FIGURE 11-1 Air monitoring should be performed prior to entering the contaminated area, during movement in the hot zone, and continuously while working in the hot zone. Dangerous gases and vapors can build up at any time due to changing conditions, such as changes in the weather, temperature, and chemical reactivity, or from container failure.

be used to make an unambiguous identification of the unknown material. Public safety sampling techniques can be used to collect samples not only for on-scene identification but also to send to outside laboratories and other agencies for further processing. We will not cover evidentiary sample collection in this chapter. Evidentiary sampling, a process related to yet very different from public safety sampling, is covered in Chapter 9, Operations Level Responders Mission-Specific Competencies: Evidence Preservation and Sampling.

Preparation and Planning

HAZWOPER (29 CFR 1910.120) requires that the incident commander identify hazards at hazardous materials incidents. The cornerstone of identification is air-monitoring equipment. Depending on the type of response your agency chooses to engage in at hazardous materials incidents, some of the most important equipment to use on scene is air-monitoring instrumentation. Air monitoring is vital for the safety of response personnel and ensuring that we have chosen the proper response zones, as well as for the protection of downwind populations. It is critical to accurately and quickly identify the hazards that airborne gases and vapors pose to the responders on scene and to the general public.

The first step is to choose what air-monitoring equipment to own and operate. This choice depends on the type of mission-specific competencies your agency has chosen to perform. For example, operations level personnel choosing to perform product control functions should have, at a minimum, a multi-gas meter that can protect them from flammable atmospheres. Operations level personnel choosing to perform clandestine drug laboratory remediation should have not only a multi-gas meter at their disposal but also a photoionization detector (PID) and **colorimetric tubes** for low-level toxic material detection and specific gas detection, respectively. Air-monitoring equipment can be expensive to acquire and has continuing annual maintenance costs that can be substantial. However, air monitoring is an essential component of hazardous materials response and must be performed in order to respond safely and effectively.

Once the air-monitoring equipment is acquired, personnel should be trained on its use, and the equipment must be maintained according to the manufacturer's recommendations. Standard operating guidelines (SOGs) should be written and made available to ensure all personnel have the ability to safely and effectively operate and interpret the air-monitoring equipment. Regular hands-on training with the equipment is an essential part of training to competency, which OSHA requires under HAZWOPER.

Every piece of air-monitoring equipment and sample identification equipment has its advantages and disadvantages. No one technology or manufacturer can do everything. Every piece of equipment will generate false positives and false negatives. As the user of air-monitoring and sample identification instrumentation, you must learn to recognize the limitations of your detection equipment in the field. In this chapter, we discuss specific air-monitoring and sample identification technology. For each technology discussed, we cover the detection method, the limitations of the technology, and the specific advantages and disadvantages of the equipment currently in use.

Air-Monitoring Procedures

Air monitoring must be conducted in a systematic fashion to ensure accurate and reliable results. Training and equipment familiarity are essential for the safety of first responders and the public. Air-monitoring equipment must be maintained on a regular basis, must be prepared for use at the site of the hazardous materials release, must be used appropriately,

colorimetric tube ■ A type of air-monitoring device that uses a chemical reaction to detect airborne contaminants. A pump is used to move a known quantity of air through a glass tube containing the detection medium.

and its results must be interpreted correctly. If air-monitoring activities are not carried out properly, the results could be disastrous.

Standard operating guidelines (SOGs) that are consistent with the manufacturer's recommendations should be written for each specific piece of air-monitoring equipment. The SOGs should cover when and how routine maintenance such as calibration is performed, how pre-entry preparations such as bump checking and **field zeroing** are carried out, how air-monitoring measurements are performed, as well as how the results are interpreted and how **correction factors** are applied. Personnel should be trained often on the use of air-monitoring equipment according to the SOGs. Any deficiencies in the SOGs that are noted during training should be promptly corrected, and personnel should be retrained on the new and improved SOGs. Improper and inaccurate air-monitoring results can lead to unnecessary exposures and even death.

ORDER OF AIR MONITORING

There is a defined order in which to monitor for hazardous materials:

1. **Radiation** (from a distance, before approaching the suspected hot zone)
2. Corrosive atmospheres
3. Oxygen
4. Flammable atmospheres
5. Toxic atmospheres

Why do we need to air monitor in this particular order? The first hazard that should be investigated is radiation, specifically gamma and neutron radiation, because neutron and gamma radiation can travel hundreds of feet in air and not be detected by any other senses until it is too late. In this day and age of terrorism and radiological dispersal devices, it may be prudent to routinely monitor for gamma radiation at any incident upon arrival. The second hazard that should be investigated is corrosivity because it will damage or destroy most air-monitoring equipment. Corrosive gases can disable sensors upon exposure, yet this can be difficult or impossible to detect in a timely fashion while in PPE.

The third hazard that should be investigated is oxygen level because combustible gas indicators typically require oxygen in order to function properly. The fourth hazard that should be investigated is flammability, after it has been determined that an appropriate oxygen level is present. And finally, we should measure toxicity after it has been determined that dangerous levels of radiation are not present, the atmosphere is not corrosive to us and our instrumentation, and we are not entering an explosive atmosphere. If we do not cover all of these bases in a systematic fashion, it is easy to overlook a dangerous hazard and become exposed and injured. In practice, we monitor for multiple hazards simultaneously because integrated multi-gas detectors expose the sensors to the atmospheric sample at the same time. Remember to make sure that corrosive monitoring is performed ahead of monitoring activities involving electronic equipment, and that flammable atmosphere monitoring is performed ahead of monitoring that uses equipment that is not intrinsically safe.

We do not always have to check for all of these hazards at every hazardous materials incident. For example, if a leaking highway cargo tanker is placarded as acetone, a flammable liquid, it is probably unnecessary to monitor for radiation and corrosive materials. We can just start with oxygen and flammability. However, if there is any question whether radioactive materials or corrosive materials could be present, start air monitoring with that hazard using the appropriate air-monitoring equipment.

field zeroing ■ The procedure of resetting an air-monitoring instrument to read zero when sensing background levels of contaminant. This must be done carefully in a known clean environment. Also known as fresh air zeroing.

correction factor ■ The value that air-monitoring readings must be multiplied by in order to get the true reading for any particular chemical. Instruments that require the use of correction factors include combustible gas indicators and photoionization detectors. Also known as relative response factor.

radiation ■ (1) Energy in the form of electromagnetic waves or particles. (2) The transfer of heat between two objects not in contact.

Based upon the chemical properties of chlorine, why is the order of air monitoring important in the incident described in the opening scenario?

Chlorine		Formula: Cl_2	CAS#: 7782-50-5	RTECS#: FO2100000	IDLH: 10 ppm
Conversion: 1 ppm = 2.90 mg/m³		DOT: 1017 124			

Synonyms/Trade Names: Molecular chlorine

Exposure Limits: **NIOSH REL:** C 0.5 ppm (1.45 mg/m³) (15-minute) **OSHA PEL:** C 1 ppm (3 mg/m³)	Measurement Methods (see Table 1): **NIOSH** 6011 **OSHA** ID101, ID126SGX

Physical Description: Greennish-yellow gas with a pungent, irritating odor.
Note: Shipped as a liquefied compressed gas.

Chemical & Physical Properties: **MW:** 70.9 **BP:** 29°F **Sol:** 0.7% **Fl.P:** NA **IP:** 11.48 eV **RGasD:** 2.47 **VP:** 6.8 atm **FRZ:** -150°F **UEL:** NA **LEL:** NA Nonflammable Gas, but a strong oxidizer.	Personal Protection/Sanitation (see Table 2): **Skin:** Frostbite **Eyes:** Frostbite **Wash skin:** N.R. **Remove:** N.R. **Change:** N.R. **Provide:** Frostbite wash	Respirator Recommendations (see Tables 3 and 4): **NIOSH/OSHA** **6 ppm:** CcS*/Sa* **10 ppm:** Sa: Cf*/PaprS*/CcrFS/GmFS/ ScbaF/SaF **§:** ScbaF: Pd,Pp/SaF: Pd,Pp: AScba **Escape:** GmFS/ScbaE

Incompatibilities and Reactivities: Reacts explosively or forms explosive compounds with many common substances such as acetylene, ether, turpentine, ammonia, fuel gas, hydrogen & finely divided metals.

Exposure Routes, Symptoms, Target Organs (see Table 5): **ER:** Inh, Con **SY:** Burning of eyes, nose, mouth; lac, rhin; cough, choking, subs pain; nau, vomit; head, dizz; syncope; pulm edema; pneu; hypox; derm; liquid: frostbite **TO:** Eyes, skin, resp sys	First Aid (see Table 6): **Eye:** Frostbite **Skin:** Frostbite **Breath:** Resp support

Solution: Chlorine is strongly corrosive and will destroy unprotected electronic equipment and **electrochemical sensors**. The first piece of air-monitoring equipment should be a wet piece of **pH paper**. It could be taped to the end of the multi-gas meter probe to serve as an early warning device if the chlorine concentration is elevated.

electrochemical sensor ■ An air-monitoring device that senses the presence of a contaminant using a chemical reduction–oxidation reaction that is converted to electronic output.

pH paper ■ Litmus paper. A colorimetric indicator used to determine the corrosiveness of an aqueous liquid. Red indicates an acid; blue indicates a base.

DATA QUALITY OBJECTIVES

It is important to consider the **data quality objectives (DQOs)** when considering air-monitoring equipment and methods. Data quality objectives refer to the ultimate goal of the air-monitoring activities. Do we just want to detect the presence of the hazardous material in the atmosphere? Or do we need to identify the hazardous material? Or do we also want to quantify the amount of hazardous material that is present? Or do we need to do a combination of all three of these? Some air-monitoring equipment can identify the type of hazardous material present; certain equipment can detect hazardous materials at low concentrations, yet other air-monitoring equipment can only identify the presence at higher concentrations. Thus, the selection of air-monitoring equipment and methods depends on the type of data needed.

For example, if the incident commander wants to downgrade the level of PPE to an air-purifying respirator (APR), he or she must know the identity of the contaminant and its concentration. Thus, air-monitoring equipment must be chosen that can

What are the data quality objectives (DQOs) in the incident described in the opening scenario?

Solution: There are several different data quality objectives (DQOs). The corrosive monitoring DQOs are to determine whether the atmospheric pH is high or low (qualitative). The chlorine level monitoring will likely be more quantitative in nature: The chlorine levels in ppm may be compared to the PEL (1 ppm) or the IDLH (10 ppm) to determine which level of protection is appropriate or to determine downwind evacuation distances.

accomplish both of these goals. We need to not only identify the contaminant but also measure its concentration at the appropriate level. Determining which air-monitoring equipment can detect the contaminant at the appropriate level can be challenging. Let's take a look at how to achieve data quality objectives with the available air-monitoring equipment.

CALIBRATION

Calibration is the process of resetting electronic air-monitoring and sample identification equipment to read a known value. The calibration process begins with a sample gas at an accurately known concentration. The air-monitoring equipment is exposed to this known concentration of gas, and we reset the instrument to this known value. For example, during the calibration of a **combustible gas indicator (CGI)**, we may have an analytical grade sample of methane at 2.5% concentration. This is 50% of the lower explosive limit (LEL) of methane. If the instrument is reading 43%, we reprogram the electronics to read 50%. Calibration is typically performed on a set schedule per the manufacturer's recommendation during routine maintenance. However, if the instrument is found to be out of calibration at a hazardous materials incident, it must be field calibrated at the emergency scene before use.

Calibration must be performed by trained personnel using high-quality calibration gases. As you can see from the preceding example, we can very easily cause an air-monitoring instrument to read incorrectly if we have old or out-of-specification calibration gas or inappropriately trained personnel performing the calibration. For example, if we mistype or misread the calibration gas that we have and reprogram the instrument to 25% rather than 50% of the LEL, the combustible gas indicator will read dramatically lower than the actual atmospheric concentration. As you can see, calibration must be performed with the utmost care!

BUMP CHECKING AND FIELD ZEROING

Air-monitoring and sample identification equipment must be prepared at the scene of a hazardous material incident before use. Two common procedures that are performed are bump checking and zeroing (or normalizing) of the equipment.

A **bump check** is a field test to verify that the instrument works properly. Either a qualitative or a quantitative bump check can be performed. A qualitative bump check exposes instrumentation to an unknown concentration of a material that the instrument is known to detect. We are simply looking for a response. This way we know that the instrument is registering change, but we do not know how accurate it is. In order to determine whether it functions accurately, we must perform a quantitative bump check, also called a bump test. This is typically done by passing the instrument's calibration gas over the sensor and comparing the instrument reading to the known concentration. It is imperative to perform a bump check before every use for your own safety. Otherwise, how do you know whether the instrument being carried into the hot zone even works?

data quality objective (DQO) ■ The degree of accuracy needed from a measurement. It may be quantitative, such as determining whether a toxic material concentration is above or below the IDLH for a PPE level decision; or qualitative, such as discerning the presence or absence of a flammable solvent to determine whether there is a container leak.

calibration ■ The procedure of resetting an air-monitoring instrument to read correctly using a known concentration of sample. The known sample is referred to as calibration gas, or cal gas, and is manufactured to rigorous technical standards.

combustible gas indicator (CGI) ■ A type of air-monitoring equipment designed to detect flammable atmospheres and read in percent of the lower explosive limit (% LEL). The CGI has a non-catalytic element and a catalytic element (which allows flammable gases to burn below their LEL). The CGI is one type of sensor typically included in multi-gas detectors.

HAZMAT HANDLE

Perform a functional check (bump check) before entering the hot zone every time you use an air-monitoring instrument!

Many air-monitoring instruments must be set to the clean, ambient **background reading** due to sensor drift. This simple procedure is variously called *field zeroing, fresh air calibration,* or *normalization.* Oxygen sensors, combustible gas indicators, carbon monoxide sensors, hydrogen sulfide sensors, and photoionization detectors (PID) all require field zeroing before use. Zeroing involves setting the sensor readings to the known ambient conditions, usually 20.9% oxygen, 0% LEL, and 0 ppm on the PID. When zeroing air-monitoring instruments, be sure you are in a clean environment, for example, not next to the exhaust pipe of your vehicle! If you field zero a multi-gas meter next to a vehicle's exhaust (which contains carbon monoxide), you will set a possibly dangerous carbon monoxide level as 0 ppm and consequently vastly underreport carbon monoxide that may be in the residence being investigated, with dire results.

REACTION TIME AND RECOVERY TIME

Air-monitoring instrumentation readings are not instantaneous. There are two primary reasons for this: First, it takes time for the gas sample to travel through the instrument, called the sample train, to the sensor. The longer the sample train, the longer it will take the sensor to be exposed to the sample and register a reading. This is extremely common during confined space operations when a 20-foot (6 m) or longer hose may be attached to the monitor. It is a good idea to factor in 1 to 2 seconds per foot of hose that is attached to the instrument. The second reason is sensor response time. Some sensors, such as photoionization detectors, are very rapid and react almost instantaneously. Other sensors, such as electrochemical sensors (for example, oxygen sensors), can take up to 20 to 40 seconds to register. These types of sensors operate using chemical reactions, which take time to complete and register electronically.

It is important to know the **reaction time** of the air-monitoring instrumentation being used because you will have to wait until the instrument has had time to respond before moving on. The T_{90} value, which indicates the time it takes to reach 90% of the full response, can be found in the instrument's manual. For example, if your instrument has a 45-second response time, and it takes 30 seconds to cross a 50-foot (15 m) room, you may be in a dangerous atmosphere before your instrument can even alert you to that fact. Due to limited response times, you must slowly and methodically carry out air-monitoring activities and not proceed too hastily.

Recovery time refers to the time it takes the sensor to be able to react to more contaminant after it has previously alarmed. Three distinct components affect recovery time: the sensor recovery time, the pump flow rate, and the sample train desorption process. All sensors use some sort of detection process that takes time. Certain processes, such as ionization, are almost instantaneous and therefore have a very short sensor recovery time. Other processes, such as electrochemical detection or surface acoustic wave detection, involve a chemical reaction or a binding process that takes seconds to minutes to complete, register electronically, and

HAZMAT HANDLE

Take your time when performing an air-monitoring survey!

revert back to baseline. Some instruments have adjustable pump speeds. The more fresh air that passes through the sample train and over the sensors, the faster the recovery time will be.

Chemicals tend to deposit, or adsorb, to the surface of various components of the sample train. The amount of chemical deposition will vary due to the chemical and physical properties of the chemical contaminant, the chemical concentration, the chemical and physical properties of the sample train components, and atmospheric temperature in relation to instrument temperature. For example, when warm atmospheric gases are pulled into a cold air-monitoring instrument, chemicals with a low condensation point will condense on the walls of the sample train. Under these circumstances, recovery time will be prolonged as the condensed liquid evaporates. Some sample train construction materials can preferentially bind certain chemicals or may be porous. Under these circumstances, recovery times may be extended as desorption processes can be extremely slow. Some detectors, such as photo-ionization detectors (PID), have very short recovery times. Other detectors, such as oxygen sensors and **surface acoustic wave (SAW) detectors**, tend to have much longer recovery times. Recovery times can be as short as milliseconds or as long as several minutes to hours.

surface acoustic wave (SAW) detector ■ A detection device that senses chemicals through a change in acoustic properties of a piezoelectric crystal after the contaminant reversibly binds to a selective matrix attached to the crystal.

CORRECTION FACTORS

As discussed earlier, air-monitoring instruments need to be calibrated in order to ensure their accuracy. Each instrument is typically calibrated to a specific gas. This does not pose problems for sensors that are specific to one gas, such as oxygen sensors. However, nonspecific sensors such as combustible gas indicators and PIDs respond differently to different chemicals. The sensors may respond more vigorously to one gas than another. If a sensor responds more vigorously to the gas being investigated at the hazardous materials incident than it does to the calibration gas, it will overreport the concentration.

On the other hand, if the air monitor responds less vigorously, it will underreport the concentration. If our instrument underreports, we might be lulled into a false sense of security and overexpose our response personnel or the public. Conversely, if it over-reports, we may take costly or dangerous actions such as unnecessary evacuations and the use of excessive PPE. Correction factors are designed to account for different instrument response efficiencies to different chemicals. The manufacturer will supply a table of correction factors (Table 11-1). Typically the instrument reading is multiplied by the correction factor to obtain the actual concentration. For example, if a chemical has a correction factor of 2.0 and the instrument reading is 100 ppm, then the actual ambient concentration of the chemical is 2 times 100 ppm, which equals 200 ppm.

ACTION LEVELS

Once we get an air-monitoring reading, what does it mean? The air-monitoring reading needs to be compared to an **action level** that will guide the hazardous materials response. Action levels are essential to air monitoring. For example, if you get a reading of 5 ppm on a photoionization detector, what does it mean? Well, that depends on how toxic the chemical is. An appropriate action level for toxic materials may be the permissible exposure limit (PEL) or the immediately dangerous to life or health (IDLH) level. The PEL may be the level at which the building may be reoccupied. If a reading is below the IDLH yet above the PEL, we may be able to use level C PPE. These are examples of how action levels guide our actions based upon real-time air-monitoring results. Table 11-2 shows common action levels for the air-monitoring equipment that may be used at hazardous materials incidents.

action level ■ The value an air-monitoring instrument reading is compared to in order to make a decision at a hazardous materials incident. An example of an action level is 19.5% oxygen, the IDLH level for oxygen, and the result of reaching this action level may be to leave the area immediately and don supplied air respiratory protection.

Air-Monitoring Equipment

Fortunately, there is a wide range of air-monitoring equipment available that has the ability to detect most common chemicals at useful concentrations. Let's explore the various types of air-monitoring instrumentation that are available in the appropriate air-monitoring order. Pay special attention to the limitations of each type of detection equipment!

TABLE 11-1	Correction Factors for a Combustible Gas Indicator (CGI) Calibrated to Methane and Various Photoionization Detectors (PIDs)			
CHEMICAL	**AREARAE CGI**	**9.8 eV RAE PID**	**10.6 eV RAE PID**	**11.7 eV RAE PID**
Acetic acid	3.4	NR	22.0	2.6
Acetone	2.2	1.2	1.1	1.4
Ammonia	0.8	NR	9.7	5.7
Benzene	2.2	0.55	0.53	0.6
Chlorine	NR	NR	NR	1.0
Diesel fuel #2	NR	1.3	0.7	0.4
Ethanol	1.7	NR	10.0	8.0
Ethyl ether	2.3	–	1.1	–
Formaldehyde	–	NR	NR	1.6
Gasoline #2	2.1	1.3	1.0	0.5
Hydrogen	1.1	NR	NR	NR
Hydrogen sulfide	interferes	NR	3.3	1.5
Iodine	NR	0.1	0.1	0.1
Methane	1.0 (cal gas)	NR	NR	NR
Methyl mercaptan	1.6	0.65	0.54	0.66
Mustard (HD)	–	–	0.6	–
Nitromethane	2.1	NR	NR	4.0
Perchloroethene	NR	0.69	0.57	0.31
Phosphine	0.3	28	3.9	1.1
Propane	1.6	NR	NR	1.8
Vinyl chloride	1.8	NR	2.0	0.6

NR = no response
Data courtesy of Rae Systems Inc.

RADIATION DETECTION

Most radiation is completely harmless, which is fortunate because radiation is all around us (Figure 4-13). Types of radiation include visible light, radio waves, microwaves, sunlight, and many others. The radiation we are talking about in the hazardous materials response context is ionizing radiation, which is extremely dangerous. Ionizing radiation cannot be seen, heard, or smelled; is capable of knocking electrons out of molecules; and causes damage by making these molecules much more reactive. In humans, this causes cellular damage, including cell death, skin damage, organ failure, and cancer.

Ionizing radiation includes alpha, beta, gamma, X-ray, and neutron radiation. (Radiation is discussed in greater detail in Chapter 4, Operations Level Responders Core Competencies: Understanding Hazardous Materials.) Each of these types of radiation is different, and a number of different detectors are specific to one or more of these types of radiation. It is critical to know what types of radiation a particular radiation detector will alarm to.

TABLE 11-2	Action Levels		
MONITORING ORDER	**DETECTION EQUIPMENT**	**LEVEL OF CONCERN**	**RATIONALE**
1. Radiation	Geiger counter Ionization detector (high level) Scintillation counter	> 1 mR/hr above background	Gamma radiation will travel hundreds of feet through air!
2. Corrosives	pH paper	Varies: < 4 or > 11 as a rule of thumb	Prolonged exposure to corrosives will destroy most of the electronic detectors.
3. Oxygen	Oxygen sensor	< 19.5% or > 23.5%	We need to know that oxygen is present at sufficient levels (> 10% as a rule of thumb) for the CGI to work.
4. Flammables	Combustible gas indicator (CGI)	> 10% of LEL or > 25% of LEL	Detect explosive atmospheres.
5. Toxic	Photoionization detector (PID) Colorimetric tubes Ion mobility spectroscopy Surface acoustic wave (SAW) detector	PEL or IDLH	Identify and determine the level of toxic materials and CBRNE agents.

Radiation detectors report their results in one of two units: **field strength** or counts per minute (cpm). Field strength, typically reported in millirem per hour (mR/hr) or in millisieverts per hour (mSv/hr), refers to a dose of radiation received. Field strength readings, with radiation detectors that are properly calibrated, should be identical from instrument to instrument. Counts per minute readings, on the other hand, vary significantly based upon the detector configuration. Counts per minute, as the name implies, is literally the number of ionizing events that a given radiation detector records. Counts per minute readings will therefore vary by detector type, configuration, and surface area (size). Field strength measurements are typically used to make hazard-risk-based assessments of the incident, such as determining exclusion zones, and establishing the length of hot zone entry times and evacuation distances. Counts per minute measurements are typically used for contamination surveys and verification of decontamination, and instrument readings must be compared to the normal background radiation level for a particular instrument in a specific area.

field strength ▪ A measure of the amount of gamma radiation at a particular location.

Follow these guidelines when using radiation detection equipment:

1. Inspect the instrument. Look for visible damage to the case, look at the detector and make sure the Mylar window is intact, and ensure that batteries are inserted.
2. Check the last calibration date of the instrument; if it is not within the manufacturer's recommended time period, use another instrument.
3. Bump check the instrument with a check source. The check source should come with the instrument. Ensure the reading is approximately what is expected (based upon the check source strength).
4. Obtain a background reading. Determine the normal background reading in the area. The normal background reading will depend upon what region of the country you are in, what altitude you are at, and what type of building you are in or near. Normal background readings vary from 5 to 50 µR/hr (microrem per hour) in the

United States. If there is a fast/slow switch, put it in slow mode to average out the natural fluctuations. Note the background reading on the air-monitoring log sheet.

5. Perform the measurement using the appropriate personal protective equipment (PPE) and record your readings.

Geiger-Mueller Tubes

Geiger-Mueller tubes are the detection technology used in Geiger counters and ionization detectors (Figure 11-2, middle). Several different manufacturers make radiation detection equipment. A ready source of radiation detection equipment is the old civil defense ionization detectors, Geiger counters, and dosimeters that your local emergency management agency probably still has (the CDV sets). Although the ionization detector is not sensitive enough to be of use in the typical hazardous material incident, the Geiger counter is ideal for radiation survey missions.

Geiger counters are probably the most useful broad-spectrum radiation detector for first responders at the operations level. Geiger counters can be configured to detect alpha, beta, and gamma radiation at higher sensitivities. They report counts per minute or field strength depending on which probe has been attached.

Ionization detectors are typically designed to detect gamma radiation at higher levels and only report field strength values. Field strength depends on the energy level of the ionizing radiation. Alpha, beta, gamma, and neutron radiation all have different energy levels. Within any individual radiation class (such as alpha), different radionuclides, or radioactive materials that emit radiation, also have different energy levels. This makes it very difficult to measure field strength across multiple radiation classes. Because gamma radiation is among the most penetrating ionization radiation, detectors are typically configured to measure gamma radiation field strength.

Scintillation Counters

Scintillation counters are designed to measure specific types of radiation and/or specific radiation energies. Scintillation counters can be configured for specificity or for sensitivity.

Geiger counter ■ Radiation detection equipment that can detect alpha, beta, and gamma radiation when properly configured.

ionization detector ■ Radiation detection equipment that can detect gamma and neutron radiation and measure field strength.

scintillation counter ■ Equipment that can be used to detect radiation. Scintillation counters can be very sensitive and radioisotope specific.

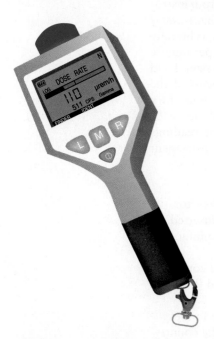
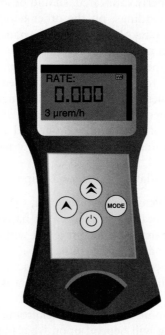

FIGURE 11-2 Common radiation detection equipment includes a general-purpose Geiger counter, which can detect alpha, beta, and gamma radiation depending on the configuration (middle); an isotope identifier, which can help identify the source (left); and a dosimeter, which is used to protect responders from unsafe levels of radiation through early detection (right).

Some common scintillating materials (phosphors) used in scintillation counters are zinc sulfide, cesium iodide, and sodium iodide. Scintillation counters are becoming much more commonly used by first responders. Your local hazardous materials team may have a radioisotope identifier that uses scintillation counter technology (Figure 11-2, left).

Dosimeters

Dosimeters are used to monitor the amount of radiation individuals receive. At hazardous materials incidents involving radiation, it is important to keep track of how much radiation each individual receives in order to compare that to the action levels recommended by OSHA, EPA, and the Department of Energy (DOE):

dosimeter ■ An instrument used to detect and measure individual radiation exposure.

Maximum annual dose for industry workers	5 rem
Max. dose for emergency responders engaged in nonlife safety	10 rem
Max. dose for emergency responders engaged in life safety	25 rem

Dosimeters can be as simple as a thermoluminescent detector (TLD) or a film badge in which cumulative radiation dose is converted to a chemical signal that can be analyzed in a laboratory later. Because the major disadvantage of these traditional dosimeters is the absence of immediate feedback, they are typically used in industries where low-level radiation exposures are relatively common and can be adequately anticipated. This is, of course, typically not the case at hazardous materials incidents. Therefore, pocket dosimeters are much more commonly used by first responders (Figure 11-2, right). Pocket dosimeters are more versatile than traditional dosimeters such as TLDs or film badges. They can measure not only cumulative dose but also field strength and/or counts per minute, and provide immediate feedback of real-time radiation levels. This information increases first responder safety tremendously.

CORROSIVE MATERIALS DETECTION

Corrosive materials include acids and bases. Table 11-3 lists some common acids and bases that you may encounter at hazardous materials incidents. Corrosive materials damage human tissue or corrode steel at a certain rate. It is important to identify the presence of a corrosive atmosphere early to prevent personal injury and damage to the air-monitoring equipment and other electronic equipment that may be carried such as radios.

The concept of pH is based upon water falling apart and re-forming. Water is formed from a proton (also called a hydrogen ion) and a hydroxide ion:

$$H_2O \leftrightarrow H^+ + {}^-OH$$

TABLE 11-3	Common Acids and Bases
ACIDS	**BASES (ALSO KNOWN AS ALKALIS OR CAUSTICS)**
Acetic acid (also flammable)	Ammonia (also flammable)
Chromic acid (also an oxidizer)	Ammonium hydroxide
Hydrogen bromide	Bleach (also an oxidizer)
Hydrogen chloride	Calcium hydroxide
Hydrogen fluoride (also highly toxic)	Lye
Hydrogen iodide	Potassium hydroxide
Nitric acid (also an oxidizer)	Potassium hypochlorite (also an oxidizer)
Phosphoric acid	Sodium hydroxide
Sulfuric acid (also an oxidizer)	Sodium hypochlorite (also an oxidizer)

This means that pH has a meaning only in aqueous solutions, or solutions that contain water. Therefore, determining the pH of anhydrous solutions such as pure (neat) solvents will give erroneous readings. Typically, these solutions will pH in the 4 to 6 range, giving the false impression of being acidic (which they generally are not).

The pH scale is measured from 0 to 14 (Figure 4-12). A pH of 7 is considered neutral, a pH below 7 is considered acidic, and a pH above 7 is considered basic. The extremes of the pH scale (0 or 14) are also the extremes of corrosiveness. The pH scale is logarithmic, which means that every change of 1 pH unit means a tenfold change in corrosiveness. For example, a solution with a pH of 4 is 1000 times as acidic as a solution with a pH of 7 because there is a difference of 3 pH units ($10 \times 10 \times 10$ equals 1000). This means that a pH change from 7 to 6 is not nearly as dangerous as a pH change from 5 to 4. Although each pH unit change of 1 is a tenfold change in corrosivity (or hydrogen ion concentration), the 5 to 4 pH change is 2 pH units lower than the 7 to 6 pH change, so it is a hundredfold greater change in hydrogen ion concentration. Practically, this means that the solution is much more acidic and dangerous, as well as significantly more difficult to neutralize.

pH paper, also known as litmus paper, is used to detect corrosive materials (Figure 11-3). When using pH paper for air-monitoring purposes, it should be wetted prior to use. Wetting the pH paper increases the sensitivity of the response, speeds up the response time of the paper, and makes the color change more permanent. This is because virtually all acids and bases are water soluble. For example, if anhydrous ammonia is released in a low-humidity atmosphere, dry pH paper will turn blue only temporarily. If you are not looking at the pH paper as you are passing through the anhydrous ammonia cloud, you may miss the response completely! In contrast, if the pH paper is wetted, anhydrous ammonia will react with the water, forming ammonium hydroxide, and turn the pH paper blue for a much longer period of time.

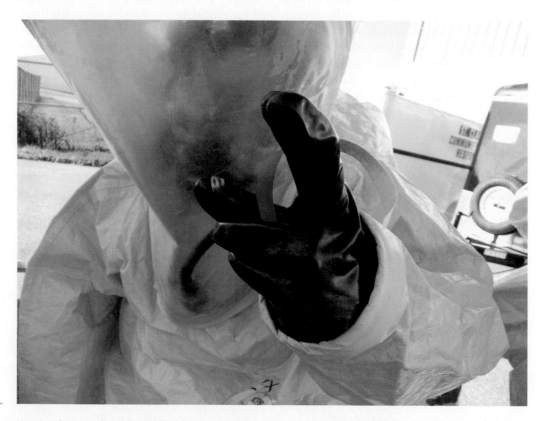

FIGURE 11-3 Using pH detection paper for air monitoring. The pH paper should be wetted before use to increase its sensitivity and to prolong the color change in dry environments. The paper turned red, indicating an acid is present.

HAZMAT HANDLE

For air-monitoring applications, wet half of the pH paper with tap water before use! This increases the sensitivity and longevity of the reading.

OXYGEN (O₂) SENSORS

Oxygen sensors operate using *electrochemistry*, which refers to the process of converting a chemical signal into an electrical signal. You are very familiar with some common electrochemistry; it is called a battery. Batteries convert a chemical reaction into electricity. Similarly, oxygen sensors convert a chemical reaction between oxygen and a metal into electricity. The oxygen oxidizes the metal, which produces ions that create an electrical current in the sensor. By measuring how much electricity is generated, how much oxygen is present can be determined. Oxygen detectors consist of two electrodes, an electrolyte solution, a battery, and electronics that can quantify the amount of signal generated by the oxidation reaction caused by the ambient atmospheric oxygen.

Electrochemical sensors such as oxygen sensors have their advantages and limitations, just like any other detection device. Advantages of electrochemical sensors include their reliability, specificity, relatively wide operational temperature range (roughly 20°F to 120°F or 5°C to 50°C), and comparatively low cost. Some limitations of electrochemical sensors include sensor readings that are affected by temperature, altitude effects due to barometric pressure, a relatively short life span (usually 6 months to 2 years), sensitivity to corrosive materials, and cross-reactivity to chemically similar compounds. For example, oxygen sensors have a marked altitude effect; are sensitive to acids because their electrolyte solution is typically alkaline; and are cross-reactive to many oxidizers such as ozone, chlorine, and other halogen gases.

A drop in oxygen level gives much more information than you may think at first. What has caused the drop in oxygen level? There are two possibilities: The first possibility is that the oxygen has been used up in a chemical reaction, such as rusting inside a closed metal tank. The second possibility is that another gas has displaced the air. In this case, we can determine the concentration of the oxygen displacing gas from the drop in oxygen concentration. First, we need to recognize that oxygen composes roughly 20% of the atmosphere and nitrogen composes roughly 80%. Thus, oxygen is roughly 1/5 of the displaced air. This means if our oxygen levels drop by 1%, there will be actually a 5% concentration of displacing gas ($5 \times 1\%$). This is because the foreign gas not only displaces the oxygen but also the nitrogen that composes 4/5 of the air.

Follow these guidelines when using oxygen sensors:

1. Inspect the instrument. Look for visible damage to the case, look at the sample train (the path the air takes inside the instrument), and ensure that batteries are inserted.
2. Check the last calibration date of the instrument; if it is not within the manufacturer's recommended time period, perform a field calibration:
   i. Enter the calibration menu of the instrument.
   ii. Ensure the calibration gas values in the instrument correspond to the available calibration gas (cal gas).
   iii. Attach the cal gas to the instrument, and perform the calibration per the manufacturer's directions.
3. Fresh air zero the instrument. Follow the manufacturer's directions to normalize the instrument to the ambient air and temperature. Make sure you are in a clean environment!

How would the chlorine from the opening scenario likely affect the oxygen sensor?

Solution: The oxygen sensor would most likely be severely damaged or destroyed (depending on the chlorine concentration and length of time it was deployed). Many oxygen sensors use an alkaline electrolyte solution, which is neutralized by the acidic chlorine gas.

4. Bump check the instrument. Apply cal gas to the instrument for a functionality check. Ensure the instrument reads close to the cal gas specifications. Close depends on your SOGs and the data quality objectives, but 10% of the value you are looking for is a good rule of thumb for reconnaissance and basic detection. A smaller percentage may be necessary for exposure level decisions and personal protective equipment (PPE) level decisions.
5. Perform the measurement using the appropriate PPE and record your results.

COMBUSTIBLE GAS INDICATORS (CGIs)

Combustible gas indicators (CGIs) are designed to detect flammable and combustible gases and vapors below the lower explosive limit (LEL) (Figure 11-4). If CGIs could detect only the level of flammable and combustible gases and vapors at or above the LEL, it would literally be too little too late.

CGIs detect the presence of flammable gases using a catalytic sensor that can burn the gases or vapors below their LEL. The chemistry involved is very similar to the catalytic converter in a car. Automotive catalytic converters are designed to burn off common by-products of incomplete combustion that would otherwise become components of smog. Combustible gas indicators have two elements: the catalytic sensor coated with platinum or palladium that allows the flammable gas to burn below its LEL and the non-catalytic reference sensor. The reference sensor is incapable of burning the flammable gas at concentrations below the LEL and does not increase in temperature when the concentration of flammable vapors is below the LEL. Therefore, the catalytic sensor increases in temperature while the reference sensor does not. The increased temperature leads to a resistance increase in only the catalytic sensor. The difference in resistance is compared and quantified by an electronic circuit called a Wheatstone bridge.

CGIs report their values as a percentage of a percentage. The LEL is a percentage: the percentage of a flammable vapor or gas in air. For example, the LEL of methane is 5%. Methane is rich enough to burn when it composes 5% of ambient air. Your CGI reports a percentage of the LEL (% LEL), or a percentage of a percentage. For example, a 1% reading on a CGI calibrated for and detecting methane (which has an LEL of 5%) is actually detecting 0.05% by volume of methane in air ($0.01 \times 5\%$). The conversion of 1% to a decimal is 0.01 (1 divided by 100). Figure 11-5 graphically illustrates this concept.

FIGURE 11-4 Internal view of the sensing components of a combustible gas indicator. The catalytic and the non-catalytic cells that form the basis of the Wheatstone bridge circuit are visible. The flame arrester is visible to the right in front of the sensor, and the permeable membrane is visible in front of the sensor to the left.

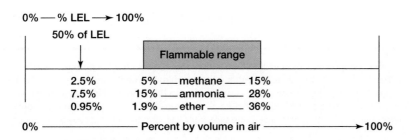

The sensitivity of combustible gas indicators varies from flammable gas to flammable gas, based primarily on the size of the gas molecule, as well as the configuration of the CGI. Typically, a diffusion barrier limits gas flow to the catalytic sensor, which means smaller molecules get to the sensor in greater numbers. Larger hydrocarbons are therefore detected less readily because it is more difficult for them to reach the catalytic sensor. This means CGIs calibrated to methane tend to underreport the concentration of most gases (which are typically larger molecules).

CGIs have advantages and limitations just like any other detection device. Their advantages include reliability and low cost, and their limitations include the need for oxygen. Depending upon the gas or vapor being measured, CGIs require at a minimum between 10% and 16% oxygen to function adequately. However, at these oxygen levels, CGIs tend to underreport the level of flammable gas. Below 10% oxygen levels, CGIs should be considered unreliable.

The catalytic sensor in CGIs can be poisoned by various chemicals including compounds containing heavy metals such as tetraethyl lead, and sulfides such as hydrogen sulfide and mercaptans. Other materials, such as silicones and halogenated hydrocarbons, can coat or otherwise affect the sensor and cause it to underreport the level of flammable gases. Sometimes these compounds can be burned off by operating the CGI for an extended period of time. Corrosive gases can corrode sensor components as well.

Some CGIs can also be operated in a supersensitive mode. In supersensitive mode the CGI can detect down to approximately 20 ppm of flammable gases, reporting the value in parts per million (ppm). Supersensitive CGIs often use a solid-state sensor in supersensitive mode, but this is not necessary. For example, a 0.1% reading on a CGI calibrated for and detecting methane (which has an LEL of 5%) is already detecting 50 ppm methane ($0.001 \times 5 \times 10,000$).

Follow these guidelines when using combustible gas indicators:

1. Inspect the instrument. Look for visible damage to the case, look at the sample train (the path the air takes inside the instrument), and ensure that batteries are inserted.
2. Check the last calibration date of the instrument; if it is not within the manufacturer's recommended time period, perform a field calibration:
   i. Enter the calibration menu of the instrument.
   ii. Ensure that the calibration gas values in the instrument correspond to the available calibration gas (cal gas).
   iii. Attach the cal gas to the instrument, and perform the calibration according to the manufacturer's directions.
3. Fresh air zero the instrument. Follow the manufacturer's directions to normalize the instrument to the ambient air and temperature. Make sure you are in a clean environment!
4. Bump check the instrument. Apply cal gas to the instrument for a functionality check. Ensure the instrument reads close to the cal gas specifications. Close depends on your SOGs and the data quality objectives, but 10% of the target value is a good rule of thumb for reconnaissance and basic detection. A smaller percentage

Would a combustible gas indicator be useful at the incident described in the opening scenario?

Solution: Yes, to check for the release of other materials that may be flammable (very likely in a train derailment). Two key points must be kept in mind when using a CGI in this incident: First, high chlorine concentrations will likely damage the CGI, even to the point of failure. Second, chlorine is an oxidizer and will contribute to combustion. Therefore, flammable materials may ignite below the LEL due to the oxidizing potential of the chlorine.

photoionization detector (PID) ■ A highly sensitive volatile organic compound detector. The detection capability of the PID depends on the ionization potential (IP) of the chemical and the bulb strength of the detector.

FIGURE 11-6 Photoionization detectors (PIDs) typically have a minimum detectable limit of 0.1 ppm; however, PIDs are available with detection limits in the low ppb.

may be necessary for exposure level decisions and personal protective equipment (PPE) level decisions.

5. Perform the measurement using the appropriate personal protective equipment (PPE) and record the results.

6. Apply the appropriate correction factor when quantitative results are needed.

TOXICS DETECTION

There is a wide range of different instrumentation available for the detection of toxic chemicals. These instruments vary in their detection methods, sensitivity, and specificity. Therefore, choosing the appropriate toxic detection device can be difficult. Here is a description of several air-monitoring instruments that detect toxic materials commonly available to hazardous materials responders at the operations level. This is a small sample of available equipment, and your agency may have other technology available that is not covered here.

Electrochemical Sensors

Electrochemical sensors can be used to detect not only oxygen, as discussed earlier, but also toxic gases such as carbon monoxide and hydrogen sulfide. Together the oxygen sensor, the combustible gas sensor, the carbon monoxide sensor, and the hydrogen sulfide sensor comprise what is often called a "four-gas meter." Four-gas meters are very common in the fire department and hazardous materials communities due to their broad utility.

Carbon monoxide (CO) and hydrogen sulfide (H_2S) electrochemical sensors both operate using the same type of chemical reaction at the sensing electrode: Carbon monoxide (or hydrogen sulfide) is reduced on the sensing electrode and generates a current. Oxygen is needed at the counting electrode to complete the circuit. Thus, oxygen is required for the operation of carbon monoxide and hydrogen sulfide sensors. The primary difference between a hydrogen sulfide and a carbon monoxide sensor is an activated carbon filter in front of the carbon monoxide sensor, which is absent from the hydrogen sulfide sensor. This explains some cross-reactivity problems these sensors have (Table 11-4). Due to its very small size, hydrogen is cross indicated on these sensors at very low concentrations.

Several dozen electrochemical sensors are available for the detection of specific toxic gases and vapors. Other electrochemical sensors include ammonia, arsine, chlorine, Freon, hydrogen chloride, hydrogen cyanide, hydrogen fluoride, phosgene, and sulfur dioxide. Any of these sensors can typically be substituted or added into a four-gas meter. In fact, many so-called four-gas meters actually have five to eight sensor ports. Therefore, if you have facilities that use some of these gases, it may be a good idea to add the appropriate sensor to your instrument. However, be aware that some of the listed gases are corrosive and may adversely affect other sensors in the instrument. That having been said, four-gas meters are comparatively cheap, especially when cost recovery ordinances are in place.

Photoionization Detectors (PIDs)

Photoionization detectors (PIDs) are designed to detect organic and inorganic gases and vapors in the low parts per million range (Figure 11-6). They were developed in the early

TABLE 11-4	Cross Sensitivity of Selected Electrochemical Sensors		
RAE SYSTEMS SENSOR TYPE	**CROSS SENSITIVITIES**		
	LARGE EFFECT (> 50%)	**MODERATE EFFECT**	**SMALL EFFECT (< 10%)**
Hydrogen sulfide	Phosphine	Methyl mercaptan Sulfur dioxide Nitrous oxide	Carbon monoxide Nitric oxide Methyl sulfide Ethyl sulfide Turpentine Ethylene
Carbon monoxide	Hydrogen Hydrogen sulfide (only with a used CO sensor or used charcoal filter) Ethylene oxide (without filter) Isobutene (with a used CO sensor)	Trichloroethylene (with a new sensor) Ethylene (especially with a used sensor)	Isobutene Isobutylene
Ammonia	Hydrazine (3:1, extremely high) Triethylamine Hydrogen sulfide Sulfur dioxide Chlorine	Nitric oxide	Hydrogen cyanide
Chlorine	Bromine Chlorine dioxide	Sulfur dioxide	Hydrogen sulfide Nitric oxide Nitrous oxide
Phosphine	Silane Diborane Germane	Sulfur dioxide Hydrogen sulfide	Hydrogen Hydrogen cyanide Ethylene

From RAE Systems Technical Note TN-114, Courtesy RAE Systems Inc.

SOLVED EXERCISE 11-5

Would electrochemical sensors be useful at the incident described in the opening scenario?

Solution: Yes, especially an electrochemical chlorine sensor. However, other electrochemical sensors (such as oxygen) may be damaged.

1970s primarily to detect volatile organic compounds such as benzene. The great advantage of the PID is that it is a reliable, low-cost, low-maintenance, extremely sensitive broad-spectrum detector. This makes PIDs ideal for hazardous materials incident decision making. A PID is often used for establishing the hot zone perimeter, initial PPE assessment for entry, spill delineation, leak detection, and decontamination verification.

ionization potential (IP) ■ The amount of energy needed to knock an electron off a chemical. In order to use a PID, we need to know the IP of a chemical and the bulb strength of the PID.

PIDs operate by ionizing the target gas using an ultraviolet (UV) lamp. A PID can detect any chemical with an **ionization potential (IP)** below the bulb strength. The ionization potential of chemicals can be found in numerous reference materials, including the NIOSH guide and the WISER program. The ionization potentials of some common gases are listed in Table 11-5. The most common PID bulb used in emergency response applications is 10.6 eV. Other available bulb strengths include 8.3 eV, 9.5 eV, 10.0 eV, 10.2 eV, and 11.7 eV. Different bulb strengths will detect different chemicals. For example, benzene, with an ionization potential of 9.24 eV, can be detected with a PID that has a 10.6 eV bulb. However, methanol, with an ionization potential of 10.84 eV, cannot be detected because its IP is greater than 10.6 eV. An 11.7 eV bulb is necessary to detect methanol using a PID. Methane has an ionization potential of 13.0 eV and cannot be detected by any PID.

Similar to combustible gas indicators, photoionization detectors detect some gases more readily than others. The sensitivity of a PID is determined by the chemistry of the target gas or vapor and the flow rate of the PID. Similar to a CGI, correction factors must

TABLE 11-5	Ionization Potentials of Selected Hazardous Materials

Red indicates chemicals that cannot be detected using a PID, whereas yellow indicates that an 11.7 eV bulb is necessary to detect the chemical (cannot be detected with a 10.6 eV bulb).

CHEMICAL	IONIZATION POTENTIAL
Acetic acid	10.66 eV
Acetone	9.71 eV
Acetylene	11.40 eV
Ammonia	10.16 eV
Benzene	9.25 eV
Carbon dioxide	13.8 eV
Carbon monoxide	14.0 eV
Chlorine	11.48 eV
Ethanol	10.47 eV
Ethyl ether	9.51 eV
Formaldehyde	10.87 eV
Hydrogen	15.43 eV
Hydrogen cyanide	13.6 eV
Hydrogen sulfide	10.45 eV
Iodine	9.4 eV
Methane	12.61 eV
Methyl mercaptan	9.44 eV
Nitrogen	15.8 eV
Nitromethane	11.02 eV
Oxygen	12.1 eV
Perchloroethene	9.32 eV
Phosphine	9.87 eV
Propane	10.95 eV
Vinyl chloride	9.99 eV
Water	12.6 eV

be applied to PID readings when gases other than the calibration gas are measured. The most common calibration gas for photoionization detectors is isobutylene, and the correction factor is supplied by the PID manufacturer.

Photoionization detectors (PIDs) are very robust, and there are no known sensor poisons, although moisture can easily damage 11.7 eV bulbs due to their construction. The biggest problem with PIDs is loss of sensitivity. This happens due to the presence of dust, smoke, humidity, and other gases such as carbon monoxide and methane that cannot be detected. Loss of sensitivity also occurs due to the presence of high concentrations of gases and vapors that are measured. In each case, the interfering gas or particulate absorbs part of the UV energy without causing any ionization. This keeps that part of the UV energy from ionizing detectable gases and causes underreporting. When the concentration of a detectable gas gets too high, typically over 2000 ppm, there is insufficient UV energy to ionize all of the gas molecules passing the sensor. This causes saturation of the detector. The linear range of most PIDs is 0 to 1000 ppm. Some manufacturers make supersensitive PIDs, such as the ppbRAE made by RAE Systems, that can detect down to the low parts per billion range.

Follow these guidelines when using a photoionization detector (PID):

1. Inspect the instrument. Look for visible damage to the case, look at the sample train (the path the air takes inside the instrument), and ensure that batteries are inserted.
2. Check the last calibration date of the instrument; if it is not within the manufacturer's recommended time period, perform a field calibration:
   i. Enter the calibration menu of the instrument.
   ii. Ensure that the calibration gas values in the instrument correspond to the available calibration gas (cal gas).
   iii. Attach the cal gas to the instrument and perform the calibration according to the manufacturer's directions.
3. Fresh air zero the instrument. Follow the manufacturer's directions to normalize the instrument to the ambient air and temperature. Make sure you are in a clean environment!
4. Bump check the instrument. Apply cal gas to the instrument for a functionality check. Ensure the instrument reads close to the cal gas specifications. Close depends on your SOGs and the data quality objectives, but 10% of the target value is a good rule of thumb for reconnaissance and basic detection. A smaller percentage may be necessary for exposure level decisions and personal protective equipment (PPE) level decisions.
5. Perform the measurement using the appropriate personal protective equipment (PPE) and record the results.
6. Apply the appropriate correction factor when quantitative results are needed.

Colorimetric Tubes

Colorimetric tubes are a low-cost, relatively specific, highly sensitive air-monitoring detection device. Colorimetric tubes can be very useful in hazardous materials emergency

SOLVED EXERCISE 11-6

Would a PID be useful at the incident described in the opening scenario?
   Solution: Yes, but not to measure chlorine levels because the ionization potential is too high at 11.48 eV for a 10.6 eV bulb (unless your PID contains an 11.7 eV bulb). However, the PID would be a good tool to detect the presence of other chemicals at low levels that may have been released.

FIGURE 11-7 Colorimetric indicators, such as colorimetric tubes, are good general-purpose, low-cost detectors capable of detecting specific chemicals—or at least chemical classes—at a wide variety of concentrations ranging from ppm to percent.

response, especially when responding to a release of a known material or when trying to verify information. The colorimetric tubes can also be helpful in the identification process. Several manufacturers make colorimetric tubes (polytest), or tube sets (simultest and CDS), that can be used to characterize an unknown contaminant into a chemical family.

Colorimetric tubes are small, chemical indicator–filled glass tubes that are sealed at both ends (Figure 11-7). The chemical indicator inside the tube changes color in response to certain chemicals or chemical families. The length of the stain indicates approximate concentration. The concentration is only approximate because colorimetric tubes can have up to a ±35% accuracy. Colorimetric tubes can be as sensitive as the parts per million range, or they can measure in the percent range. Various manufacturers have more than 100 different colorimetric tubes on the market for different applications in various industries.

The first step when using colorimetric tubes is to check their expiration date, appropriate operation range (sensitivity), and tube integrity (including the tube itself and the colorimetric indicator inside). Using expired tubes can be dangerous because the colorimetric indicating reagents lose their effectiveness over time. The second step is to check the accuracy of the pump (leak check). A known quantity of air, usually 100 cubic centimeters (cc), is pumped through the glass tube using either a piston pump or a bellows pump. A bellows pump can be checked by depressing the bellows and placing an unopened tube on the end. After waiting 5 minutes, the pump should remain almost completely depressed, which means that there is no leak. Once the equipment passes muster, a measurement can be taken.

When using colorimetric tubes, it is extremely important to completely read and understand the directions supplied with the tubes. At a minimum, the following information should be noted:

- The preparation of the tube (if any).
- The number of pump strokes needed per measurement, and how long this is expected to take. (For example, sulfuric acid tubes require 100 pump strokes, and the overall measurement takes approximately 75 minutes. This is obviously a very important piece of information to know before entering the hot zone in full PPE! It should be noted that tubes requiring 100 pump strokes are extremely rare. Most tubes require 10 or fewer pump strokes.)
- The interval between pump strokes (if any).
- **Cross sensitivities** the tube may exhibit.
- The color change to be expected with the chemical of interest and any chemicals the tube is cross sensitive to.
- Tube detection range and accuracy.
- Any special instructions (such as taking 10 pump strokes, then breaking an ampule, and taking 10 more pump strokes)

Generally, follow these guidelines when using colorimetric tubes:

1. Inspect the pump and perform a leak test. Look for visible damage to the pump, such as cracking or visible leaks.
2. Check the expiration date of the colorimetric tubes.

cross sensitivity ■ The ability of a detection instrument to detect other chemicals for which it was not designed. Cross sensitivity usually results in a false positive; in other words, the chemical to which the detection equipment is cross sensitive is incorrectly assumed to be the chemical of interest.

3. Ensure that the sensitivity range is appropriate (for example, ppm versus %).
4. Read the instructions carefully, and note the number of pump strokes, the color change to be expected, and any cross sensitivities.
5. Score and break open both ends of the colorimetric tube, and insert with the arrow pointing toward the pump (in the direction of airflow). Ensure that the lowest numbers are farthest away from the pump!
6. Carefully take a pump stroke, and wait for the pump to completely reinflate. Repeat this procedure for the required number of pump strokes.
7. Read the length of stain along the tube. If the staining is uneven, take the most conservative reading (highest number).

The advantages of colorimetric tubes include comparatively low cost, ease of use, and minimal maintenance and upkeep of equipment. The limitations of colorimetric tubes include a limited shelf life, the detection tubes should be refrigerated, relatively poor accuracy, and cross sensitivities. The shelf life of most colorimetric tubes is 1 to 3 years, depending on the reagent used in the colorimetric reaction. Most colorimetric tubes will last longer stored in a refrigerator because the reagents break down more slowly at lower temperatures. Some tubes may even be reused if no color changes occurred. However, be careful to read the directions carefully and ensure that the tube is not saturated.

Colorimetric tubes are generally not chemical specific, but rather they react to a class of chemicals. As has been discussed, this leads to cross sensitivities, or cross-reactivities, otherwise known as false positives. For example, MSA makes an aromatic hydrocarbon colorimetric tube that detects benzene, toluene, and xylene. Dräger makes a benzene-specific colorimetric tube with a pre-layer that filters out the other aromatics (such as toluene and xylene). However, if the pre-layer becomes saturated, the readings will be inaccurate.

ion mobility spectroscopy (IMS) ■ A detection device that ionizes contaminants and measures the drift time of the ions in order to identify the material.

Ion Mobility Spectroscopy (IMS)

Ion mobility spectroscopy (IMS) is commonly used to detect chemical warfare agents, explosives, and illegal drugs. It can also be used to discover toxic industrial chemicals (TICs) and toxic industrial materials (TIMs). The most common IMS instrument used by first responders is the APD 2000 made by Smiths Detection (Figure 11-8).

IMS technology uses a radioactive isotope to ionize the gases and vapors it is trying to identify. The ions are then separated by their size and mobility in a drift tube. At the end of the drift tube is a collector electrode where the ions are collected and produce a signal. The type and amount of ions, combined with the time it takes them to move down the drift tube, provide a unique pattern or signature that identifies the unknown material. Some IMS detectors have large databases and good sensor technology, whereas others do not. IMS detectors have been notorious in the past for the large amount of false positives, especially from such common materials as diesel fuel, vehicle exhaust, household cleaners, solvents, and even wintergreen oil. Nevertheless, IMS technology is a good tool to have in the toolbox.

FIGURE 11-8 Chemical warfare agent detectors often use ion mobility spectroscopy (IMS) to detect substances often.

HAZMAT HANDLE

Whenever air-monitoring instrumentation is used to make a presumptive identification, at least three different detectors should give consistent results to avoid false positives.

SOLVED EXERCISE 11-8

What type of air-monitoring/detection equipment could be used at the incident described in the opening scenario? What are the advantages and disadvantages of each?

Solution: Wetted pH paper may be used to detect the corrosive atmosphere created by the chlorine cloud. The advantage is that it is cheap and will not be damaged by the chlorine; the disadvantage is that it will detect chlorine only at relatively high levels.

Colorimetric tubes may be used to detect chlorine. The advantage is that they are relatively low cost and will not be damaged by the chlorine. The disadvantage is that they are not continuous air-monitoring devices and so are much more time consuming to use. In addition, a new tube must be used at each location.

An electrochemical chlorine sensor, which may be part of a multi-gas meter, may also be used. The advantage is that it is quantitative and will detect chlorine at relatively low levels (ppm). However, the elevated chlorine levels may damage other sensors in the multi-gas meter, including the oxygen sensor, the combustible gas indicator, the CO sensor, and the hydrogen sulfide sensor.

FIGURE 11-9 General sample collection activities at a clandestine laboratory.

Sample Collection of Unknown Materials

Often samples of suspected hazardous materials must be collected for presumptive field identification, definitive laboratory verification, or evidentiary purposes (Figure 11-9). This section covers ways to collect and process gas, liquid, and solid samples of possible hazardous materials. Sample collection for evidentiary purposes was covered in Chapter 9, Operations Level Responders Mission-Specific Competencies: Evidence Preservation and Sampling.

Evidence may be collected only by sworn law enforcement officers and fire investigators, or by other first responders under their direct supervision. Evidence may also be collected only after probable cause has been established or a warrant has been issued. Therefore, if a fire department–based hazmat team collects a sample as "evidence" at an emergency incident, that sample may be thrown out of court due to improper search and seizure. Because the hazmat team was called to an emergency, probable cause may not exist for a law enforcement operation, which evidence collection is by definition. Non-law-enforcement agencies should therefore always call their sampling activities "public safety sampling" instead of "evidentiary sampling." Nevertheless, agencies' public safety sampling protocols should adhere to high standards of documentation, certified tools and containers, and good sampling protocols that minimize cross contamination. Such guidelines will not only give more reliable field test results and lab analysis but also allow law enforcement to seize these public safety samples into evidence at a later time, after a warrant has been obtained. So in practice, the performance of public safety sampling and evidentiary sampling looks very similar; the difference is intent. The intent of a public safety sample is to identify a substance to mitigate an emergency, whereas the intent of an evidentiary sample is to solve and prosecute a crime.

Sample collection protocols are designed to ensure the consistency and reproducibility of sampling protocols, to ensure the acquired sample actually contains the material of interest, to minimize the likelihood of cross contamination of samples, and to prevent external contamination of the samples.

Sampling equipment includes compatible receiving containers, tools to acquire the samples such as spatulas, pipettes and swabs, and equipment to prevent cross contamination such as towels, wipes, and surface covers. Whenever possible, sampling equipment should be certified clean per ASTM methods. Try to maximize the compatibility of the sampling equipment with the sample, especially the receiving container. Chemicals are generally collected in glass containers, whereas biological samples are collected in plastic containers. However, some exceptions do apply. For example, hydrofluoric acid is incompatible with glass and must be collected in a plastic container.

Three types of samples are typically collected: the unknown sample, a control sample, and an equipment sample. The unknown sample contains the material of interest. A control sample is taken from a clean area near the sampling point using the identical sampling method. The control sample is designed to ensure that the sampling technique itself does not introduce contamination and should be taken after the unknown samples have been collected. An equipment sample is a complete, duplicate unused set of all of the equipment used to obtain the unknown sample. The equipment sample is designed to ensure that the equipment being used for sampling has not been contaminated prior to use.

The way the sampling points are determined can affect the outcome of the sampling mission. There are several ways to collect samples: a systematic method, a judgmental method, and a random method. The systematic sampling method involves setting up a grid and collecting samples at predetermined intervals along that grid. Its advantages are that samples are taken at regular intervals, and it is the most thorough and unbiased sampling method. Its disadvantages are that it is time and labor intensive, and requires a substantial amount of equipment to collect the required number of samples.

The judgmental sampling method involves determining the sampling points based upon the observations and expertise of the sampling team. Its advantages are that a large number of samples are not required, and there is a higher likelihood of getting a sample of interest with an experienced sampling team. Its disadvantage is that the material of interest may not be obtained if the judgment of the sampling team is faulty.

The random sampling method involves taking samples at random within a predetermined area, and it is used when resources are limited and there is no immediate indication of which location to sample. It is the least desirable of the three sampling methods. Whichever sampling method you choose, always be prepared to justify your actions!

SAMPLING PROCEDURES: TWO-PERSON TECHNIQUE

The two-person sampling technique is designed to minimize cross contamination of samples and was specifically designed for evidentiary sample collection (Figure 11-10). However, this technique also ensures that public safety samples are as pristine as possible, which will lead to better sample identification results and laboratory test results.

The two-person sampling technique uses a minimum of two people per team. The attendant handles only clean sampling tools and equipment as well as the double-bagged sample of interest. The sampler collects the sample of interest using the opened tools handed to him or her by the attendant. All sampling tools are disposable and are used

FIGURE 11-10 Demonstration of the two-person sample collection technique. This system is designed to minimize the possibility of cross contaminating samples during evidence collection.

for one sample only. The sampler wears a minimum of two layers of gloves. The outer layer will always be changed between sampling points to eliminate any chance of cross contaminating the sample.

Documentation is very important during sampling activities. The names of the attendant and sampler, the time of sampling, the sample location, and the sample tracking number should all be noted for each individual sample. Sometimes public safety samples unexpectedly become evidence. When the sampling techniques and sampling personnel are well documented, the samples will be much more likely to stand up in court.

Gases

Gas samples may include the contents of abandoned cylinders, air samples within buildings or open areas, headspace samples of containers, atmospheric samples, gaseous contents of bags, and so on. Sampling gases can be a difficult process. Hazardous materials technicians or environmental contractors may use pumps or vacuum canisters (such as SUMMA canisters), but these are typically not readily available to personnel operating at the operations level. A neat trick is to use a photoionization detector (PID) as a pump: PIDs typically have a threaded gas outlet, and most manufacturers supply a nipple that can be screwed into the outlet and connected to a gas bag. Remember, PIDs use a nondestructive form of ionization in the detection process, and the sample train is usually made of stainless steel and glass, which minimally absorb chemicals.

Liquids

Liquid samples may include contents of abandoned containers, freestanding liquids, surface waters, liquid runoff, sewage, liquids found in clandestine laboratories, pastes, and so forth. Liquids are typically sampled using plastic or glass pipettes (Figure 11-11, left). A cotton or Dacron swab can be used to sample small amounts of liquid, or liquids in hard-to-reach places (Figure 11-11, right). Pastes and gels can be scraped up using a spatula or other device.

Solids

Solid samples may include pellets, powders, dusts, granules, gels, sediments, soils, plants, residues, clothing, and many other solid materials. Solids are typically sampled using a spatula or cotton swab (Figure 11-12). Some solids may be fine powders, which aerosolize

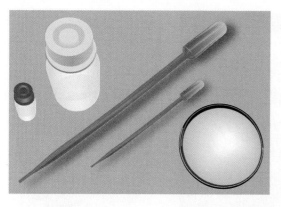
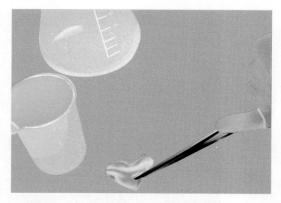

FIGURE 11-11 Liquid sample collection equipment (left). Collection of a liquid sample using a swab (right).

FIGURE 11-12 Solid sample collection equipment.

easily and may be a potential inhalation hazard. Typically, solids do not have a high vapor pressure and do not pose a significant respiratory hazard. However, under certain circumstances, such as white powder incidents or when fine particulates are known to be involved, the highest level of respiratory protection may be warranted.

Solid and Liquid Sample Identification

Solid and liquid field sample identification equipment has become much more common in recent years thanks to the increased funding provided by the U.S. Department of Homeland Security (DHS). The two most common types of sample identification equipment are Raman spectroscopy–based technologies and Fourier transform infrared (FTIR)

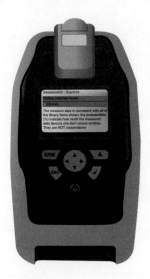

FIGURE 11-13 Solid and liquid sample identification equipment. The AhuraFD model of Raman spectrometer (left). The Hazmat ID model of an FTIR spectrometer (middle). The TruDefender FT model of FTIR spectrometer (right). *Drawings on left and right are courtesy of Thermo Fisher Scientific*

Raman spectroscopy ■ A method of sample identification that uses near-infrared single wavelength scattering and that can identify most covalently bonded compounds when they are present in sufficient concentration.

covalent bond ■ A shared pair of electrons between two atoms that binds them. Covalent bonds can be seen by Raman and FTIR spectrometers.

ionic bond ■ The electrostatic force of attraction between oppositely charged ions that binds them. An example of ionic bonding is sodium chloride.

fluorescence ■ The intense production of a broad range of electromagnetic radiation by a substance in response to the input of energy. Infrared fluorescence sometimes interferes with Raman spectroscopy.

spectroscopy (Figure 11-13). These complementary technologies are very effective for presumptive identification of unknown liquid and solid materials.

RAMAN SPECTROSCOPY

Raman spectroscopy uses a near-infrared laser to excite the **covalent bonds** within an unknown material to produce a pattern of scattered light. The scattered light is measured and generates a spectrum, which can be compared to a known database of chemical spectra, similarly to the way a fingerprint is compared to a database of known fingerprints.

Raman spectroscopy can be used to identify the material through translucent containers. Most glass containers, plastic containers, and thinner envelopes are translucent enough to make field identification possible without even sampling the contents. However, many containers are not translucent, and the contents must be sampled.

Raman spectroscopy has advantages and disadvantages. Just like any other technique, Raman spectroscopy is not a magic bullet that can do everything, and it has limitations. It can be used to identify materials that contain covalent bonds. Fortunately, most chemicals do contain covalent bonds and can be readily identified using Raman spectroscopy. However, materials that contain only **ionic bonds**, such as table salt, or sodium chloride, cannot be identified using Raman spectroscopy. A major advantage of Raman spectroscopy is that it cannot identify water. Water is completely Raman silent. Raman spectrometers are therefore more sensitive (about tenfold more sensitive) to hazardous materials dissolved in water than other liquid and solid sample identification technologies such as Fourier transform infrared spectroscopy.

Fluorescence can interfere with Raman spectrometers. Fluorescence occurs when a material emits a shower of infrared light upon laser excitation. Fluorescence does not offer any specific information about the material; it just interferes with the Raman signal needed for identification. This interference occurs in the same way that an oncoming vehicle with its high beams on interferes with your night vision while driving.

The following procedure should be used to identify an unknown solid or liquid material using Raman spectroscopy in point-and-shoot and vial mode:

1. Inspect the instrument. Look for visible damage to the case, make sure the laser pathway is not heavily contaminated or obstructed, and ensure that batteries are inserted.
2. Turn the instrument on.
3. SAFETY WARNING: Never scan dark materials or suspected explosives in their original containers or in sample vials! An explosion causing serious injury or death may otherwise occur. If scanning is absolutely necessary, take a small sample (the size of a grain of sand) and scan it on a noncombustible surface away from the rest of the suspect material.
4. When using in point-and-shoot mode: Position the sample at the focal point of the laser. This position is critical to ensuring accurate and effective identification and is the most common reason why sample identification fails. Also, be aware of the eye hazard that the external laser may pose.
5. When using in vial mode: Ensure the vial is at least one quarter full.
6. The instrument will automatically perform a library search. Matches with a 0.95 confidence factor or above are considered reliable by most manufacturers. However, the unknown spectrum should always be compared to that of the library entry to make sure that the peaks match well.
7. If there are any doubts, most manufacturers offer some sort of reach-back service that should be accessed.

FOURIER TRANSFORM INFRARED (FTIR) SPECTROSCOPY

Fourier transform infrared (FTIR) spectroscopy exposes unknown chemicals to a broadband infrared source and measures which wavelengths of infrared light are absorbed by the covalent bonds of the sample. The absorbed wavelengths are measured and generate a spectrum, which can be compared to a known database of chemical spectra, similarly to the way a fingerprint is compared to a database of known fingerprints. In this way, FTIR spectroscopy is conceptually similar to Raman spectroscopy.

FTIR spectroscopy can be used to identify solids and liquids (as well as gases with the appropriate equipment). Unlike Raman spectroscopy, FTIR spectroscopy requires direct contact with the unknown material being identified. This is because it is an absorbance-based technique, and due to engineering limitations, fieldable FTIR equipment uses the attenuated total reflectance (ATR) illumination mode.

FTIR spectroscopy has advantages and disadvantages just as any other technique does. Similar to Raman spectroscopy, FTIR spectroscopy can be used to identify materials that have covalent bonds. FTIR spectroscopy cannot identify chemicals that contain only ionic bonds. FTIR and Raman spectroscopy differ in the types of covalent bonds that each technique is able to identify. Polar covalent bonds absorb infrared energy particularly well, whereas nonpolar covalent bonds scatter infrared energy particularly well. Thus, FTIR spectrometers see polar covalent bonds very well, whereas Raman spectrometers see nonpolar covalent bonds particularly well. That having been said, both types of spectrometers identify the vast majority of chemicals very well. The identification capabilities of Raman and FTIR spectrometers are summarized in Table 11-6.

The following procedure should be used to identify an unknown solid or liquid material using FTIR spectroscopy:

1. Inspect the instrument. Look for visible damage to the case, make sure the diamond surface is not contaminated, and ensure that batteries are inserted.
2. Turn the instrument on.
3. Collect a background reading.

Fourier transform infrared (FTIR) spectroscopy ■ A method of sample identification that uses infrared absorption and can identify most covalently bonded compounds when found in a sufficient concentration.

TABLE 11-6	Identification Capabilities and Limitations of Raman and FTIR Spectrometers	

RAMAN SPECTROSCOPY	FTIR SPECTROSCOPY
CAPABILITIES	**CAPABILITIES**
Compounds containing covalent bonds can be identified (most chemicals)	Compounds containing covalent bonds can be identified (most chemicals)
Minimum detection threshold in water is approximately 1%	Minimum detection threshold in water is approximately 10%
Hazardous material does not contact Raman spectrometer (less decon needed)	
Ability to identify unknown material through clear containers and standard envelopes	
LIMITATIONS	**LIMITATIONS**
Purely ionic compounds not seen (NaCl, HF, etc.)	Purely ionic compounds not seen (NaCl, HF, etc.)
Dark-colored materials may ignite or explode	Need to sample unknown material and have direct contact with instrument
Highly fluorescent materials	Water has a strong FTIR signal and makes it more difficult to detect material of interest

4. Place a small amount of the sample on the diamond. Solid materials need to be compressed onto the diamond surface. Most instrument manufacturers include a tool to facilitate this procedure. Liquids need only a thin coating on the diamond surface.
5. The instrument will automatically subtract the background reading from the unknown spectrum and perform the library search. Matches with a 0.95 confidence factor or above are considered reliable by most manufacturers. However, the unknown spectrum should always be compared to that of the library entry to make sure that the peaks match well.
6. If there are any doubts, most manufacturers offer some sort of reach-back service that should be accessed.

Documentation and Data Sharing

Documentation is a critical aspect of air monitoring and sample identification. Good documentation of air-monitoring readings and sample identification results will allow others to use and interpret those results to protect other first responders and the public. The documentation you keep can also convince others of the accuracy of your instrument readings and procedures.

The air-monitoring and sample identification results that you obtain will also be of interest to other agencies. It is very important to share your results with other appropriate agencies to maximize the safety of other first responders and the public. These presumptive field identifications will also be of interest to the laboratory performing a definitive analysis. You and your agency may be interested in comparing the laboratory results with your field identification results. Any discrepancies between your results and the laboratory results may indicate that your standard operating guidelines or training program needs to be improved.

Summary

Performing proper air monitoring at hazardous materials incidents is critical for the safety of both first responders and the public. Air monitoring should be performed in a defined order to systematically rule out hazards and ensure the chemical sensors will operate within design parameters. Generally, air monitoring should begin with gamma radiation detection from the cold zone, followed by corrosive atmosphere detection, oxygen measurements, flammable gas and vapor detection, and finally toxic atmosphere detection. It is vital to know the capabilities and limitations of the air-monitoring equipment your agency uses. The only way to do this is to train regularly with the equipment under conditions that are similar to the types of responses you can expect in your jurisdiction. Remember that the HAZWOPER standard requires that you train to competency! Often it will be necessary to collect samples and perform field identification. Raman and FTIR spectroscopy are excellent tools that provide semiautomated sample identification in the field and are simple enough to operate under emergency conditions.

Review Questions

1. What is the appropriate order of air monitoring and why?
2. What is the difference between calibration and a bump check?
3. What is a correction factor and why is it necessary?
4. Why should pH paper be wetted prior to air-monitoring applications?
5. What units does a combustible gas indicator (CGI) read out in? Why is this significant?
6. Explain how the two-person sample collection technique works.

Problem-Solving Activities

1. Review your agency's standard operating guidelines (SOGs) covering air-monitoring activities at hazardous materials incidents or incidents involving weapons of mass destruction (WMD).
   a. Determine when air monitoring is required to be performed at hazardous materials incidents.
   b. List the air-monitoring equipment your agency uses.
   c. Determine the detection limits of this equipment.
   d. Determine under what conditions false positives and/or false negatives may be expected.
   e. What chemicals is the equipment cross-reactive to?
   f. Perform a functional check on all air-monitoring equipment that you may use on a hazardous materials incident.
2. You are operating a combustible gas indicator (CGI) that has been calibrated to methane. As you approach an overturned tanker that is placarded with UN 1090, the CGI starts to alarm at 10%. Using Table 11-1, determine what the actual percent LEL is for this particular chemical. What should your actions be?
3. You are operating an oxygen sensor that is indicating 20.5% oxygen at a cryogenic argon release. The instrument was properly calibrated, field zeroed, and bump checked prior to use. Background oxygen levels are 20.9%. Based upon these meter readings, what is the ambient concentration of argon at your location? You are wearing level C PPE. Is this appropriate? Why or why not?
4. Review your agency's standard operating guidelines (SOGs) covering sampling activities at hazardous materials incidents or incidents involving weapons of mass destruction (WMD).
   a. Determine when sampling activities will be performed at these incidents.
   b. List the sampling equipment your agency uses.

5. Review your agency's standard operating guidelines (SOGs) covering liquid and solid sample identification activities at hazardous materials incidents or incidents involving weapons of mass destruction (WMD).
   a. Determine when liquid and solid sample identification will be performed at hazardous materials incidents.
   b. List the sample identification equipment your agency uses.
   c. Determine the detection limits of this equipment.
   d. Determine under what conditions false positives and/or false negatives may be expected.
   e. Identify three materials that you can find in your kitchen or garage using this equipment.
6. Locate your air-monitoring log sheets and your sample identification log sheets, and familiarize yourself with them. Fill out the sample identification log sheet with one of the three materials you identified in Problem-Solving Activity 5.

References and Further Reading

Centers for Disease Control and Prevention. (2007). *NIOSH Pocket Guide to Chemical Hazards*. Washington, DC: U.S. Government Printing Office.

Hawley, Christopher. (2007). *Hazardous Materials Air Monitoring & Detection Devices* (2nd ed.). Florence, KY: Delmar Cengage Learning.

Maslansky, Carol J., and Steven P. Maslansky. (1993). *Air Monitoring Instrumentation*. New York: John Wiley & Sons.

National Fire Protection Association. (2008). NFPA 472, *Standard for Competence of Responders to Hazardous Materials/Weapons of Mass Destruction Incidents*. Quincy, MA: Author.

Occupational Safety and Health Administration. (1990). 29 CFR 1910.120. *Hazardous Waste Site Operations and Emergency Response (HAZWOPER)*. Washington, DC: U.S. Department of Labor.

Rae Systems. (2005). *Application & Technical Notes Guide* (3rd ed.) San Jose, CA: Author.

Weber, Chris. (2007). *Pocket Reference for Hazardous Materials Response*. Upper Saddle River, NJ: Pearson/Brady.

Operations Level Responders Mission-Specific Competencies: Victim Rescue and Recovery

confined space, *p. 322*
victim retrieval device (VRD), *p. 334*

OBJECTIVES

- List 11 factors affecting the feasibility of victim rescue.
- List four victim considerations that should be taken into account when mobilizing for victim rescue.
- Describe the essential components of hot zone medical treatment.
- Describe the 3/30 rule of 2003 and its implications for victim rescue and recovery at WMD incidents.
- Describe four ways emergency decontamination may be performed.
- Describe four methods of removing a victim from the hot zone.
- Describe three victim retrieval devices and their use.

You Are on Duty! Solvent Explosion at the Manufacturing Facility

DANVERS, MA

A 911 call is received at 2:45 A.M. on November 22, 2006, that a large explosion has occurred. The caller is unsure of the exact location. The explosion knocked houses off their foundations and could be felt 45 miles (72 km) away. Other 911 calls pour in and the explosion is determined to have occurred at a local paint, solvent, and ink manufacturer. The explosion causes massive destruction in the immediate area, including the destruction of 24 houses and 6 businesses. Ten people are eventually hospitalized. The cause of the explosion was determined by the U.S. Chemical Safety Board (CSB) to be a heated 2000-gallon (7570-L) tank containing highly flammable liquid with a valve left open overnight that generated an ideal fuel–air mixture inside the sealed building that spontaneously ignited. Other chemicals that were involved in the ensuing fire were over 50,000 pounds (22,680 kg) of nitrocellulose and thousands of gallons of flammable liquids. An unconscious victim is found near the flammable liquid tank. He is covered in liquid and appears to have significant traumatic injuries.

- How would you conduct search and rescue operations in the building of origin and the blast-affected area?
- What concern do you have regarding the hazardous materials that are stored at the facility and caused the explosion?
- What special precautions do you need to take when entering the area?
- What special equipment and techniques will be required to effect rescues?

Let's try to answer these questions in this chapter.

Operations level responders can have one of the biggest impacts on the outcome of life-threatening hazardous material and weapons of mass destruction (WMD) incidents by rapidly performing victim rescue and recovery. It typically takes hazardous materials teams from 20 minutes to an hour to arrive and set up for hot zone entry operations. For injured victims, whether the incident is a simple medical emergency or a result of chemical exposure, this is an eternity (Figure 12-1).

When operations level responders are able to enter the hot zone and perform victim rescue and recovery quickly, they will save lives. For example, imagine that a simple medical emergency such as an asthma attack occurs at a hardware store. In the process of collapsing, the patient knocks down a shelving unit full of solvent bottles, some of which break. The store is evacuated, and the fire department and EMS are called. If the patient could be removed from the solvent spill within minutes, quickly decontaminated,

FIGURE 12-1 Multiple victims in the hot zone. Rescuing and rendering medical care to victims at hazardous materials incidents can be a difficult and dangerous operation. *Art by David Heskett*

and given advanced life-support treatment, he would have a good chance of survival. If we were to wait 45 minutes to get hazardous materials technicians dressed out in full personal protective equipment (PPE), it would be a body recovery. For a quick snatch and grab in this scenario, structural firefighter protective equipment would be more than adequate to make a rescue. In this chapter, we discuss how your agency can prepare and plan for such rescues, as well as which techniques and safety precautions can be implemented to tip the hazard-risk assessment process in favor of rapid victim rescue and recovery.

Preparation and Planning

Preparation and planning are essential at the agency level and should include detailed planning for fixed site facilities within your jurisdiction. At this level, preparation should include an analysis of available personal protective equipment (PPE) and rescue equipment that may be utilized for victim rescue and recovery in light of chemical hazards. In addition, agencies should have detailed standard operating guidelines (SOGs) spelling out the necessary PPE, equipment, rescue techniques, and training. Fixed site facilities that store, use, process, or generate hazardous materials need to have detailed pre-incident plans in place that specifically outline situations and locations in which victim rescue and recovery will be attempted. These pre-incident plans should be reviewed and trained on annually, at a minimum. Throughout the planning phase, a comprehensive hazard-risk assessment should be conducted and documented.

FEASIBILITY

The feasibility of performing victim rescue and recovery operations should be analyzed at the pre-incident planning stage as well as at the start of any hazardous materials incident. Some of the factors that should be taken into account are:

- Chemical hazards
- Facility layout
- Facility safety equipment and procedures
- Available PPE
- Available rescue equipment (such as victim recovery devices [VRD])
- Availability of sufficient trained personnel
- Location of the victims
- Nature of victim injuries
- Victim exposure versus victim contamination
- Available decontamination
- Available medical treatment in the hot zone, in the warm zone during decontamination, during transport, and at the definitive care location

When looking at the feasibility of victim rescue and recovery, we are trying to determine the benefit to the victim and to balance that with the risk to the responders. For example, if there will be little risk to the responders performing victim rescue, the patient outcome will likely be positive, and the appropriate PPE and rescue equipment are available, a victim rescue should almost certainly be attempted. On the other hand, if the risk to the responders is high and the patient outcome is highly doubtful, a victim rescue should almost certainly not be attempted because the risk–benefit analysis is not in our favor. These highly subjective decisions will be based largely upon agency preparation, the incident commander's experience and training, and evaluation of tactical considerations of the hazardous materials incident.

What are some of the considerations regarding the feasibility of a rescue operation to retrieve the victim next to the tank in the opening scenario?

Solution:
1. Is this a body recovery or a rescue?
2. How much flammable liquid has been spilled?
3. What danger does the flammable liquid pose during patient packaging and extrication?
4. What PPE is available for rescuers? Is it adequate?
5. How will decontamination be accomplished?
6. Can warm zone treatment be started?
7. How will the patient be transported?

SAFETY PROCEDURES

If victim rescue and recovery is determined to be feasible, solid safety procedures should be implemented. These measures include using a NIMS-compliant incident command system, appropriately trained personnel, suitable personal protective equipment, and effective rescue equipment; using a properly trained and equipped backup team or rapid intervention team; and having transport-capable advanced life-support EMS services available on scene. Special conditions at the hazardous materials incident may necessitate additional safety precautions. These conditions may include, but are not limited to, the nature of the chemical release, quantity of the release, location of the release, location and condition of the victim, weather conditions, likelihood of secondary devices or booby traps, or any other unusual agency response considerations.

The Buddy System

As has been discussed, at least two personnel should always enter the hot zone together. This is referred to as the buddy system, which is mandated by HAZWOPER for any personnel entering the hot zone. Personnel should *never* enter the hot zone alone. The incident commander and others may be held criminally liable if an injury or death occurs due to a violation of this rule. The buddy system is also mandated by the respiratory protection standard, and specifically refers to personnel using breathing apparatus. Using the buddy system increases the safety of personnel entering the hot zone. If one person suffers a medical emergency, has an equipment failure, or is overcome by the chemical, the other person may be able to effect a rescue and summon outside help in the form of the backup team.

Backup Teams

A backup team, or rapid intervention team (RIT), is also required by HAZWOPER when personnel enter the hot zone. In the respiratory protection standard, this is referred to as the two-in/two-out rule. There is an exception to this rule for the rescue of confirmed victims. However, a backup team is an essential safety net for rescue crews entering the hot zone, and the incident commander must immediately request additional resources if a rescue team is entering the hot zone without a backup team.

The backup team serves as a rescue team should anything happen to the entry team in the hot zone. The backup team members should be ready to enter the hot zone at a moment's notice. They should have donned their PPE—at least at the same level of PPE as the entry team—and be ready to go on air. Appropriate tools such as rescue tools need to be at their disposal. These include tools to remove the entry team in case they become incapacitated, such as Sked boards or stretchers (see "Victim Considerations"). Remember, it is difficult to remove people dressed in full PPE while you yourself are in PPE. This action requires practice and proficiency with the rescue equipment. Also, the backup team must be vigilant to avoid the same hazard that caused the entry team to become incapacitated.

What safety procedures should be in place before a rescue is attempted in the incident described in the opening scenario?

Solution:

1. A backup team must be in place.
2. Personal protective equipment (PPE) must protect adequately against the flammable and chemical hazards (and others based upon research).
3. The rescue team must have continuous air-monitoring capabilities, especially a combustible gas indicator (CGI).

Safety Briefing

Typically, before anyone enters the hot zone, a thorough hazard assessment must be conducted to evaluate the hazards and institute appropriate countermeasures. After a site safety plan has been formed, the entry team, backup team, and decontamination team must be briefed before anyone enters the hot zone. This briefing alerts the team members to the dangers present in the hot zone and the actions they are expected to perform. When victim rescues must be rapidly conducted, the safety briefing will be concise and focus only on the essentials:

- Health hazards or suspected health hazards
- PPE being used and its limitations in the current situation
- Site layout (if available)
- Rescue objectives (triage versus victim recovery)
- Radio frequencies
- Stay times
- Emergency signals
- Decontamination line location
- Emergency decontamination procedures

The safety briefing should be carried out in the presence of all personnel entering the hot zone, the decontamination line personnel, and the group leaders. Typically, the incident commander, hazmat branch director, research group leader, or safety officer conducts the safety briefing. Who carries out the briefing will depend on the size of the incident, its complexity, and the technical expertise of the relevant personnel.

RESCUE TEAM ORGANIZATION

Hazardous materials incidents and weapons of mass destruction incidents often pose grave risks to would-be rescuers. Therefore, any rescue operation should be well organized: Rescue team positions should be well defined, and personnel assigned to these positions must be very familiar with their roles and the main functions they are to carry out. Here follows a look at three different types of hazardous materials incidents that may require rescue operations.

As with any hazardous materials incident, a NIMS-compliant incident command system must be used for victim rescue operations. At a minimum, an incident commander and a safety officer must be appointed. Any time entry operations are commenced in the hot zone, a backup team should be on standby to make a rescue. The entry team and backup teams must consist of a minimum of two people each under emergency response situations such as victim rescue per 29 CFR 1910.134 (Respiratory Protection Standard).

Resource
Central

See OSHA Respiratory Standard (29 CFR 1910.134) from the Code of Federal Regulations for more information.

Industrial Facility Emergencies

Industrial facility emergencies involving chemical releases can be challenging to hazardous materials operations level responders. Industrial facilities are usually sizable and complex, and contain a large amount of dangerous industrial equipment that may be energized. Industrial equipment may have electrical hazards such as high voltages and amperages, pneumatic hazards, hydraulic hazards, and mechanical hazards. Preplanning will be an essential component for successful industrial facility victim rescue. When victim rescue operations involve confined spaces, it gets even more complicated.

Confined Space Incidents

Confined space incidents add a whole other level of complexity to hazardous materials incidents. OSHA defines confined spaces as any space that is large enough for a person to enter and perform work, has limited or restricted means of entry and egress, and is not designed for continuous occupancy (Figure 12-2). Examples of confined spaces include tanks, silos, vessels, storage bins, hoppers, vaults, pits, and utility openings such as sanitary sewers and storm sewers. You may have noticed that these spaces are often designed to contain hazardous materials!

OSHA regulates permit-required confined space entry operations in 29 CFR 1910.146. This regulation requires, among other things, that team positions and responsibilities be defined, that all members are trained and competent, that air monitoring for hazardous materials is carried out, and that key members are proficient in the use of rescue equipment and rope rescue systems. The details of confined space rescue will not be covered in this chapter; that topic can fill another book. Here is a brief look at the confined space entry team positions and their responsibilities. "Entrants" enter the confined space; "attendants" monitor the entrants from outside the confined space; and the "entry supervisor" manages the confined space entry, evaluates hazards, and maintains the overall safety of the operation. Personnel performing any of these functions must be appropriately trained in confined space operations. Rescuers, including hazardous materials operations level personnel, must adhere to the same confined space entry criteria or face civil and/or criminal penalties.

confined space ■ Per OSHA, an area not designed for continuous human occupancy that has limited means of entry and egress and may contain a hazardous atmosphere. Examples of confined spaces include sewer systems and tanks with manhole access.

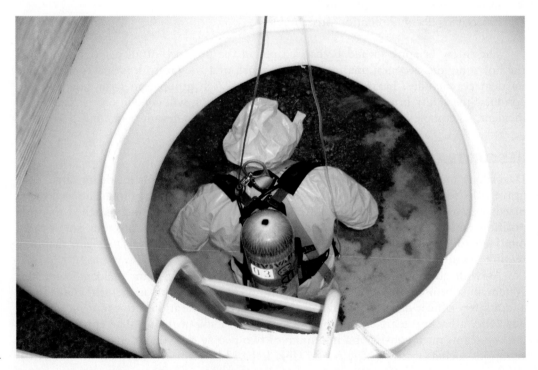

FIGURE 12-2 Confined space rescue operations involving a hazardous materials release.

Terrorist and Weapons of Mass Destruction (WMD) Incidents

Hazardous materials incidents that are terrorist incidents or involve weapons of mass destruction (WMD) are even more dangerous due to the intentional nature of the hazardous materials release. In this case, the hazardous materials incident is compounded by the potential presence of terrorists or other perpetrators who are intent on harming you; the possibility of secondary devices and booby traps; the increased toxicity or hazard posed by the weapon of mass destruction; and the nature of the release itself, which is intended to be significant in scope and harmful to humans.

Law enforcement should always be involved early in terrorist incidents for a number of reasons. Before entering the hot zone on these types of incidents, the area must be screened for secondary devices, which is the job of the explosive ordnance disposal (EOD) team, or bomb squad.

Joint training during the preplanning stages will be essential for rapid victim rescue and recovery. If this training does not occur before the terrorist incident, the fire department will likely not enter the hot zone due to the possible presence of perpetrators and secondary devices; and the police department will likely not enter due to the possible presence of hazardous materials or weapons of mass destruction. So who goes in? Nobody does in a timely fashion. Needless to say, the public will be the loser in this type of standoff!

INCIDENT RESPONSE CONSIDERATIONS

The following considerations should be addressed in standard operating guidelines (SOGs) in the form of tactical guidance the incident commander can utilize in difficult hazardous materials situations. Victim rescue versus victim recovery decisions can be some of the most stressful decisions an incident commander has to make. Do I place my own at risk, or do I let a civilian die? Preparation and planning will help reduce the stress on the incident commander and greatly increase the survivability of victims of hazardous materials incidents.

Victim Considerations

One of the primary incident response considerations should be the victims themselves. What is the victims' condition? What is their location? These two questions, along with the available agency resources, will determine the feasibility of a victim rescue operation. There are four victim possibilities:

1. The victim is ambulatory, and you have line-of-sight contact with him or her.
2. The victim is non-ambulatory, but you have line-of-sight contact with him or her.
3. The victim is ambulatory, but you do not have line-of-sight contact with him or her.
4. The victim is non-ambulatory, and you do not have line-of-sight contact with him or her.

These facts, when other conditions are factored in, determine whether this will be a rescue operation or a recovery operation. The difference is that rescue operations are time sensitive and must be performed as quickly and safely as possible, whereas body recoveries are typically not time-critical operations. For rescue operations, some risk based upon the perceived benefit to the victim will be incurred. For body recoveries, no appreciable risk should be incurred.

When we are faced with ambulatory patients, the benefits of rescue are high. Depending on how long the victim has been in the chemical environment, something is known about the danger of that hazardous material. The higher the exposure level and the longer the patient remains viable, the less danger that hazardous material poses in the short term. Conversely, when we are faced with non-ambulatory patients, the benefits of rescue may be lower. It is not known whether the patient's condition is due to an unrelated emergency such as a preexisting medical condition or trauma unrelated to the chemical exposure. However, we must assume that the chemical may have caused the injuries and is hazardous. Non-ambulatory patients also require more personnel and equipment to make a rescue. For example, a Sked board or stretcher must be used for each victim removed from the hot zone, which will require at least two rescuers per patient. Conversely, ambulatory patients can usually self-extricate from the hot zone, which requires just one rescuer with a bullhorn.

The complexity and risk of a rescue is much lower when we can see the victim, that is, when we have line-of-sight contact. This makes assessment of the patient easier, makes communication with the patient easier if he or she is conscious, and permits the command staff and backup team to have contact with the rescue team. Line-of-sight victim rescues are much safer operations than non-line-of-sight rescue operations. When we cannot see the victim, patient assessment and communications become much more difficult, and rescue operations are more dangerous. These factors must be considered in the hazard-risk assessment. Depending on the situation, non-ambulatory victims with whom we do not have line-of-sight communication may be deemed recovery operations.

Hot Zone Treatment

Basic medical treatment of victims should begin in the hot zone due to the length of time rescues and decontamination take at hazardous materials incidents. Basic medical treatment refers to the ABCs: airway, breathing, and circulation. Imagine a victim has a simple medical emergency at an industrial facility, such as an asthma attack, and she is not breathing. What would be the patient outcome if it took 10 minutes to package the patient and move her to the decontamination line, 10 more minutes to decontaminate the patient, and then we finally start medical treatment? Obviously the patient cannot survive 20 minutes without breathing. Therefore, in the hot zone we must immediately initiate the ABCs. If trauma is involved, do not forget c-spine considerations.

Even in full PPE, responders can make some basic assessments of the ABCs. Breathing can be assessed by watching the chest rise. Circulation can be assessed using the capillary refill technique or a pulse oximeter. If the patient does not have a viable airway, an attempt must be made to open it using an appropriate method such as the head-tilt, chin-lift method. If the patient is still not breathing, you must immediately start artificial respirations using a bag valve mask (BVM). If capillary refill or the pulse oximeter indicates the patient does not have viable circulation, start cardiopulmonary resuscitation (CPR). Starting hot zone emergency medical treatment is an essential lifesaving operation for victims of hazardous materials incidents and weapons of mass destruction. Of course, advanced life support (ALS) treatment, such as intravenous fluids (IVs) and intubations, should not be started until sufficient decontamination has been completed in the warm zone.

Personal Protective Equipment

A key consideration when performing victim rescue and recovery is the available personal protective equipment (PPE). Determining whether available personal protective equipment is appropriate or not is not an easy task. If the chemical hazard has not been identified, the worst-case scenario must always be assumed. In this case, firefighters should wear their structural firefighter protective equipment and self-contained breathing apparatus (SCBA) at a minimum. This will provide them with the highest level of respiratory

SOLVED EXERCISE 12-3

When the rescue team arrives at the victim in the opening scenario, team members notice his breathing is shallow and 5 breaths per minute. What type of hot zone treatment should this victim receive?
Solution: Assisted respirations with a bag valve mask (BVM).

protection but virtually no chemical splash protection. If firefighters use this type of PPE ensemble, they must be very careful to avoid direct contact with any hazardous materials. Because this may nevertheless happen inadvertently, chemical protective clothing should be strongly considered.

Some of the PPE considerations will be chemical compatibility as well as the condition of the victim. For example, if it takes 10 minutes for the agency to respond to the incident, and the victim is fully alert despite exposure or contamination, the immediate risk to the responders is probably fairly low. However, keep in mind that some chemicals may have long-term or chronic effects, such as causing cancer. Firefighters are very familiar with this hazard-risk assessment: When performing vehicle extrication with a leaking gas tank, firefighters do not think twice about donning their structural firefighter protective gear and removing the victim. Of course, they have a hose line in place in case of ignition of the flammable gasoline vapors and may even have air monitoring in place as well. They have taken care of the immediate risk of flammability, but what about long-term effects? Gasoline contains benzene, a known carcinogen. The firefighters performing the vehicle extrication have taken this into their risk–benefit analysis and determined that a victim rescue could be performed in structural firefighter protective gear. Rescuing a victim of a different hazardous materials emergency is no different. Evaluation of the available personal protective equipment, including structural firefighter protective gear, is a cornerstone of a good hazard-risk assessment.

Rescue and the 3/30 Rule of 2003

The 3/30 rule, developed in 2003 by the U.S. Army Soldier and Biological Chemical Command (SBCCOM), is quite controversial. The 3/30 rule refers to using firefighter protective ensembles (FFPE) with self-contained breathing apparatus (SCBA) to perform rescues at chemical weapons release incidents. Research indicates that turnout gear affords adequate protection for a 3-minute reconnaissance if viable (living) victims are not immediately visible, and a 30-minute rescue mission if viable victims are present at least 10 minutes after agent release. Nerve agents and mustard agents were specifically studied. Keep in mind that FFPE plus SCBA provides much greater protection against nerve agents than against mustard agent. Blister agent symptoms may be delayed for up to 18 hours. This may initially lull victims and responders into a false sense of security, increasing their exposure. The 3/30 rule applies only to rescue operations, *not* to agent detection or mitigation efforts. Rescue operations may be conducted for 30 minutes if:

1. Living victims are present.
2. Rescue occurs after vapor concentration has peaked (10 minutes after release).
3. Mustard agent (HD) is not suspected (in this case, only a 2-minute reconnaissance is recommended).
4. Chemical agent detectors and certified chemical protective clothing are not available.
5. Medical assistance is available at the scene.
   - Reconnaissance must be limited to 3 minutes when living victims are not visible.
   - Exit the area immediately if chemical contamination without viable victims is present.
   - Exit the area immediately if signs of mustard agent are present (oily liquid, garlic odor, skin reddening, or blistering in victims).
   - After exiting in either case, victims and/or rescuers must undergo immediate emergency decontamination with soap and water.

Positive-pressure ventilation is very effective at lowering vapor concentrations (40% to 75% percent reduction within 10 minutes of operation). Care should be taken not to spread the contamination to other occupied portions of the building or to people gathered outside. Ensure that these areas are properly isolated before beginning ventilation.

Should the rescue in the opening scenario be carried out in turnout gear or chemical protective clothing? What are your considerations in making this decision?

Solution: In order to effect rapid victim removal, turnout gear with SCBA would most likely be the optimal choice. Turnouts offer excellent protection against the flammability hazards and respiratory hazards, but poor protection against skin absorption hazards. Therefore, the rescue should be performed as quickly as possible (a snatch and grab). If further research indicates this is an extremely hazardous substance, chemical protective clothing with flash protection may be indicated.

Vermiculite and/or foam can be helpful in reducing the vapor hazard during the search and rescue operation. Hazmat teams performing agent detection and mitigation efforts must wear appropriate chemical protective clothing (CBRN rated).

Departments intending to adopt the rule should read the report in its entirety, (SBCCOM, Chemical Command, "Risk Assessment of Using Firefighter Protective Ensemble with Self-Contained Breathing Apparatus for Rescue Operations During a Terrorist Chemical Agent Incident," August 2003).

Emergency Decontamination

Any time first responders at the operations level enter the hot zone, even for a victim rescue and recovery, a decontamination plan should be in place. This may be as simple as pulling a hose line off the first due engine, as in a vehicle extrication with a leaking gas tank (emergency decontamination) (Figure 5-24). Or it may be more complex and include technical decontamination with soap and water and mechanical removal of chemical contamination. Technical decontamination is covered in greater detail in Chapter 8, Operations Level Responders Mission-Specific Competencies: Technical Decontamination.

Operations level responders may set up emergency decontamination for victims at hazardous materials incidents and weapons of mass destruction incidents. This is one of the most important functions operations level responders can perform to ensure public safety. The purpose of emergency decontamination is to prevent further injury to the patient from the hazardous material.

Emergency decontamination is very simple and fast to set up. It should be conducted in the warm zone, upwind, uphill, and upstream of the incident. How emergency decontamination is carried out is up to your imagination and may include:

- Charged hose line in the field
- Water or foam and water fire extinguisher
- Shower facilities in a nearby building
- Emergency shower and eyewash at industrial or research facilities

When life safety is involved, the EPA has stated that runoff does not have to be collected. However, as soon as resources permit or the life safety priorities have been taken care of, all runoff from decontamination efforts must be collected and treated as hazardous waste until proven otherwise.

The faster we can get the hazardous material off the patient, the better his or her prognosis will be. All of a patient's clothing must be removed for decontamination to be effective. This includes underwear and jewelry. If these articles are not removed, chemicals can be trapped underneath. Visible contamination can be brushed off or padded off, and the remaining contamination can be washed off.

Some chemicals can be removed with just water, but many chemicals cannot. Therefore, the most common and effective decontamination solution is soap and water.

What type of decontamination plan should you have in place for the incident described in the opening scenario?

Solution: Because the exact identity of the liquid is unknown, the best option to decontaminate the victim is with a soap and water solution. Flammable liquids are often not very soluble in water; therefore, soap is an essential component. A CGI and PID should be used to verify that decontamination was successful. Dry decon may be used to decontaminate the rescuers.

However, be aware that not all chemicals can be removed even with soap and water. Yet if water is the only decontamination solution available, use it! Emergency decontamination should be followed with technical decontamination or patient decontamination unless complete decontamination can be verified. The decontamination process and procedures are covered in more detail in Chapter 8, Operations Level Responders Mission-Specific Competencies: Technical Decontamination.

Victim Rescue Techniques and Equipment

The victim rescue techniques and equipment employed will depend primarily on the location and condition of the victim. The type of personal protective equipment used will depend mainly on the chemical and physical characteristics of the hazardous material and the nature of the release.

VICTIM CONDITION

The condition of the victim will dictate how difficult and time consuming the rescue will be. If there are many victims—in other words, if it is a mass casualty incident—the situation becomes orders of magnitude more difficult. We discuss how to help ambulatory victims ("green" patients) help themselves out of the hot zone without contaminating you and others in the process. We also discuss equipment and techniques to remove non-ambulatory patients from the hot zone, including setting up an assembly line using ropes and a mechanical advantage system.

Ambulatory Victims

Ambulatory patients are among the easiest to rescue. The victim rescue operation may be as simple as getting on a bullhorn or public address system and telling people to evacuate the immediate area of the hazardous materials release and meet at a triage point, or it may require personnel to enter the hot zone and assist them (Figure 12-3). The triage point should be located out of the immediate area of the chemical release at the edge of the hot zone where the victims can be triaged to determine whether they need to be decontaminated and what medical treatment is required.

A triage system such as START can be very effective in mass casualty incidents (Figure 12-4 and Box 12.1). It is very important to ensure that victims remain in the immediate area and do not self-transport to their homes, medical clinics, or hospitals. If the victims are not corralled at the hazardous materials release site, the risk of cross contamination is greatly enhanced.

Non-ambulatory Victims

Conversely, non-ambulatory patients are the most difficult to rescue and require resource-intensive victim rescue and recovery operations. Non-ambulatory patients will need stretchers, Sked boards, or gurneys to be transported out of the hot zone (Figure 12-5). This will require a minimum of two rescuers per patient and, depending on the size and condition of the victim, possibly more. Victim extrication can be simplified using mechanical advantage systems such as ropes and pulleys or roller systems. These

FIGURE 12-3 The walking wounded may need to be assisted out of the immediate release area and guided toward the decontamination line.

systems are especially useful when dealing with multiple non-ambulatory victims. During line-of-site victim rescue operations, mechanical advantage systems such as the Z-rig (a 3:1 rope and pulley system) can drastically reduce the number of personnel required to make multiple rescues. However, these systems may not always be feasible depending on the location of the victims. For example, if the victims are deep within the building out of sight and around corners, it will be almost impossible to set up a mechanical advantage system. Without this aid, the operation will be very labor and resource intensive, and may affect the decision between calling it a rescue operation versus a body recovery operation.

VICTIM REMOVAL EQUIPMENT AND TECHNIQUES

Non-ambulatory victims, especially when they are unconscious, will be extremely difficult to remove from the hot zone wearing full PPE. It is therefore important to use equipment whenever possible.

Lifts, Carries, and Drags

Unaided lifts using PPE will be difficult and dangerous, and the chance of compromising your PPE are very high. Therefore, only selected lifts will be discussed. Any lifts that involve holding, carrying, or handling the victim in close contact and direct pressure on your PPE for a prolonged period of time increase the chances of tearing or otherwise damaging your chemical protective clothing or causing the hazardous chemical to permeate through your PPE more quickly (such as turnout gear if using the 3/30

FIGURE 12-4 START triage is an effective method to determine how to assign limited resources at a mass casualty incident. Mass casualty hazmat and WMD incidents are very stressful to manage due to the large number of resources and personnel that are needed to help victims, especially non-ambulatory victims, in the hot zone.

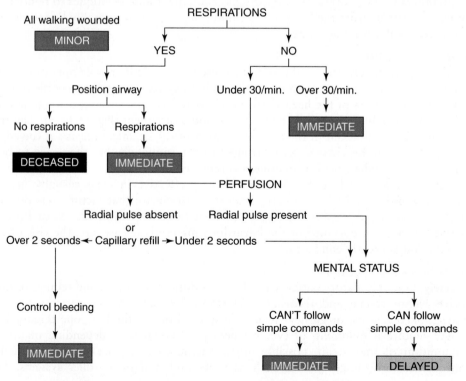

BOX 12.1 START TRIAGE

Triage is a system of prioritizing the order of patient care when available resources are limited. This most commonly occurs during mass casualty incidents. A mass casualty incident cannot be defined simply by a certain number of patients because the resources of different departments, agencies, and areas of the country differ dramatically. A mass casualty incident is any incident that overwhelms the resources of the responding agency.

START, which stands for "simple triage and rapid treatment," is a triage system based upon three quickly and readily identifiable signs: rate of breathing, pulse rate, and mental status. Depending on these three indicators, victims are sorted into four categories: green, yellow, red, and black (Figure 12-4). Ambulatory patients that have only minor injuries and can usually self-evacuate are triaged "green" and may even be used to help treat other patients (such as applying pressure to another patient's arterial bleed). Patients that are breathing, have a good pulse, and can follow simple commands are considered "yellow" and can wait for treatment. Patients that are having difficulty breathing, have a weak pulse, or cannot follow commands are considered "red" and need immediate treatment. Patients that are not breathing or are barely breathing (fewer than 10 breaths per minute), have no pulse, or have a very weak pulse and cannot follow commands are considered "black," or unsalvageable. In large mass casualty incidents, these patients will not be treated and will die due to a lack of resources.

Notice that, during a mass casualty incident, patients that in a typical situation would be aggressively treated are essentially left to die. The reasoning behind this system is that these patients would use too many resources and cause many more, less severely injured patients to die. In a mass casualty incident, we cannot save everyone, and we must use the available resources to save the most people possible. This may seem harsh, but it is necessary. Mass casualty incidents are some of the most difficult incidents for responders.

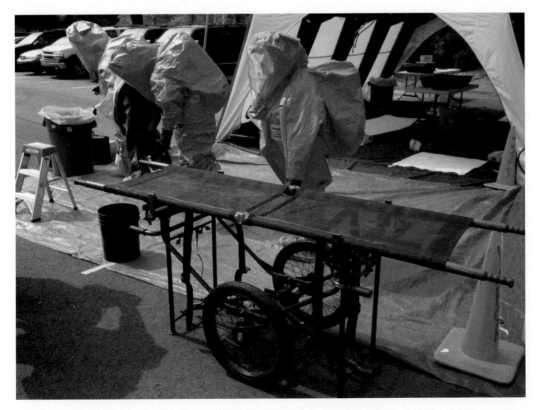

FIGURE 12-5 Heavy-duty stretchers with large wheels allow rescuers to navigate rough terrain, in some cases including steps.

rule guidelines). If this type of lift is your only choice, you must weigh the hazards and risks versus the benefit to be gained very carefully. Generally, lifts that require victim–rescuer contact should be used only with victims that you have reason to believe are not contaminated (such as with a gas or vapor exposure). Remember, if many people have been incapacitated by the chemical in a short amount of time, it is probably an extremely toxic and dangerous material. The indirect lifts discussed here will minimize the chances of damaging your PPE and becoming a victim yourself.

The amount of effort that you exert using lifts in full PPE will be tremendous. The chances of suffering from heat stress, including the potentially deadly heatstroke, are extremely high without the proper precautions. Personnel should wear some type of cooling vest, be well hydrated before entering the hot zone, and be cycled through short work shifts. The length of time a rescuer will be able to spend in the hot zone, or the number of victims a rescuer will be able to remove, will depend on environmental conditions, rescuer fitness, and the length and terrain of the rescue. Personnel must be monitored very closely, and the backup team should be ready to intervene in case a rescuer becomes incapacitated due to heat exhaustion or heatstroke. It is essential to practice these lifts and carries under controlled conditions, in full PPE, with safety systems in place so that rescue limitations can be accurately determined and unrealistic expectations are not formed by the command staff.

Clothing Drag   The clothing drag is an effective way to quickly remove a non-ambulatory victim that is wearing normal clothing. The victim is essentially grabbed by the collar and dragged out (Figure 12-6). The advantages of the clothing drag are that it is simple and requires no extra equipment. The disadvantage of this drag is that the victim's body creates friction, which makes victim removal more difficult.

Blanket Drag   The blanket drag is an effective way of removing a victim that is wearing flimsy clothing, such as a robe or nightgown (Box 12.2). A blanket or tarp is placed

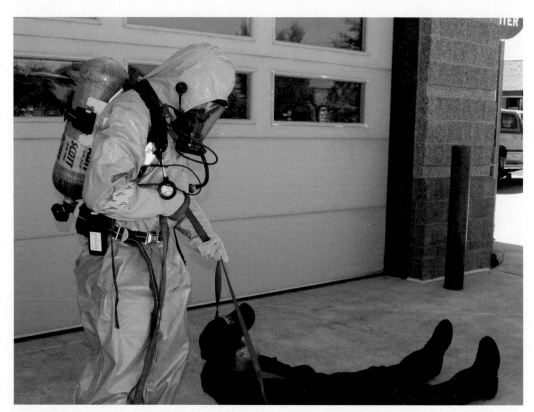

FIGURE 12-6 The clothing drag is an efficient victim removal method, but it affords the victim little protection. If the drag is performed poorly, the victim may have difficulty breathing or sustain additional trauma.

BOX 12.2 BLANKET DRAG

1. Place a blanket, tarp, or other suitable surface next to the victim.
2. If the victim is on his or her back, first roll the victim on his or her side, and then place the tarp underneath the victim's back.
3. Roll the victim onto the tarp and center the victim.
4. Drag the tarp from the end nearest the victim's head.

underneath the victim, and the blanket is used as a support to drag the victim out (Figure 12-7). The advantages of the blanket drag are that it is simple, and a blanket or other suitable material can usually be readily found. The disadvantage of this drag is that the victim's body on the blanket creates friction, which makes victim removal more difficult. However, if the "blanket" is chosen wisely, such as a plastic tarp, the friction can be reduced dramatically. In addition, most fire department engine companies and truck companies have tarps readily available. These tarps can be quickly folded to a suitable size or cut into appropriate sizes when many rescues need to be made.

Sling Drag The sling drag is another way to remove a victim that is wearing flimsy clothing. A sling of webbing, rope, or long clothing placed underneath the armpits of the victim is used to drag the victim out (Figure 12-8 and Box 12.3). The wider the sling, the more comfortable for the victim and the less chance of causing injury or aggravating a preexisting injury. Of course, if the victim is in an IDLH atmosphere, minor secondary injuries are of minimal concern. The advantage of the sling drag is that it can be quickly and easily performed with minimal equipment. The disadvantage of this drag is that the victim's body creates friction, which makes victim removal more difficult.

Chair Carry In a chair carry, the victim is placed in a chair, and one rescuer grabs the legs of the chair at the feet of the victim while the second rescuer grabs the back of the chair near the victim's torso and head (Figure 12-9). The victim may have to be restrained in the chair using rope or webbing to keep from falling off, depending on the type of chair, the victim's physical appearance, and the danger of the evacuation route. The advantages are that contact with the victim is minimized, the victim can be carried up and down stairs, and chairs are typically readily available. The disadvantage of the chair carry is that it requires two people.

Victim Retrieval Devices (VRD) and Rescue Equipment

One of the biggest disadvantages of lifts and drags is the amount of energy needed to perform them. When strenuous effort is needed in level A or level B PPE, heat-related emergencies in the first responder quickly become a significant concern. Using transfer devices to remove victims can dramatically reduce the amount of effort rescuers must expend

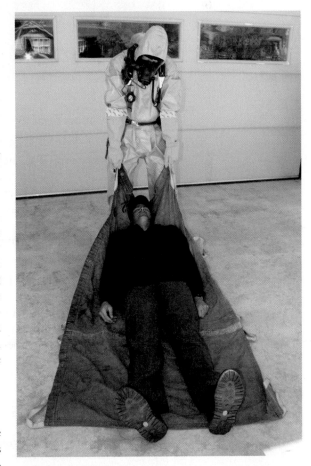

FIGURE 12-7 A blanket drag can be accomplished with any flat and flexible material, such as a blanket, tarp, shower curtain, large towel, or even a door.

FIGURE 12-8 Sling drag using a firefighter's emergency escape webbing.

to remove victims from the hot zone. The disadvantage is that this equipment must be on hand and available for use.

Patient Backboard Patient backboards can be used to remove victims from the hot zone. The victim can be strapped to the backboard and carried out by two people, or dragged by one person. Using a rigid backboard has the advantage of reducing the friction to a small surface area where the backboard edge contacts the ground. The advantage of using backboards is that most ambulances, medic units, and fire departments carry two or more backboards at any given time. The disadvantages of backboards are that they are bulky, and strapping the patient onto the backboard takes time.

Litter or Stokes Basket Litters and Stokes baskets have advantages and disadvantages similar to those of patient backboards. A Stokes basket is essentially a reinforced wire basket capable of supporting the weight of a person. Stokes baskets are often used for vertical lifting and lowering operations. The victim is strapped into these devices using webbing or rope and can be carried by two or more people or dragged by a single person

BOX 12.3 SLING DRAG

1. Place the webbing across the victim's chest and underneath the armpits.
2. Grab the webbing from over the victim's chest, and tuck the free ends through the loop.
3. Carefully tighten the loop around the back of the neck. CAUTION: The loop must be on the back and through the armpits to avoid victim strangulation!
4. Pull the victim using the sling. Position your hands to cradle the victim's head while moving him or her.

How would you remove the victim in the opening scenario from the hot zone?
Solution: There are many options, but a Sked board with c-spine precautions is one possibility.

(Figure 12-10). One advantage they offer is an attachment point for ropes and lifting, which may be necessary for complicated rescues at hazardous materials incidents. The disadvantage of these devices is the amount of time it takes to strap the person into the litter or basket.

Sked Board   Sked boards are often used in confined space rescues due to their narrow profile and lifting capabilities (if so rated) (Figure 12-11). These boards are also extremely useful at hazardous materials and WMD incidents. The Sked board is made of pliable plastic, yet is rigid enough to easily support the weight of the victim. The plastic also provides much lower friction than the victim's clothing, a blanket, or a Stokes basket when dragged. The Sked board also provides a point to tie off a rope, allowing the victim to be dragged from a distance when there is a clear line of sight. In fact, the victim can even be dragged using a simple mechanical advantage system, and the boards can be attached to multiple attachment points on a rope to remotely evacuate multiple patients rapidly.

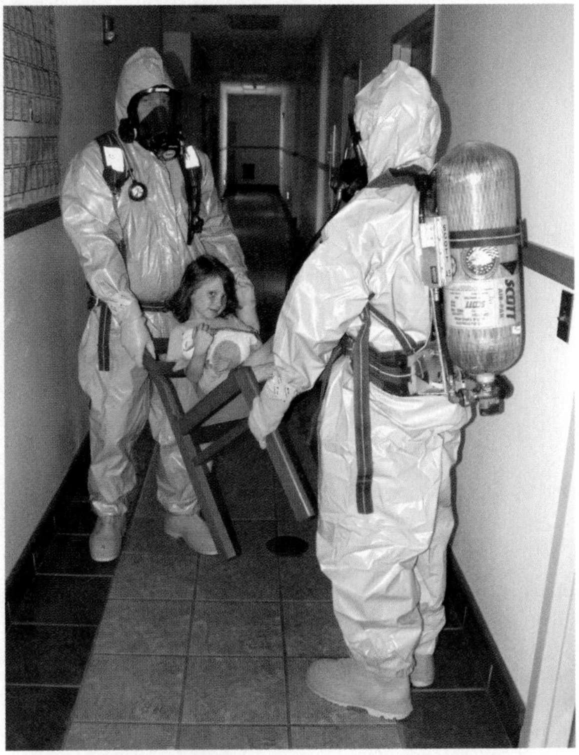

Assembly-Line Hot Zone Victim Removal Using Mechanical Advantage Systems
Sked boards, litters, and Stokes baskets are designed for use in combination with ropes and a mechanical advantage system to evacuate victims from heights, down the sides of a building, or from depths such as confined spaces. During mass casualty incidents involving hazardous materials or WMDs, ropes and mechanical advantage systems can

FIGURE 12-9 The chair carry is a readily available method to move victims and typically requires two rescuers. However, the chair could be dragged by a single rescuer when conditions permit (in the absence of stairs or other obstacles).

be used to ease the workload of first responders, thereby reducing heat-related injuries and strain- and sprain-type injuries, and speeding up the horizontal evacuation process. This section covers the application of ropes and mechanical advantage systems. In order to perform raises and lowers, you will need to have formalized technical rescue training due to the extreme dangers in which these maneuvers place both the victims and the rescuers.

Ropes, Knots, Carabiners, and Pulleys   The ropes used for a horizontal type of rescue may be rated rescue rope, water rescue rope (typically polypropylene), or utility rope of sufficient strength. Almost any type of rope may be used because the mechanical advantage system will be applied to an essentially *horizontal surface*, where rope failure will not result in further movement or otherwise endanger the victim. If the rope will be used for vertical raises and lowers or for bringing patients up or down inclines, the rope must be rescue rated, and all personnel constructing the raise/lower system must have formal technical rescue training. No exceptions!

Any secure knot will do the job for a horizontal type of rescue. However, the figure-eight follow through and the butterfly knot are very useful in this situation. The figure-eight follow through is a useful knot to tie a rope to an anchor, or to attach a carabiner to the

FIGURE 12-10 Stokes baskets, which are often used for high-angle rescues, can be used to move victims out of the hot zone.

FIGURE 12-11 A victim being evacuated using a Sked board. Depending on the type of surface, the transfer distance, and the size of the victim, one or two rescuers can effectively move a victim using a Sked board.

victim retrieval device (VRD) ■ Equipment used to rescue non-ambulatory victims from the hot zone.

end of the rope. The butterfly knot can create an attachment point for a carabiner in the middle of the rope. This knot can be used to attach **victim retrieval devices (VRDs)** along the length of the rope in the hot zone; these devices can then be quickly removed in the warm zone as the patients are removed from the line for decontamination, secondary triage, and treatment.

Simple 3:1 Mechanical Advantage System (Z-Rig) A mechanical advantage (MA) system is easy to construct and can significantly reduce the workload that rescuers need to perform. For example, a 3:1 mechanical advantage system reduces a 150-pound (68-kg) load to 50 pounds (23 kg) (not including friction). All that is needed to construct a Z-rig is a long piece of rope, two pulleys, and two carabiners. The Z-rig takes its name from the shape of the rope when it is laid out and in use.

Assembling a Z-rig style 3:1 mechanical advantage system for a horizontal (or nearly horizontal) pull:

1. Lay a rope out in a Z configuration, and assign one end of the rope to be the load end and the other to the haul end. One end will be attached to the victim (the load); the other will be used for hauling the victim out of the hot zone using the victim retrieval device (VRD). The load end should point toward the hot zone; the hauling end should point toward the cold zone. Note: The rope should be at least three times as long as the distance to be covered in the haul. In other words, if 150 feet (46 m) must be traversed, the rope should be 450 to 500 feet (137 to 152 m) long.
2. Place a strong knot (such as a figure-eight follow through) at the load end of the rope. This will be attached to the victim retrieval device (such as a Sked board).
3. Place a pulley in each of the two bights (bends in the Z) of the rope. One pulley should point toward the hot zone, the other pulley toward the cold zone.
4. Attach the pulley pointing toward the hot zone to the victim retrieval device, a few inches away from the rope (to avoid rubbing), using a carabiner.
5. Attach the pulley pointing toward the cold zone to a sturdy anchor using a carabiner and webbing or short piece of rope.
6. The mechanical advantage system is ready for use on a horizontal surface with straight, unobstructed access to the victim loading area (where the haul will begin). To facilitate the return of the victim retrieval device, a second rope may be placed on the other side of the VRD. The hot zone personnel can use this rope to retrieve the empty VRD.
7. Note: This simple type of mechanical advantage setup should *not* be used for lowering or raising systems. A lowering or raising system is defined as any situation in which letting go of the rope (haul line) would cause the load to move on its own. Safety systems are not built into the preceding mechanical advantage system!

SkyHook Continuous Loop Rescue System   A commercial product that is excellent for quickly retrieving multiple victims from the hot zone in an assembly-line fashion is the SkyHook Continuous Loop Rescue System. This device can be rapidly set up and consists of a rope with attachment sites for victim retrieval devices (VRDs), a hoist, and an optional motor. The occupied VRDs can be quickly retrieved from the hot zone, and the empty VRDs can be quickly returned without detaching VRDs from the rope.

SOLVED EXERCISE 12-7

Write the safety briefing you will give to the rescue team in the opening scenario.

Solution: You will be entering the hot zone from the north to perform a rescue of the unconscious victim next to flammable liquid tank #5 located on the north side of the facility. You will use the Sked board to package the victim in a snatch-and-grab rescue and the medical bag to perform any immediately lifesaving measures such as artificial respirations using the BVM or chest compressions. You will be wearing full structural firefighting protective gear with SCBA, which will provide only very limited chemical protection. You will wear chemical protective gloves underneath your leather gloves. Spend the least amount of time possible in the hot zone, and do not kneel in or otherwise come in contact with the product. The chemical hazard is posed by a mixture of solvents, including MEK, acetone, methanol, and toluene, which are of varying toxicity and are very flammable liquids. In case you do come into contact with the material, a decontamination line using dry decon for the PPE and wet decon using soap and water as necessary has been set up on the north side of the hot zone. Contamination will initially be detected, and decontamination will be verified, using a photoionization detector (PID), which can detect all four flammable liquids involved in the release. You will perform continuous air monitoring for oxygen and combustible gases using the RKI Eagle monitor while downrange. Calibration has been verified through quantitative bump checking, and the monitor has been fresh air zeroed. A backup team will be standing by with a charged hose line and rescue equipment. Other hazards include debris from the explosion. Have you looked at the site map, and do you understand the area you will be working in? We will have radio communications on our main fire frequency. Do you have any questions?

Summary

Due to the rapid arrival of responders at the operations level at hazardous materials and weapons of mass destruction incidents, many lives could be saved by quickly performing rapid and effective victim rescue operations. A high level of preparation and planning are required to safely and effectively perform victim rescue operations. Agencies need to acquire the appropriate personal protective equipment and rescue equipment before attempting these operations. Agencies must also train their personnel in the appropriate victim rescue techniques and safety procedures. Incident commanders must perform a rapid and effective hazard-risk analysis before initiating victim rescue operations to ensure the safety of all responders on scene while maximizing the chance of survival of trapped victims. When all the pieces are in place, hazardous materials responders at the operations level can make a real difference when they are capable of performing victim rescue and recovery operations.

Review Questions

1. What are the advantages of early victim rescue and recovery operations at hazardous materials and weapons of mass destruction incidents?
2. What are 11 factors to consider when determining the feasibility of victim rescue and recovery operations?
3. Name six safety procedures that should be in place before victim rescue and recovery operations are commenced.
4. What are the four tactical conditions that you may find victims in at the scene of hazardous materials or weapons of mass destruction incidents?
5. What confined space entry team positions are defined in the confined space regulations?

Problem-Solving Activities

1. Using a SARA Title III facility in your jurisdiction, determine under what conditions rescue and recovery operations should be carried out. What factors did you consider when making this decision? Take into account the training and equipment available to you and your agency.
2. Using one of the conditions from Problem-Solving Activity 1, determine how you would perform the victim rescue. Take into account the training and equipment available to you and your agency.

References and Further Reading

Byers, M, M. Russell, and D. J. Lockey. (2008). Clinical care in the "Hot Zone." *Emergency Medicine Journal* 25:108-112.

Michigan State Police Emergency Management and Homeland Security Training Center. (2008). *Hot Zone Rescue for the Operational Level Responder* (Student Manual). Lansing, MI: Author.

National Fire Protection Association. (2008). NFPA 472, *Standard for Competence of Responders to Hazardous Materials/Weapons of Mass Destruction Incidents.* Quincy, MA: Author.

National Fire Protection Association. (2008). NFPA 473, *Standard for Competencies for EMS Personnel Responding to Hazardous Materials/Weapons of Mass Destruction Incidents.* Quincy, MA: Author.

Occupational Safety and Health Administration. (1984). 29 CFR 1910.134, *Respiratory Protection Standard.* Washington, DC: U.S. Department of Labor.

Occupational Safety and Health Administration. (1990). 29 CFR 1910.120, *Hazardous Waste Site Operations and Emergency Response (HAZWOPER).* Washington, DC: U.S. Department of Labor.

Weber, Chris. (2007). *Pocket Reference for Hazardous Materials Response.* Upper Saddle River, NJ: Pearson/Brady.

13

Operations Level Responders Mission-Specific Competencies: Illicit Laboratory Incidents

KEY TERMS

acid gas generator, *p. 364*

binary weapon, *p. 350*

condenser, *p. 342*

culture media, *p. 352*

distillation, *p. 340*

fractional distillation, *p. 342*

incubator, *p. 352*

precursor, *p. 339*

reagent, *p. 339*

reflux, *p. 340*

salting out, *p. 345*

scrubber, *p. 343*

simple distillation, *p. 342*

SLUDGEM, *p. 351*

solvent, *p. 339*

teratogen, *p. 358*

tissue culture, *p. 354*

tweaker, *p. 343*

vacuum filtration, *p. 345*

yeast extract, *p. 353*

OBJECTIVES

- Describe the types of chemicals that may be found in illicit laboratories.
- Describe the type of equipment that may be found in illicit laboratories.
- Describe the dangers that illicit laboratories pose to the first responder.
- Define the unique indicators of a clandestine methamphetamine lab.
- Compare and contrast illegal drug, chemical warfare agent (CWA), biological warfare agent (BWA), explosive, and radiological dispersion device (RDD) labs.
- Describe the types of booby traps and secondary devices that may be found in illicit labs.
- List the safety considerations before entering and operating in illicit labs.
- List four possible dangers from handling containers in illicit labs.

You Are on Duty! Meth Lab Discovered at the Local Motel

Kissimmee, FL

You are dispatched to a report of an explosion at one of the motels by the highway. It is a cool January evening, and upon arrival smoke and fire are showing from a third-floor unit. As fire crews gain access, they notice that the room's door has been blown off its hinges. The fire is quickly extinguished, and during overhaul what appear to be makeshift laboratory equipment and chemicals are found. This seems to be some sort of clandestine lab. Firefighters and hotel employees speculate that this may be a meth lab.

■ What do you do?

Let's try to answer this question in this chapter.

Illicit, or clandestine, laboratories can be some of the most dangerous hazardous materials incidents to which you can respond (Figure 13-1). By definition, these are not only hazardous materials incidents but also crime scenes. As such, you must be extremely careful of not only the hazardous atmospheres and the chemicals and equipment that are present, but also the presence of booby traps, secondary devices, and armed and dangerous suspects. Therefore, responses to clandestine laboratories are typically multiagency coordinated responses.

FIGURE 13-1 An example of an illicit methamphetamine laboratory that contains a wide variety of laboratory and homemade equipment, chemicals, and precursors.

NFPA 472 (2008) recommends training to the awareness and operations levels with the additional Mission-Specific Competency in Personal Protective Equipment (Chapter 6). Due to the complexity and dangers involved in clandestine laboratory response, it is highly recommended to also have the Mission-Specific Competency in Air Monitoring and Sampling (Chapter 11) as well as the Mission-Specific Competency in Evidence Preservation and Sampling (Chapter 9). If you have a thorough understanding of air monitoring and sampling, you will be less likely to overlook critical safety hazards. If you have a thorough understanding of evidence preservation and sampling, you will be much less likely to destroy evidence, which may make criminal prosecution impossible.

In this chapter, we cover how to recognize and identify different illicit laboratory types, discuss the key response considerations and necessary tools needed for clandestine laboratory, or clan lab, response, as well as the basis of sound tactical decision making for responding to illicit laboratories.

Illicit Laboratory Recognition and Identification

The many different types of illicit laboratories may produce many different final products. Recognition of the type of laboratory is essential for your own safety, the safety of the public, notification of the appropriate authority having jurisdiction (AHJ), and successful evidence collection and ultimately prosecution. Currently, the most common types of laboratories are clandestine drug laboratories, especially methamphetamine-producing labs.

CLANDESTINE LABORATORY CONFIGURATION

Clandestine labs may be stationary or mobile. Stationary illicit laboratories may be found just about anywhere: hotel and motel rooms, homes, storage units, apartments, rural areas, and urban areas. Mobile illicit laboratories may be found in cars, trucks, vans, RVs, campers, and boats. Due to the illicit nature of the clandestine laboratory operation, most clan lab operators will try to find a relatively secluded area to run their operation.

Clandestine laboratories are designed to produce a final product. The laboratories may be set up to extract a product from raw material, convert an unrefined product to a refined product, or synthesize the final product from several raw materials. The complexity of the laboratory will increase as it progresses from extraction to conversion, to synthesis operations. However, a relatively simple extraction laboratory may be just as dangerous as a complex synthetic laboratory, depending on the chemicals present, how the equipment is set up, the expertise and housekeeping skills of the operator, and any booby traps or secondary devices that may be present.

Chemicals

A wide range of chemicals and other hazardous materials may be present at illicit laboratories. Each of these chemicals plays a vital role in generating the final product. The most important chemicals are the **precursors**, which are the raw materials that will be converted into the final product. Without the appropriate precursor in a sufficient quantity, the clandestine laboratory will not be able to produce a final product. A common example of a precursor is pseudoephedrine in the synthesis of methamphetamine. **Reagents** are the chemicals that facilitate the conversion of the precursor to the final product. Reagents may or may not become part of the finished product. An example of a reagent is lithium metal in the synthesis of methamphetamine. Some reagents are catalysts, which do not become part of the finished product; consequently, catalysts are required in small amounts because they are not used up during the chemical reaction. An example of a catalyst is red phosphorus in the synthesis of methamphetamine. **Solvents** are chemicals used to dissolve the reagents and precursors so that the chemical reaction can occur. Solvents are used in large amounts and are usually the cause of fires and explosions in clan labs. Solvents are one of the primary reasons it is essential to use a combustible gas indicator

precursor ■ A key reagent that is incorporated into the final product.

reagent ■ A substance that is used up in a chemical reaction and often becomes part of the final product.

solvent ■ A liquid capable of dissolving another material (solid, liquid, or gas) to form a mixture (a solution).

What are some of the chemical hazards you may expect in the illicit lab incident described in the opening scenario?

Solution: Flammable liquids (ethers, alcohols, etc.), corrosives (lye, sulfuric acid, HCl, etc.), water-reactive materials (sodium and lithium metals), and toxic materials (phosphine, etc.) can all be found in clandestine drug labs. Clandestine explosives labs will have oxidizers (peroxides, nitric acid, perchlorates, etc.) in addition to flammable liquids and other corrosives. Chemical warfare agent (CWA) labs have many other toxic compounds along with the aforementioned items. Biological warfare agent (BWA) labs may initially look similar to what a home brewer or home winemaker may have set up.

reflux ■ The process of carrying out a chemical reaction with the addition of heat.

distillation ■ The process of separating a mixture of liquids by heating the mixture to the boiling point of the individual substances, starting with the lowest boiling point liquid, and subsequently condensing the purified substance in a condenser. Each substance is purified in turn by primarily converting the most volatile liquid in the mixture into a vapor at any one time, and not the other liquids in the mixture.

(CGI) during any illicit laboratory reconnaissance, during the chemical and hazard assessment phase, while gathering evidence, and during the remediation phase. Otherwise, the clan lab response can turn deadly!

Glassware and Equipment

Illicit laboratories may contain a lot of exotic glassware and equipment, or they may consist of a number of normal items hooked up into funny-looking contraptions (Figure 13-2). All of this equipment is designed to facilitate one or more chemical reactions. Exotic glassware and equipment is commercially designed for use in industrial laboratories or research laboratories, on either a small scale or a large scale, depending on the particular use. The funny-looking contraptions constructed of normal household items are improvised laboratory glassware and equipment, which are often used in mom-and-pop labs.

Let's explore some of the chemical reactions that you might run across in illicit laboratories. **Reflux** reactors are used to heat a chemical reaction over a period of time (Figure 13-3). Heat increases the speed of the reaction. The reflux reactor is designed

FIGURE 13-2 Examples of laboratory and homemade glassware.

to limit the evaporation of the solvent or other volatile chemicals used in the reaction. The reflux reaction used in illicit laboratories is similar in concept to simmering a stew on your stove using a covered pot. The steam condenses on the lid and drips back into the stew to continue the cycle.

Distillation reactions are used to separate chemicals. Typically, they separate the final product or intermediate from the reagents, solvents, and precursors (Figure 13-4). The reaction vessel at the bottom is heated to volatilize the components. In a **fractional distillation** apparatus, the distillation column is used to remove less volatile components, which allows the more volatile components to separate and continue to rise and move into the **condenser**, or condensation column. A **simple distillation** apparatus does not have the vertical distillation column and is used to separate materials that have markedly different boiling points (Figure 13-4, left). A fractional distillation apparatus is used to separate materials that have similar boiling points (Figure 13-4, right).

It is important to know what different glassware setups are designed to do so we may understand the hazards the glassware system could pose, and to be able to determine the most effective sampling points for evidence collection and sample identification. Both reflux apparatus and distillation apparatus must remain open to the atmosphere so that temperature changes will not cause an explosion or implosion. Typically, a tube will lead to a **scrubber**, which removes any harmful vapors (Figure 13-5). A scrubber typically contains a corrosive solution, either an acid or a base, which reacts with the harmful vapors and inactivates them. When investigating clandestine labs, be extremely careful not to disconnect any scrubbers, because you may inadvertently create an extremely hazardous, even deadly,

FIGURE 13-3 A reflux reactor is used to produce a final product from reagents and precursors over a period of time with the addition of heat. A reflux reactor is analogous to a slow cooker used in cooking. In fact, slow cookers have been discovered in illicit labs being used as reflux reactors to produce meth.

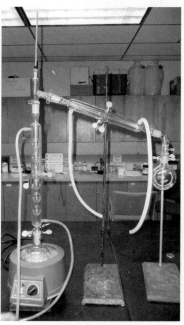

FIGURE 13-4 A simple distillation setup is shown on the left. A fractional distillation setup is shown on the right. Distillation is used to separate liquids from one another based upon their relative volatility.

What are some of the glassware and equipment hazards you may find in the meth lab described in the opening scenario?

Solution: A wide variety of equipment and glassware could be found in this type of lab including glassware to make the meth (such as round-bottom flasks, plastic bottles, cooking pots, etc.), heating equipment (hot plates, etc.), cooling equipment (such as an ice bath), gas generators (such as gasoline cans or soda bottles with a hose connected, producing hydrogen chloride gas), compressed gas cylinders (such as propane-type cylinders containing anhydrous ammonia), and glassware for evaporation (such as baking dishes).

fractional distillation ▪ The process of separating a mixture of liquids with similar boiling points by heating the mixture using a vertical fractionating or distillation column above the distillation vessel to increase the purity of each substance. Fractional distillation is used to separate materials with boiling points that are generally less than 50°F (25°C) apart. Also known as fractionation.

condenser ▪ A device used to transform a gas or vapor into a liquid while carrying out a chemical reaction or purifying a mixture. Condensers may be part of the equipment at an illicit lab. Also known as condensation column.

simple distillation ▪ The process of separating two or more liquids with boiling points at least 50°F (25°C) apart from each other through a gradual heating process.

atmosphere. Also be aware that, even if there is a scrubber, it may be set up incorrectly and not effectively remove dangerous contaminants. Air monitoring will let you know how dangerous the atmosphere is.

METHAMPHETAMINE LABS

Methamphetamine is an extremely addictive stimulant that can be easily produced from ephedrine or pseudoephedrine in one chemical step. This is one reason there are thousands of small mom-and-pop meth labs all across the country. In addition, there are also super labs run by large drug cartels. Each type of lab poses its own unique hazards.

Small mom-and-pop meth labs typically have smaller quantities of chemicals on hand than the super labs. Because the people that produce methamphetamine, known as "cooks," typically also use methamphetamine, they can be extremely unpredictable and

FIGURE 13-5 A chemical vapor scrubber. The liquid in the bottom of the Erlenmeyer flask on the right is used to absorb or neutralize the offending vapor or gas.

violent. This is especially true if they are in the tweaking stage of methamphetamine addiction (see Box 13-1). Small mom-and-pop labs may be protected by booby traps and other secondary devices designed to injure other meth cooks or users whom they believe are out to steal their finished product (see "Booby Traps and Secondary Devices" and Figure 9-4). However, booby traps have injured many law enforcement officers and clan lab responders. If you have a suspect in custody, he or she will often reveal the presence and location of booby traps because first responders are usually not the primary target. Therefore, it is very important to use good interrogation techniques at suspected meth labs.

scrubber ■ Equipment used to remove harmful compounds from the vapor phase of a chemical reaction.

BOX 13.1 METHAMPHETAMINE ADDICTION

Methamphetamine—also known as meth, speed, crank, and crystal—is an extremely addictive central nervous system stimulant. Meth causes its high, intense pleasure and euphoria by releasing a surge of dopamine, the primary pleasure neurotransmitter, in the brain. With prolonged use, meth destroys the dopamine receptors, making it very difficult to feel pleasure and eventually causing depression to set in.

There are several stages of methamphetamine abuse.

STAGE 1: THE RUSH

This is the initial intensely pleasurable response the user feels, caused by the release of adrenaline (epinephrine) throughout the body and dopamine in the brain. Characterized by a soaring pulse, blood pressure, and metabolism, the rush phase can last 5 to 30 minutes.

STAGE 2: THE HIGH

The high makes the user feel superior to others, and he or she becomes argumentative and often interrupts other people and finishes their sentences. The high phase can last 3 to 24 hours.

STAGE 3: BINGING

Binging is a continuation of the high. The user tries to maintain the high by smoking or injecting more methamphetamine, but each time the high decreases until it eventually stops altogether. Binging can last 2 to 15 days.

STAGE 4: TWEAKING

Tweaking occurs when the high can no longer be maintained with additional doses of methamphetamine. Tweaking is the most dangerous stage for first responders in contact with meth abusers. Symptoms include:

- Anxiety and insomnia
- Irritability and frustration
- Aggressiveness and psychotic behavior
- Paranoia
- Hallucinations
- Fatigue
- Depression
- Intense methamphetamine cravings

The extremely aggressive and paranoid behavior makes meth addicts in this phase, called **tweakers**, extremely unpredictable and dangerous. Loud noises and voices, uniforms, and bright lights can cause spontaneous aggressiveness. As a first responder, be extremely cautious when approaching, interviewing, treating, detaining, or apprehending these individuals!

tweaker ■ A methamphetamine addict in the most dangerous stage of the addictive cycle.

STAGE 5: THE CRASH

The user becomes extremely tired and listless, sleeping tremendous amounts. This phase, called "couch surfing," lasts 1 to 3 days.

STAGE 6: "NORMAL"

The user appears normal after the methamphetamine has left the system. This phase may last 1 to 14 days, but chronic users usually skip this stage and almost immediately seek more meth.

STAGE 7: WITHDRAWAL

After a lengthy time, maybe months, the user loses the ability to experience pleasure and becomes clinically depressed. Symptoms of depression include lethargy, a lack of energy, and sleeping a lot. The user may even feel suicidal. Due to severe depression, users in the withdrawal stage almost invariably begin using meth again.

The first three stages (the rush, the high, and binging) produce an intense euphoria and may last from several hours to weeks (with repeated doses). This euphoric phase becomes increasingly difficult to attain and maintain as the dopamine receptors are destroyed, leading to shorter highs and the use of higher meth doses. The last stages, including "tweaking," as the meth leaves the body, lead to agitation, which is often accompanied by violent behavior.

Visible signs of chronic meth use include severe acne caused by the addicts themselves scratching the skin trying to get at the "crank bugs" they think are crawling underneath. These sores take longer to heal because of tissue destruction and poor blood perfusion. Chronic meth addicts appear years to decades older than they actually are due to poor diet, wrinkles from the skin's losing its elasticity and luster, tooth decay from the corrosive nature of the meth and poor oral hygiene ("meth mouth"), and visible sores from scratching themselves.

Methamphetamine may be found in two distinct forms in any given clan lab. Methamphetamine hydrochloride is water soluble and is formed under acidic conditions (low pH). Methamphetamine base, or meth oil, is not water soluble and is formed under alkaline conditions (high pH). For example, methamphetamine hydrochloride may be dissolved in drinking water and transported covertly. Conversely, methamphetamine oil may be dissolved in a nonpolar solvent such as Coleman fuel and transported covertly. Be sure to pH any solutions that you come across in an illicit laboratory. Any acidic or basic solutions may contain methamphetamine and so must be sampled and possibly quantified for evidentiary purposes and then properly disposed of.

There are several different common methamphetamine synthesis routes that you may encounter. We explore each of these in turn. The type of illicit meth lab you will most commonly encounter will depend on the area of the country in which you work (Figure 13-6).

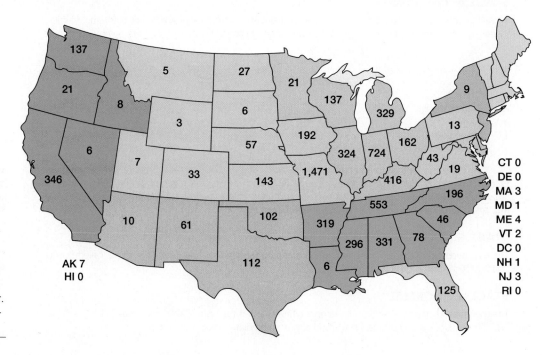

FIGURE 13-6 Methamphetamine laboratory distribution across the United States in 2010, which saw an increase in methamphetamine activity over the previous year. *Courtesy of the U.S. Drug Enforcement Administration*

Ephedrine and Pseudoephedrine Reduction Method

Methamphetamine can be made from either ephedrine or pseudoephedrine in numerous ways. The recipes involve a chemical conversion of either ephedrine or pseudoephedrine to methamphetamine. Through a single chemical reaction, called a reduction reaction, the precursor is converted to the final product. Recipes typically contain more than one step because of additional separation and purification processes. Let's explore some of the common methamphetamine synthesis routes.

Each ephedrine or pseudoephedrine lab will typically start with a tablet or liquid extraction step using a solvent and/or acids and bases (depending on the precursor source). You may find coffee grinders, mixers, or mortars and pestles to grind up the tablets, and hot plates, a stove, or heating mantles to evaporate the solvent. A filtering system is used to remove the undissolved solids. Filters may be as simple as a coffee filter and funnel to more complex glassware such as a Buchner funnel **vacuum filtration** system (Figure 13-7). The reflux reaction and meth oil production are then carried out according to each individual process (as described later). After the meth oil is produced, it is dissolved in a readily available solvent such as Coleman fuel or ether.

Methamphetamine is subsequently salted out of solution using a hydrogen chloride (HCl) gas generator (Figure 9-6). **Salting out** is the process that converts the meth oil to a meth salt (methamphetamine hydrochloride). Common constituents of HCl generators are sulfuric acid, muriatic acid, rock salt, and aluminum foil. If you find a gas can with a long hose attached, be especially careful because highly corrosive hydrogen chloride gas may be emitted unexpectedly. Even if the HCl generator does not look active, moving the container may mix the reagents inside and restart the hydrogen chloride generation process. Because the HCl generator is often reused, the tip of the hose is an especially good place to find residual methamphetamine that can be used as evidence, even if the bulk of the methamphetamine that was produced has already been removed from that location. After salting out, the solid methamphetamine final product is then filtered and possibly washed with clean solvent.

The most common chemicals found in methamphetamine labs are:

Denatured alcohol
Red phosphorous
Iodine
Anhydrous ammonia
Ammonium nitrate
Lithium metal
Sodium metal
Isopropyl alcohol (IPA)
Coleman fuel
Hydriodic acid (HI)
Sulfuric acid
Muriatic acid (HCl)
Rock salt
Aluminum foil
Methanol (HEET)
Ether
Sodium hydroxide (lye)

Other chemicals you may find in alternate methamphetamine recipes are:

Phenyl-2-propanone
Methylamine

vacuum filtration ■ The use of a negative pressure to separate solids from a liquid. A Buchner funnel, side arm flask, and vacuum source can be used to perform a vacuum filtration.

salting out ■ The process of separating two liquids by making one insoluble thereby selectively causing a phase change and turning it into a solid.

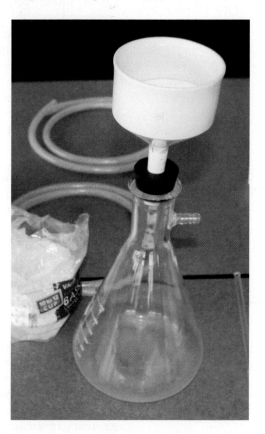

FIGURE 13-7 Vacuum filtration equipment. A Buchner funnel is used to remove a solid from a liquid.

Formic acid
Mercuric chloride
1,3-Dimethylurea
Lithium aluminum hydride
Sodium borohydride
Perchloric acid
Palladium
Barium sulfate
Phosphorus pentachloride
Thionyl chloride
Hydrogen
Benzyl chloride
Magnesium
Acetaldehyde
Phenylalanine
Benzyl chloroformate
p-Toluenesulfonyl chloride
Various solvents
Various acids and bases

Anhydrous Ammonia or "Nazi" Method

This methamphetamine production method uses anhydrous ammonia and sodium or lithium metal as unique ingredients. It also begins with a tablet extraction as mentioned earlier. In the reduction reaction, anhydrous ammonia and lithium or sodium metal are combined with the precursor. The anhydrous ammonia dissolves the lithium metal and forms a powerful reducing agent. After the reaction is complete, water is used to convert the unreacted lithium metal to lithium hydroxide, forming an alkaline reaction mixture. The meth oil will be floating on top of the alkaline reaction mixture. A solvent, such as Coleman fuel, is used to remove the meth oil, which is then converted to the final product as mentioned earlier.

The presence of a large number of damaged or destroyed lithium-ion batteries is a strong indication of the existence of an active methamphetamine lab that is using the Nazi or ammonia methamphetamine production method. Unique hazards of this method result from the storage of anhydrous ammonia in propane tanks, which may lead to tank failure. Anhydrous ammonia, a very corrosive material, attacks the brass fittings of the valve system and corrodes the interior of the tank. A visual indication that a propane tank contains, or may have contained, anhydrous ammonia is a bluish-green discoloration of the valves (Figure 9-6). Other hazards are lithium and sodium metal, which are water-reactive materials that spontaneously combust in the presence of moist air or water. This is why these metals are typically stored underneath kerosene or another suitable anhydrous solvent.

Red P Method

This methamphetamine production method uses red phosphorus and iodine as unique ingredients. It also begins with a tablet extraction as mentioned earlier. In the reduction reaction, red phosphorus and iodine are combined with the precursor in a reflux reaction. Following the cooking process, the final product is separated from the red phosphorus using filtration. A strong base, such as sodium hydroxide or lye, is used to separate the methamphetamine from the acidic reaction mixture, thus creating meth oil, which is then converted to the final product as mentioned earlier.

The presence of a large number of unused matchbooks is a strong indication of the existence of an active methamphetamine lab that is using the red P methamphetamine production method. Unique hazards of this method include the production of deadly phosphine gas when the red phosphorus is overheated. There have been numerous incidents of cooks being overcome by phosphine fumes and dying when the chemical reactions were improperly carried out and the toxic vapors were not vented. Another hazard is iodine. Although iodine is a solid, it sublimates, which means that it produces very toxic vapors.

One Pot Method

The one pot method differs from the preceding recipes in that all the ingredients are combined in one container and left to often react in a remote area. This method uses similar ingredients to the anhydrous ammonia/Nazi method described earlier, with ammonium nitrate replacing the anhydrous ammonia. The cooking process may be left unattended, and children may be injured when they unexpectedly come across the unattended one pot method. This method also has greater danger because a large amount of pressure can be generated when the reagents are combined in improper amounts. After a specified amount of time, the cooks come back to recover their reaction vessel, produce the meth oil, and salt out the final methamphetamine product as described earlier.

Phenyl-2-Propanone (P2P) Reduction Method

This methamphetamine production method uses phenyl-2-propanone (P2P) as the precursor instead of ephedrine or pseudoephedrine. This method has shown a resurgence since sales restrictions were instituted for ephedrine and pseudoephedrine. The key precursors are phenyl-2-propanone and methylamine; other ingredients in the process may be mercuric chloride, aluminum, sodium metal, palladium oxide, 1,3-dimethylurea, formic acid, ethanol, and/or methanol.

P2P is also a controlled substance. Therefore, the cooks may be making P2P from phenylacetic acid, which is a precursor in the production of P2P. This acid has a very unique odor, smelling strongly of cat urine. When you are in this type of lab, you will likely know it if you do not have the proper respiratory protection!

OTHER ILLEGAL DRUG LABS

Some of the most dangerous drug laboratories you can run into are PCP, LSD, and fentanyl laboratories. We will not go into great detail as to which chemicals are used in which processes. If you suspect you have encountered one of these laboratories, it is imperative that you exit the laboratory immediately, notify the DEA, and consult with technical experts as soon as possible.

Ecstasy (MDMA) and MDA

Ecstasy, or 3,4-methylenedioxymethamphetamine (MDMA); 3,4-methylenedioxyamphetamine (MDA); and 3,4-methylenedioxyethylamphetamine (MDEA) are in the amphetamine and methamphetamine class of drugs. Laboratories processing these drugs will contain typical chemical laboratory glassware and improvised glassware as mentioned earlier. The unique hazards you will encounter are highly flammable materials such as methylamine, ethylamine, and ether.

Chemicals you may encounter in MDMA (ecstasy), MDA, or MDEA laboratories are:

Isosafrole
Hydrogen peroxide
Formic acid
Ammonium formate
Muriatic acid
Piperonal
Nitroethane
Lithium aluminum hydride
Hydrobromic acid
Ammonia
Methylamine
Aluminum foil

Mercuric chloride
Mercuric bromide
Sodium hydroxide
Sodium cyanoborohydride
Methylenedioxyphenyl-2-propanone
Isopropanol
Ether

Fentanyl

Fentanyl is a powerful narcotic. Fentanyl laboratories will contain typical chemical laboratory glassware and improvised glassware as mentioned earlier. The unique hazards you will encounter are highly toxic and flammable materials such as acetonitrile and ether.

Chemicals you may encounter in a fentanyl laboratory are:

Acetonitrile
Propionic anhydride
Toluene
Ether
Muriatic acid
Potassium iodide
Sodium carbonate
Phenyl bromoethane

PCP

Phencyclidine, called PCP or angel dust, is a powerful hallucinogenic drug. PCP laboratories will contain typical chemical laboratory glassware and improvised glassware as mentioned earlier. The unique hazards you will encounter are highly flammable materials such as ether, and very toxic materials such as cyanide salts (potassium and sodium cyanide).

Chemicals you may encounter in a PCP laboratory are:

Piperidine
Muriatic acid
Cyclohexanone
Sodium cyanide
Potassium cyanide
Phenylmagnesium bromide
Sodium bisulfite
Ether
Isooctane
Ammonium chloride
Ammonium hydroxide
Methanol

GHB

γ-Hydroxybutyric acid (GHB) is a quick-acting depressant, also known as the date rape drug, that causes euphoria and memory loss. GHB laboratories will contain typical chemical laboratory glassware and improvised glassware as mentioned earlier.

Chemicals you may encounter in a GHB laboratory are:

Butyrolactone
Sodium hydroxide
Potassium hydroxide
Alcohol
Ethanol
Methanol

What clues could lead you to the type of illicit lab that was operating in the motel room described in the opening scenario?

Solution: Typically, no single component will be a smoking gun; but a combination of key precursors, other reagents, recipes, equipment, and the final product are a strong indication of what the clandestine lab is producing. For example, most of us have many of the components of a meth lab in our houses on any given day: a box of cold medication in the bathroom medicine cabinet, a can of starter fluid in the garage, drain cleaner under the kitchen sink, baking dishes and cooking pots in the kitchen. The difference is that in a meth lab the precursors will likely be found in unusually large quantities, and most of the components will be found near each other (for convenience).

CAT

Methcathinone (CAT) is a stimulant closely related to methamphetamine. CAT recipes start off with the same precursors as methamphetamine recipes, namely pseudoephedrine or ephedrine. CAT laboratories will contain typical chemical laboratory glassware and improvised glassware as mentioned earlier. The hazards you will encounter are similar to those in the methamphetamine laboratories. One unique hazard you may find at CAT labs is the very toxic chemical sodium dichromate, a known human carcinogen.

Chemicals you may encounter in a CAT laboratory are:

Pseudoephedrine
Ephedrine
Sulfuric acid
Sodium dichromate
Potassium permanganate
Acetic acid
Sodium hydroxide (lye)
Magnesium sulfate
Acetone
Toluene

CHEMICAL AGENT LABS

Fortunately, chemical agent laboratories are extremely uncommon. Yet they can be deadly due to the extreme toxicity of the final product being produced. Therefore, you should have a basic understanding of the types of chemical agents and their hazards as well as the laboratories you may encounter that produce these agents. We will not go into great detail as to which chemicals are used in which processes for security reasons. If you suspect you have encountered one of these laboratories, it is imperative that you exit the laboratory immediately, notify the FBI, and consult with technical experts as soon as possible. The state WMD civil support teams (WMD-CSTs) are composed of very well-trained hazardous materials responders and are technical experts in CWAs. They are a vital resource to have on site when confronted with this type of laboratory.

Blood Agents

Blood agents are a broad class of chemical agents that in some way affect the ability of the body to transport or use oxygen. Some blood agents affect the red blood cells and the protein within them that carries oxygen, called hemoglobin. Other blood agents, most notably hydrogen cyanide and cyanogen chloride, primarily affect the cells' ability to use oxygen. They bind to a protein called cytochrome oxidase and inactivate it, thereby preventing the cells from using oxygen to produce energy.

Blood agents are among the simplest chemical agents to produce, increasing the likelihood of your running across a rudimentary blood agent laboratory. Hydrogen cyanide is the simplest blood agent to produce, requiring only a cyanide salt and an acid. These chemicals may also be configured as a **binary weapon** that produces a blood agent when the two individual chemicals are mixed upon activation. Cyanide may also be produced from natural sources such as fruit seeds through a distillation process. This very laborious technique does not yield large quantities of hydrogen cyanide.

binary weapon ■ Munitions composed of two chemical precursor parts that when mixed form a chemical warfare agent.

Blood agent laboratories will contain typical chemical laboratory glassware and improvised glassware as mentioned earlier. A key piece of unique glassware that you may encounter is called a cold finger, which may be used to solidify the final product using a dry ice and acetone cold bath. The unique hazard you will encounter is primarily the highly volatile and dangerous final product: hydrogen cyanide.

Selected chemicals you may encounter in a blood agent laboratory are:

Sodium cyanide
Potassium cyanide
Potassium ferrocyanide
Sodium ferrocyanide
Various acids
Dry ice
Acetone

Vesicants

Vesicants, or blister agents, are a broad class of chemical agents that cause skin lesions and blistering. Most vesicants are slow acting and can take hours to manifest signs and symptoms. Early recognition and use of appropriate personal protective equipment is therefore extremely important when dealing with vesicants and vesicant laboratories. The common slow-acting vesicants are the sulfur mustards and nitrogen mustards. Lewisite, an arsenic-based vesicant, is fast acting. All of the vesicants will cause extreme skin damage, including blistering, and carcinogenic and teratogenic effects.

Vesicant laboratories will contain typical chemical laboratory glassware and improvised glassware as mentioned earlier. There is a wide variety of chemical synthesis routes for the common vesicants, including sulfur mustard, nitrogen mustards, and lewisite. The unique hazards you will encounter are the delayed onset of symptoms when exposed to the final product; the highly damaging, carcinogenic, and teratogenic nature of the vesicants; and the use of several highly toxic compressed gases such as hydrogen chloride, ethylene oxide, and chlorine.

Selected chemicals you may encounter in a vesicant laboratory are:

Thiodiglycol
Hydrogen chloride gas
2-Chlorohydrin
Sodium sulfide
Hydrogen sulfide gas
Ethylene oxide gas
Ethanol
Calcium hypochlorite
Thionyl chloride
Chlorine gas
Sulfur
Sulfur monochloride
Ethane
Activated charcoal
Nitrogen gas
Acetylene gas
Arsenic trichloride

2-Chloroethanol
Various acids and bases
Various solvents

Nerve Agents

Nerve agents are a broad class of chemical agents that interfere with nerve signaling and typically cause death by asphyxiation when the muscles of the diaphragm fail to function properly. Specifically, nerve agents inactivate the protein acetylcholinesterase, which breaks down the neurotransmitter acetylcholine, ending nerve signaling under normal conditions. But nerve agents inactivate acetylcholinesterase, and key muscles continue to contract. When the diaphragm continues to contract, the lungs cannot take in fresh air. When the muscles in the eyes continue to contract, it leads to pinpoint pupils, the classic nerve agent poisoning symptom. When secretory muscles continue to contract, they continue to excrete fluid. These effects are what define the primary signs and symptoms of nerve agent poisoning: **SLUDGEM** stands for salivation (drooling), lacrimation (tearing), urination, defecation, gastrointestinal upset, emesis (vomiting), and miosis (pinpoint pupils).

Nerve agent laboratories will contain typical chemical laboratory glassware and improvised glassware as mentioned earlier. There is a wide variety of chemical synthesis routes for the common nerve agents, including sarin, VX, tabun, and soman. The unique hazards you will encounter are the extremely toxic nature of nerve agents, the skin absorptive properties of nerve agents, the possible use of very corrosive and toxic chemicals such as hydrogen fluoride, and typically comparatively complicated chemical synthesis routes using a lot of glassware and equipment.

Selected chemicals you may encounter in a nerve agent laboratory are:

Phosphorous trichloride
Aluminum trichloride
Oxygen gas
Methyl chloride
Methylene chloride
Methanol
Hydrogen fluoride
Sodium metal
Ammonia
Isopropanol
Bromoacetic acid
Sulfur dioxide gas
Phosphorous oxychloride
Potassium fluoride
Triethyl phosphate
Potassium bifluoride
Ammonium bifluoride
Sodium fluoride
Various solvents
Various acids and bases
Various alcohols (precursors)

SLUDGEM ■ The signs and symptoms of nerve agent poisoning. SLUDGEM stands for salivation (drooling), lacrimation (tearing), urination, defecation, gastrointestinal upset, emesis (vomiting), and miosis (constricted or pinpoint pupils).

BIOLOGICAL AGENT LABS

Biological agent (BWA) laboratories, although uncommon, are easy to set up and operate. Anyone that has brewed beer at home has set up a biological laboratory, namely, growing the yeast that produces the alcohol in the beer. Furthermore, biological agent laboratories and the associated reagents and equipment are extremely common in industry

How would your approach differ if the incident described in the opening scenario were a chemical agent lab?

Solution: Safety and crime scene management are the biggest differences between a clandestine drug lab and a clandestine CWA lab. Chemical agents and their precursors are much more toxic than illegal drugs. The chemical processes (such as reactions, distillations, and purification) will be more complicated than most illegal drug manufacturing processes. Any personnel entering the lab must be in suitable PPE and use proper entry techniques (such as air monitoring). The clandestine lab would become a federal crime scene under the jurisdiction of the FBI because CWAs are considered weapons of mass destruction (WMD). It is very important that the crime scene be disturbed as little as possible (ideally only for life safety purposes) in order to permit clean evidence collection.

incubator ■ The equipment used to grow biological agents at a controlled temperature.

culture media ■ The food sources for growing microorganisms in fermentors as well as animal, plant and human cells in tissue culture.

and academia, due to the commercial and scientific usefulness of advanced molecular biological techniques. Biological agent laboratories may be set up using commercially and industrially available reagents and equipment, or relatively simple homemade or homebrew type of equipment.

Generally, the most difficult part of setting up a biological agent laboratory is obtaining a starter culture of the target organism. Many biological agents are weaponized, or converted to a form and size that is easily dispersed and inhaled. If you suspect you have encountered one of these laboratories, it is imperative that you exit the laboratory immediately, notify the FBI, and consult with technical experts as soon as possible. The state WMD civil support teams (WMD-CSTs) are composed of very well-trained hazardous materials responders and are technical experts in BWAs. They are a vital resource to have on site when confronted with this type of laboratory.

Three primary types of biological agents are commonly produced: bacteria, viruses, and toxins. Each of these different production methods will use different reagents and equipment, which allows them to be distinguished from one another with a little practice and experience.

Biological laboratories use a lot of unique glassware and equipment that is designed to keep the bacteria and viruses alive and thriving. Arguably the most important of this equipment and glassware is the **incubator**. Incubators come in many different sizes and shapes, depending upon the type of organism they are designed to culture, and the amount of product they are designed to produce (Figure 13-8). Small-scale incubators can be as simple as an Erlenmeyer flask containing **culture media** in a temperature-controlled orbital shaker. Large-scale incubators are fancy contraptions with many different inlets and outlets that permit the addition of essential gases such as oxygen and carbon dioxide, the addition of culture media and other nutrients, sampling points, and harvesting. The type of culture media that is used will help us to distinguish the different classes and types of organisms being grown.

Several different steps are necessary for successful biological agent production:

1. Bioprospecting
2. Isolation of the organism of interest
3. Characterizing the bacterial culture
4. Long-term storage of the bacterial culture

FIGURE 13-8 Incubator that could be used in a biological agent lab. *Photo by author. Courtesy of Boekel Scientific*

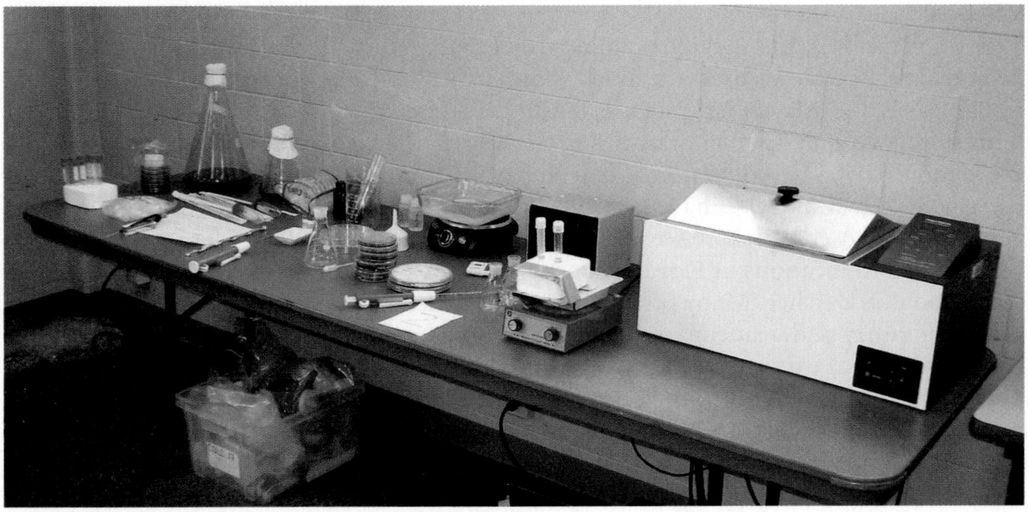

FIGURE 13-9 Bio-
logical agent laboratory
equipment.

5. Fermentation (small scale or large scale)
6. Product recovery
7. Weaponization

Each of these steps will be represented by unique reagents and equipment in a biological agent laboratory. Figure 13-9 illustrates the laboratory equipment you may find in use during these different stages.

Bacteria

Bacterial biological warfare agents include anthrax, plague, brucellosis, tularemia, cholera, and glanders. Many of these agents, such as anthrax, plague, and cholera, are endemic throughout different parts of the world and can be relatively easily bioprospected. Bioprospecting is the process of cultivating a biological organism from the wild. However, stocks of many biological warfare agents are commercially available because they are used in industrial and academic labs throughout the world for research and vaccine development.

Unique hazards of biological agent laboratories may be the presence of weaponized final product that poses a grave inhalation hazard. Generally, the types of chemicals used in bacterial laboratories are less hazardous than those found in drug, chemical warfare agent, toxin, or explosives laboratories.

Common reagents you may find in biological agent laboratories that are growing bacteria:

Yeast extract
Bacto™ agar
Casein
Powdered milk
Beef extract
Fish meal
Dextrose
Corn syrup
Glucose
Sucrose
Magnesium salts
Nitrogen gas
Disinfectants
Ethanol
Bleach

yeast extract ■ A nutrient source for the growth of bacteria used in illicit biological agent laboratories.

Viruses

Viral biological agents include smallpox, Marburg virus, Ebola virus, and other viral hemorrhagic fevers. Viruses are much more difficult to bioprospect than bacteria. For example, smallpox was eradicated in the 1970s and is now theoretically located in long-term storage in only two laboratories in the world (the Centers for Disease Control and Prevention [CDC] in Atlanta, GA, and Vector Institute in Russia). Terrorists seeking to use smallpox will have to have contacts within the biowarfare community, which is a very real fear. A comparable scenario happened in the fall of 2001 when anthrax letters were mailed to locations in Florida; Washington, DC; and New York allegedly by a U.S. bioweapons researcher. Many other dangerous viruses are commercially available and are used in industry and academia for research and vaccine production.

A unique hazard of viral agent laboratories may be the presence of weaponized final product that poses a grave inhalation hazard. Generally the types of chemicals used in viral production laboratories are less hazardous than those found in drug, chemical agent, toxin, or explosives laboratories.

Common reagents you may find in biological agent laboratories that are growing viruses:

tissue culture ■ A method used to grow mammalian cells that often serve as a growth medium for viruses.

Eggs
Culture media
Carbon dioxide gas
Growth factors
Tissue culture media
Disinfectants
Ethanol
Bleach

Toxins

Toxins are poisonous materials derived from living organisms. Toxins may be small molecules, similar in size to chemical agents, or large proteins such as ricin. Toxin manufacture may have two distinct production phases: the growth of the biological agent, and the isolation and purification of the toxin itself. Thus, toxin laboratories may have characteristics of both biological agent laboratories and chemical agent laboratories. They may contain biological production equipment and reagents as well as many chemicals and glassware. As with all of the laboratories that have been discussed, the glassware may be commercially available laboratory glassware or made from homemade, improvised materials.

Growth of the biological agent will be similar to the viral or bacterial processes described earlier. Isolation of the toxin will involve opening the biological agent to release its contents, which contain the toxin. This may be the final step for some small molecule toxins and may be used as a crude, dirty mixture. Other toxins, especially proteins, are typically purified using the cycles of a combination of salting out and filtration.

Common reagents you may find in biological agent laboratories that are producing toxins:

Rancid meat
Castor beans
Ammonium sulfate
Sodium sulfate
Sodium hydroxide
Acetone
Heptane
Carbon tetrachloride
Hexamethaphosphate (HMP)
Casein
Powdered milk

How would your approach differ if the incident described in the opening scenario were a biological agent (BWA) lab?

Solution: Safety and crime scene management are the biggest differences between a clandestine drug lab and a clandestine BWA lab. Weaponized biological agents can be extremely hazardous by inhalation and will not produce immediate signs and symptoms (with the exception of certain toxins). A BWA lab will look substantially different from a meth lab. Personnel entering the lab must still be in suitable PPE and use proper entry techniques (such as air monitoring). Other detection equipment, primarily handheld assays and polymerase chain reaction (PCR) assays, will be necessary to evaluate the suspected lab. The clandestine lab would become a federal crime scene under the jurisdiction of the FBI because BWAs are considered weapons of mass destruction (WMD). It is very important that the crime scene be disturbed as little as possible (ideally only for life safety purposes) in order to permit clean evidence collection.

Yeast extract
Peptone
Glucose
Corn syrup
Sodium citrate
Sodium benzoate
Sodium azide
Various acids and bases
Various solvents

EXPLOSIVES LABS

Explosives are the most commonly used weapons of mass destruction (WMD) in the United States and worldwide. Traditionally, commercially available and easily produced explosives such as ammonium nitrate–fuel oil mixtures (ANFO) have been used. For example, this explosive was used in the Alfred P. Murrah Federal Building bombing in Oklahoma City in 1995. Recently, terrorists have been producing organic peroxide–based explosives such as triacetone triperoxide (TATP) and hexamethylene triperoxide diamine (HMTD). TATP is an explosive that is commonly used by al-Qaeda operatives. We will not go into great detail as to which chemicals are used in which processes for security reasons. If you suspect you have encountered one of these laboratories, it is imperative that you exit the laboratory immediately, notify the FBI and your local bomb squad, and consult with technical experts as soon as possible.

Explosives laboratories contain typical chemical laboratory glassware and improvised glassware as mentioned earlier. There is a wide variety of chemical synthesis routes for the common explosives, including ANFO, TNT, RDX, PETN, urea nitrate, TATP, and HMTD. The unique hazards you will encounter are the extremely sensitive nature of the final product, strong oxidizers, corrosives, and flammable liquids. Common names associated with oxidizers that you may find in an explosives laboratory are:

Peroxide (such as hydrogen peroxide or methyl ethyl ketone peroxide)
Nitrate (such as nitric acid, ammonium nitrate and potassium nitrate, or saltpeter)
Perchlorate (such as ammonium perchlorate)
Chlorate (such as sodium chlorate)
Hypochlorite (such as calcium hypochlorite, or bleach)
Permanganate (such as potassium permanganate)
Chromate (such as sodium chromate and chromic acid)
Dichromate (such as potassium dichromate)
Iodate (such as lead iodate)
Periodate (such as potassium periodate)

Powdered metals, such as aluminum and magnesium, may also be used as fuels in an improvised explosive.

Nitrated Explosives

You often will encounter highly corrosive nitric acid in laboratories that are producing nitrated (nitro group–based) explosives such as TNT, picric acid, RDX, HMX, PETN, and nitroglycerin. Nitric acid is an oxidizing acid and can donate a nitro group to the final product, the nitrated explosive. The presence of nitric acid, a strong oxidizer, will be a good indication that you are dealing with a nitrated explosives lab.

Selected chemicals you may encounter in a nitrated explosives laboratory are:

Ammonium nitrate
Potassium nitrate
Fuel oil (or related)
Nitric acid
Mercury
Ethylene glycol
Methanol
Picric acid (trinitrophenol)
Phenol
Pentaerythrite
Glycerin
Toluene
Hexamethylenetetramine
Urea
Sugar
Charcoal
Cotton
Aluminum powder
Sulfur powder
Sulfuric acid
Various acids

Peroxide Explosives

Peroxide-based explosives have been widely used recently, especially in the Middle East by terrorist organizations. Several plots have been disrupted in the United States and the United Kingdom in recent years involving the synthesis and use of this type of explosive. Given the relative ease with which these ingredients can be acquired, this is the most likely explosives lab you will encounter. Be careful when investigating any illicit laboratory, and do not assume it is a drug lab. The presence of hydrogen peroxide, a strong oxidizer, will be a good indication that you are dealing with a peroxide explosives lab.

Selected chemicals you may encounter in a peroxide explosives laboratory are:

Acetone
Hydrogen peroxide
Methyl ethyl ketone
Toluene
Hexamine
Citric acid
Sulfuric acid
Hydrochloric acid (muriatic acid)
Various acids

Radiological Dispersal Devices (RDDs)

Radiological dispersal devices (RDDs) have been in the news recently as common explosive devices with radioactive material attached to, or incorporated into, them. The

How would your approach differ if the incident described in the opening scenario were an explosives lab?

Solution: Safety and crime scene management are the biggest differences between a clandestine drug lab and a clandestine explosives lab. Explosives and their precursors are much more unstable than illegal drugs or their precursors. Some explosives manufacturing processes are exceedingly simple and may be initially overlooked. The presence of unique precursors such as oxidizers, including peroxides, nitrates, and perchlorates, is a key indication of an explosives lab. Personnel should not enter this type of lab without advice from technical experts such as an EOD team. The clandestine lab would then become a federal crime scene under the jurisdiction of the FBI because explosives are considered weapons of mass destruction (WMD). It is very important that the crime scene be disturbed as little as possible (ideally only for life safety purposes) in order to permit clean evidence collection.

greatest hazard of RDDs remains the explosives. Most of the casualties will result from the energy released by the explosive. However, the radioactive material is a powerful psychological weapon and creates a sense of worry among the public in the general vicinity of the detonation site. Common radioactive materials that you may encounter in an RDD laboratory are radioactive americium, cesium, thorium, strontium, cobalt, or other common medical or research isotopes such as radioactive iodine, phosphorus, or sulfur. The unique hazard of RDD laboratories will be the presence of radiation from poorly shielded radioactive materials, which can be detected only by using radiological detection instruments such as a Geiger counter.

Preparation and Planning

There are three distinct phases in which an illicit laboratory may be discovered:

1. Operational (active)
2. Nonoperational (inactive)
3. Boxed (inactive)

Operational labs are fully functional and actively producing materials, nonoperational labs are fully or partially set up but not active (pre- or postproduction), and boxed labs are in storage. As you can imagine, each of these phases is not created equal. The most dangerous phase to find an illicit lab in is the functional and actively producing phase because all the necessary chemicals, reagents, solvents, precursors, and products are present. During this phase, most of the equipment is energized and poses significant hazards such as a potential ignition source, electrocution hazard, thermal hazard, or explosion hazard. Actively producing laboratories must be treated with the utmost care! Inactive labs are generally the least dangerous type of lab to encounter. However, also be cautious around the inactive labs, even the boxed ones, because dangerous chemical residue may remain on the glassware and equipment.

The common hazards that illicit laboratories may share are:

- Flammability
- Toxic chemicals
- Corrosive materials
- Dangerous glassware and equipment
- Electrical hazards
- Mechanical hazards
- Thermal hazards
- Pressurization hazards
- Secondary devices and booby traps

Many chemicals, such as benzene, carbon tetrachloride, and chloroform, are carcinogens, or cancer-causing agents. Yet other chemicals, such as benzene, chloroform, phenylacetic acid, and iodine, are **teratogens**, which cause birth defects.

HAZWOPER (29 CFR 1910.120) addresses how to control hazards at hazardous materials incidents. Hazards should first be addressed using engineering controls, followed by administrative controls, followed by personal protective equipment (PPE). What does this mean? Engineering controls generally use technology to control or eliminate a hazard. For example, rather than using personal protective equipment and entering a flammable atmosphere, first use the engineering control of ventilation, which can significantly reduce or eliminate the threat posed by the flammable atmosphere. When no feasible engineering controls to eliminate a hazard exist, then apply administrative controls.

Administrative controls do not address the hazards directly, but rather focus on how to make the way we work around those hazards as safe as possible. For example, when dealing with high heat in an open area, an administrative control procedure would be to shorten the duty cycle of entry team members from 45 minutes per person to 20 minutes. Finally, when engineering controls and administrative controls have proven to be insufficient to eliminate the hazard, then use personal protective equipment. For example, even though engineering controls and administrative controls have been applied, the characterization of a clandestine laboratory still poses significant hazards. Therefore, entry team members also wear the appropriate level of personal protective equipment. Typically, all three hazard control methods are applied at hazardous materials incidents. But always remember that the use of personal protective equipment is the hazard control method of last resort!

Response

During a response to any illicit laboratory, the following must be accomplished:

1. Coordinate crime scene operations with the law enforcement agency having jurisdiction.
2. Ensure that the law enforcement agency having jurisdiction secures and preserves the crime scene.
3. Perform the first entry and reconnaissance jointly with SWAT, EOD, and hazardous materials teams to avoid booby traps, secondary devices, and hazardous atmospheres.
4. Perform continuous air monitoring during the operation.
5. Have a decontamination plan in place that can effectively and safely deal with all of the deployed assets, including SWAT, EOD, and K-9 units.
6. Mitigate the immediate hazards while preserving evidence.
7. Coordinate post–crime scene remediation operations.
8. Thoroughly document all completed operations.

Before entering the illicit laboratory, conduct a thorough hazard assessment based upon available facts such as dispatch information, interviews of witnesses and neighbors, a visual overview of the exterior, and thermal imaging data from the exterior. Always be aware of the potential for booby traps inside the building itself as well as in the exterior yard and grounds.

BOOBY TRAPS AND SECONDARY DEVICES

Booby traps and secondary devices are a very real risk at illicit laboratory operations. These devices come in many forms, including explosives, chemicals, firearms, and mechanical devices (Figure 3-37). Common booby traps include containers full of hazardous chemicals propped above doorways that fall on anyone entering, fishhooks strung at eye level both indoors and outdoors, shotguns rigged with a pull string tied to doorknobs, lightbulbs partially filled with gasoline or gunpowder inserted into light sockets, and secondary devices constructed from explosives. It is therefore very important to watch where you walk, be careful opening doors, and not to turn on light switches. Although most of these booby traps are designed to prevent the theft of illegal drugs, they are indiscriminate. Often the cooks will disclose the location of these booby traps to law enforcement officers to avoid stiff penalties should they injure or kill a first responder. But, certainly, do not count on it! In many cases booby traps and secondary devices have not been disclosed.

RECONNAISSANCE AND INITIAL LABORATORY CHARACTERIZATION

Before entering the immediate area of the illicit laboratory, a site safety plan should be in place that includes all of the following:

1. Atmospheric monitoring
2. Selection of appropriate respiratory protection and PPE
3. Decontamination plan

Air Monitoring

Atmospheric monitoring should be performed for the five types of hazards that may be encountered: radiation, corrosive atmospheres, oxygen-deficient or -enriched atmospheres, flammable atmospheres, and toxic atmospheres (Figure 13-10). Radioactive materials may be encountered in certain illicit drug laboratories, and most certainly in radiological dispersal device (RDD) laboratories. Geiger counters can be used to monitor for radiation (Figure 13-11). Corrosive atmospheres may be encountered in just about any of the laboratories that have been discussed. Wetted pH paper can be used to monitor for corrosive atmospheres. Flammable atmospheres may also be encountered in just about any of the labs that have been discussed. We must monitor for oxygen content to ensure that the combustible gas indicator (CGI) will function correctly. Therefore, most

FIGURE 13-10 Remote air-monitoring activities at a clandestine laboratory incident using a robot.

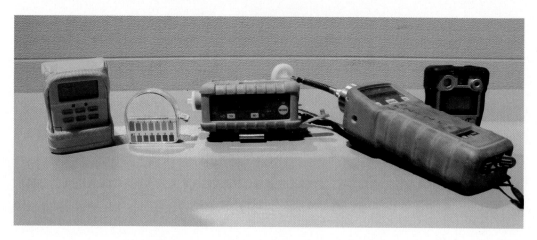

FIGURE 13-11 A variety of air-monitoring instrumentation. From the left is a Geiger counter, pH paper, a multi-gas monitor including oxygen sensor and combustible gas indicator (CGI), a photoionization detector (PID), and an ammonia detector. This equipment allows the hazmat responder to monitor for all five atmospheric hazards: (1) radiation, (2) corrosive materials, (3) oxygen, (4) flammable materials, and (5) toxic materials.

instrumentation, such as multi-gas detectors, that monitors for flammable atmospheres will also contain an oxygen sensor. Toxic atmospheres may also be encountered in just about any of the labs that have been discussed. Many different types of detectors are appropriate for low-level detection of toxic materials. Some of these are broad-range

SOLVED EXERCISE 13-7

What type of air monitoring would you do before continuing overhaul operations at the incident described in the opening scenario?

Solution: Before overhaul continues, atmospheric monitoring should be performed for all five hazards in the appropriate order: Monitor for gamma radiation from a distance (to rule out a dirty bomb manufacturing facility). As the lab is approached, corrosive atmospheres should be ruled out using wetted pH paper by placing it just in front of the other air-monitoring equipment. Oxygen and flammable vapors should be measured using a multi-gas meter, followed by the measurement of low-level toxic materials. These can be measured in a targeted fashion using a colorimetric indicator (tubes, paper, or tape) based upon the suspected hazard (phosphine, iodine, etc., in a meth lab); or a general survey instrument such as a photoionization detector (PID) could be used. If any hazards are found, the appropriate safety measures must be taken before continuing overhaul operations.

SOLVED EXERCISE 13-8

As a law enforcement officer investigating the incident in the opening scenario, what type of air monitoring would you do?

Solution: Before the lab is processed, atmospheric monitoring should be performed for all five hazards in the appropriate order: Monitor for gamma radiation from a distance (to rule out a dirty bomb manufacturing facility). As the lab is approached, corrosive atmospheres should be ruled out using wetted pH paper by placing it just in front of the other air-monitoring equipment. Oxygen and flammable vapors should be measured using a multi-gas meter, followed by the measurement of low-level toxic materials. These can be measured in a targeted fashion based upon the suspected hazard (such as colorimetric tubes for phosphine and iodine in a meth lab); or a general survey instrument such as a photoionization detector (PID) could be used. If any hazards are found, the appropriate safety measures must be taken before continuing the investigation.

detectors such as photoionization detectors (PIDs); others are very specific such as hydrogen sulfide sensors and colorimetric tubes. Air-monitoring instrumentation and strategies are discussed in greater detail in Chapter 11, Operations Level Responders Mission-Specific Competencies: Air Monitoring and Sampling.

Personal Protective Equipment

Respiratory protection and personal protective equipment must be carefully selected when responding to illicit laboratory incidents (Figure 13-12). Initial characterization of an unknown illicit laboratory should be conducted in level A personal protective equipment. If there are some initial indications that the type of laboratory is known, the initial characterization may be carried out in level B PPE. For example, if you have been dispatched to the suspected methamphetamine laboratory, level B PPE may be appropriate for an initial characterization. However, keep in mind that if unexpected hazards are encountered, you may have to exit the laboratory and upgrade your PPE. Always be acutely aware of flammability hazards, especially when wearing chemical protective clothing (CPC) that does not offer flash protection. Exit the area immediately if flammable vapors are discovered. This is why a combustible gas indicator (CGI) must be used when responding to illicit labs.

Illicit laboratory processing is often carried out in level B or level C PPE, depending on the hazard assessment. Once the hazards and chemicals have been characterized during the initial reconnaissance using air monitoring, the level of PPE is often downgraded. Always make sure that the PPE you are using is compatible with the chemicals found in the laboratory. Remember, we always want to be more cautious in unknown situations! Selection and use of personal protective equipment is discussed in greater detail in Chapter 6, Operations Level Responders Mission-Specific Competencies: Personal Protective Equipment.

FIGURE 13-12 Common personal protective equipment used at illicit laboratory incidents.

What type of PPE is appropriate at the incident described in the opening scenario? Why?

Solution: Turnout gear may be appropriate to complete overhaul if no skin absorptive chemical atmospheric hazards are found. If chemical hazards are found, chemical protective clothing (CPC) with flash protection may be necessary to complete overhaul.

Law enforcement officers may need to process the lab in level A protection if extensive atmospheric and chemical hazards are found, or they may require as little as level C or even level D protection if engineering controls and good work practices can be safely employed at the clandestine lab location. Continuous air monitoring must be carried out to ensure the continued safety of the workplace.

Decontamination

A decontamination plan must be in place before anyone enters an illicit laboratory per HAZWOPER (29 CFR 1910.120). At a minimum, this means a source of water must be available to remove contaminants from entry personnel. Typically, this requires that a technical decontamination line be in place that includes the following: a source of water; the ability to use soap; tools and equipment to perform decontamination; a secondary containment system for the used decontamination solution; and a disposal method for disposable PPE, used decontamination solution, and other decontamination waste (Figure 13-13). Personnel must be trained in proper technical decontamination of entry personnel. Technical decontamination is discussed in greater detail in Chapter 8, Operations Level Responders Mission-Specific Competencies: Technical Decontamination.

LABORATORY PROCESSING

During the initial reconnaissance, the area should be thoroughly photographed and videotaped before anything is touched or moved to give investigators a clear indication of what the laboratory looked like before intervention. Also during this overview stage, try to determine what type of illicit laboratory you are dealing with and how the chemical

FIGURE 13-13 A decontamination line being used at a clandestine methamphetamine lab investigation and cleanup. *Courtesy of Mike Becker, Longmont (CO) Fire Department*

FIGURE 13-14 Processing of clandestine laboratories. The hazmat responder is identifying, photographing, and logging the chemicals and glassware found at the site.

process is laid out within the lab. It is essential to determine these two pieces of information for a safe and successful illicit laboratory response.

Two common methods for processing illicit laboratories are the "process-in-place" method and the "process outside" method (Figure 13-14). The process-in-place method refers to characterizing, sampling, and dismantling the laboratory right where the laboratory is found. In this case, the analytical work that must be performed is done inside the lab before the chemicals are removed and placed in the chemical staging area. This method requires continuous air monitoring and ventilation during laboratory processing. An advantage of this method is that it is easier to understand the laboratory process under investigation. A major disadvantage of this method is that you are working around numerous hazardous materials in a cramped, confined space with poor lighting and other hazards. The process outside method refers to characterizing and sampling the laboratory components outside or away from the illicit laboratory itself. The advantage of this method is that a minimum amount of time is spent in the primary hazard zone, the illicit laboratory itself. The disadvantage of this method is that once the laboratory has been dismantled, it may be difficult or impossible to reconstruct the laboratory process, if it was not thoroughly documented during the reconnaissance phase of the operation.

Illicit laboratories should be processed systematically. Always work from one side of the lab to the other, or have another orderly work method in place. Following a systematic method will make the laboratory processing operation safer, make it easier to write the incident report, and make it easier to testify in court. Remove any easily accessible, nonessential materials that may become trip hazards first. Typically, next remove the highest hazards first to increase the safety of the processing operation. Then work on identifying, characterizing, and sampling the components of the laboratory that are essential to the process. Finding the final product is the ultimate goal for law enforcement purposes. The final product should be sampled, identified, quantified, and collected for evidence. After the essential components of the laboratory have been processed, the rest of the reagents can be identified and disposed of properly. Always keep in mind that you may find the final product in unexpected places later during the processing operation.

Container Handling

The initial characterization of an illicit laboratory should be conducted in a minimum of level B protection. During this stage, unknown containers will be visually examined, handled, and sampled. After all the containers have been thoroughly characterized and we are ready to remediate the lab, we may downgrade to level C protection based upon a thorough hazard assessment.

The first step before handling the container is a damage assessment. This assessment includes looking for distended or bulging containers, indicating a possible pressure buildup inside; leaking or weeping container seams; or other visible damage to containers such as cracks, dents, or corrosion. Also make sure that the container closures are present and not damaged. For example, caps on corrosive containers often are cracked and leak, whereas gaskets on solvent containers often are corroded or brittle and leak. For your safety, all of these issues should be addressed before handling these containers in any way.

Distended or bulging containers are often the most dangerous due to the possibility of pressure buildup. If possible, these containers should be moved out of the lab using remote handling before personnel continue with laboratory characterization or remediation. If this is not possible, a decision needs to be made whether the bulging container is stable and other containers can be characterized and removed while it is still in place, or whether this container needs to be dealt with first. Consult an expert before making these extremely important decisions!

Some common dangerous containers found in illicit laboratories include:

■ **Acid gas generators** (Figure 9-6)
■ Anhydrous ammonia containers (Figure 9-6)
■ Hot glassware (Figures 13-3 and 13-4)
■ Corroded gas cylinders (Figure 9-6)

Containers that require the use of self-contained breathing apparatus (SCBA) should be addressed and disposed of first. Always use the buddy system. When handling any container, use dry gloves; hold the container with both hands; and do not handle the container by its cap, neck, or any other delicate area. Move the containers out of the immediate work area, and make sure to segregate the chemicals by compatibility. For example, do not place strong acids and strong bases together, and do not place oxidizers and fuels (solvents) together in the same immediate area. Chemical staging areas should have good ventilation, plenty of room, and easy and unhindered access to avoid accidents. Before placing chemicals in the staging area, lay down a sheet of plastic as a secondary containment area in case any of the containers leak.

Sample Identification

Before sampling a container, ensure that it is safe to do so! Keep in mind that some containers may be booby-trapped, other containers may be over pressurized or weak, and yet other containers should not be opened because you are in the inappropriate level of PPE or ventilation is not sufficient.

Once it is determined safe to handle and open a container to perform sample identification, choose the appropriate tools and receiving containers. Generally, chemicals should be placed in glass containers, and materials of biological origin should be put in plastic containers. Liquids are typically sampled using pipettes, whereas solids are usually sampled using spatulas. Keep in mind that other techniques might be appropriate, such as swabbing. When dealing with liquids, carefully examine the container to determine whether multiple layers are present. If they are, be sure to sample each layer. For example, the final product such as meth oil may be a thin layer atop an aqueous solution. This layer may be very easy to miss, and it is, of course, one of the most important samples to recover. Depending on your jurisdiction's sampling protocols, you may be able to place multiple layers in a single

acid gas generator ■ The container and chemical contents that are designed to release an acidic gas and are often used in clandestine laboratories. These contraptions can be very dangerous to first responders if handled improperly. The most common acid gas that is generated is hydrogen chloride in methamphetamine labs.

container (as you found them), or you may have to place each individual layer in a separate container. Consult the analytical laboratory for the appropriate collection method in your area. Also sample any sediment found at the bottom of containers.

Next, consider how to identify the material, choosing among the different techniques available. Air-monitoring equipment may be used to assess the vapors that the material is emitting by carefully monitoring the headspace of the container. The liquid itself may be assessed using pH paper, oxidizer paper, and M8 or M9 papers (Figure 13-15). Field identification techniques include Raman spectroscopy, Fourier transform infrared (FTIR) spectroscopy, gas chromatography/mass spectrometry, and X-ray fluorescence (Figure 13-16). Not all agencies have access to this type of advanced equipment. Therefore, we will ultimately collect a sample to send to an analytical laboratory, especially if we suspect we are dealing with the final product. Sample identification is discussed in greater detail in Chapter 11, Operations Level Responders Mission-Specific Competencies: Air Monitoring and Sampling.

Evidence Collection

Evidence collection should be prioritized in the order of final product followed by essential precursors. The other chemicals—reagents and wastes—may or may not be considered evidence. However, consult with the local crime lab to determine the appropriate procedures. Remember, there is limited space in the evidence locker! Do not forget to collect an equipment blank and a sample blank, collect photographic documentation, and seal the sample container with evidence tape. Evidence collection is discussed in greater detail in Chapter 9, Operations Level Responders Mission-Specific Competencies: Evidence Preservation and Sampling.

FIGURE 13-15 The hazmat responder is determining the pH of a liquid at a clandestine laboratory.

REMEDIATION PLAN

Once all of the chemicals have been characterized, your job is not complete. After the chemicals have been characterized at a clandestine lab, we must dispose of them properly. This requires handling the containers so we can sort them by compatibility, take samples for laboratory analysis, and over pack leaky containers or those in poor condition.

SOLVED EXERCISE 13-10

You find several intact containers with solids and liquids in the illicit drug laboratory described in the opening scenario. How would you handle them as a firefighter? As a law enforcement officer?

Solution: As a firefighter performing overhaul operations, leave the containers for the hazardous materials team or the meth lab response team to identify and handle properly.

As a law enforcement officer, these containers must be properly documented for evidentiary purposes. The containers should be photographed before manipulation, the contents must be identified using appropriate equipment or reagents, the containers and their contents may be collected as evidence (such as the final product, meth), or the identified material must be disposed of properly in accordance with environmental regulations.

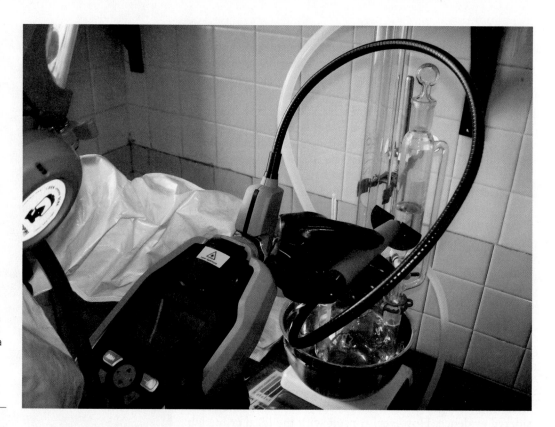

FIGURE 13-16 Using a Raman spectrometer at an illicit laboratory to identify the contents of a reaction flask. *Photo by author. Pictured is the AhuraFD, courtesy of Thermo Fisher Scientific*

The Resource Conservation and Recovery Act regulates hazardous waste from cradle to grave. The authority having jurisdiction is considered the generator of the hazardous waste and is mandated by law to dispose of it properly. This means finding a responsible hazardous waste contractor, filling out the hazardous waste manifest, and making sure that the return copy of the manifest is received once hazardous waste is disposed of according to law (Figure 13-17). Failure to do so can result in significant civil, and possibly criminal, penalties depending on the severity of the infraction.

SOLVED EXERCISE 13-11

You turn the incident described in the opening scenario over to a private cleanup contractor. Who is responsible for the hazardous waste that was generated in the lab?

Solution: The law enforcement agency having jurisdiction is responsible for the cradle-to-grave management of the hazardous waste, according to RCRA. The law enforcement agency must retain documents indicating the proper disposal for at least 30 years. In many cases, the DEA or the state police will assume responsibility for drug labs, especially in smaller jurisdictions, due to the complexity and cost associated with meth lab remediation.

Summary

Illicit laboratories can be some of the most dangerous hazardous materials incidents for first responders. These laboratories usually have chemical hazards, mechanical hazards, and electrical hazards, and often contain booby traps and violent offenders. Before entering a suspected clandestine laboratory, make sure you have law enforcement support. Never enter an illicit laboratory without the proper personal protective equipment and air-monitoring instrumentation. Ensure

that decontamination is available on scene. Depending on the type of laboratory, contact the appropriate agency having jurisdiction. This may include the Drug Enforcement Administration (DEA), the Bureau of Alcohol, Tobacco, Firearms and Explosives (ATF), or the Federal Bureau of Investigation (FBI). Never forget that illicit laboratories are both hazardous materials incidents and crime scenes.

Review Questions

1. What is the most common illicit laboratory encountered in the United States and Canada?
2. What are the key differences between a biological warfare agent laboratory and a chemical warfare agent laboratory?
3. How do organic peroxide–based explosives differ from other common industrial and military explosives?
4. Name nine common hazards of illicit laboratories.
5. Describe the importance of air-monitoring operations at illicit laboratory incidents.
6. What are four common dangerous containers at illicit laboratories?

Problem-Solving Activities

1. Review your agency's standard operating guidelines (SOGs) for response to illicit laboratory incidents.
2. Review your agency's personal protective equipment plan.
3. If your agency carries personal protective equipment, don and doff it according to the SOG.
4. If your agency carries decontamination equipment, set up a decontamination line you may use at an illicit laboratory incident.
5. You come across a clandestine laboratory that contains anhydrous ammonia, lithium batteries, and pseudoephedrine. What is it probably manufacturing?
6. You come across a clandestine laboratory that contains red phosphorus, iodine, and ephedrine. What is it probably manufacturing?
7. You come across a clandestine laboratory that contains isosafrole. What is it probably manufacturing?
8. You come across a clandestine laboratory that contains butyrolactone. What is it probably manufacturing?
9. You come across a clandestine laboratory that contains pseudoephedrine, sodium dichromate, and potassium permanganate. What is it probably manufacturing?
10. You come across a clandestine laboratory that contains thiodiglycol and an acid. What is it probably manufacturing?
11. You come across a clandestine laboratory that contains incubators, yeast extract, and casein. What is it probably manufacturing?
12. You come across a clandestine laboratory that contains hydrogen peroxide and acetone. What is it probably manufacturing?

Please print or type. (Form designed for use on elite (12-pitch) typewriter.) Form Approved. OMB No. 2050-0039

UNIFORM HAZARDOUS WASTE MANIFEST	1. Generator ID Number	2. Page 1 of	3. Emergency Response Phone	4. Manifest Tracking Number

5. Generator's Name and Mailing Address   Generator's Site Address (if different than mailing address)

Generator's Phone:

6. Transporter 1 Company Name	U.S. EPA ID Number

7. Transporter 2 Company Name	U.S. EPA ID Number

8. Designated Facility Name and Site Address   U.S. EPA ID Number

Facility's Phone:

9a. HM	9b. U.S. DOT Description (including Proper Shipping Name, Hazard Class, ID Number, and Packing Group (if any))	10. Containers		11. Total Quantity	12. Unit Wt./Vol.	13. Waste Codes
		No.	Type			
	1.					
	2.					
	3.					
	4.					

14. Special Handling Instructions and Additional Information

15. GENERATOR'S/OFFEROR'S CERTIFICATION: I hereby declare that the contents of this consignment are fully and accurately described above by the proper shipping name, and are classified, packaged, marked and labeled/placarded, and are in all respects in proper condition for transport according to applicable international and national governmental regulations. If export shipment and I am the Primary Exporter, I certify that the contents of this consignment conform to the terms of the attached EPA Acknowledgment of Consent.
I certify that the waste minimization statement identified in 40 CFR 262.27(a) (if I am a large quantity generator) or (b) (if I am a small quantity generator) is true.

Generator's/Offeror's Printed/Typed Name   Signature   Month Day

16. International Shipments ☐ Import to U.S. ☐ Export from U.S. Port of entry/exit:____
Transporter signature (for exports only):   Date leaving U.S.:

17. Transporter Acknowledgment of Receipt of Materials

Transporter 1 Printed/Typed Name   Signature   Month Day

Transporter 2 Printed/Typed Name   Signature   Month Day

18. Discrepancy

18a. Discrepancy Indication Space ☐ Quantity ☐ Type ☐ Residue ☐ Partial Rejection ☐ Full Reje
Manifest Reference Number:

18b. Alternate Facility (or Generator)   U.S. EPA ID Num
Facility's Phone:

18c. Signature of Alternate Facility (or Generator)

19. Hazardous Waste Report Management Method Codes (i.e., codes for hazardous waste treatment, disposal, and recy

1.	2.	3.	4.

20. Designated Facility Owner or Operator: Certification of receipt of hazardous materials covered by the manifest except as noted in Item 18a

Printed/Typed Name   Signature   Mon

EPA Form 8700-22 (Rev. 3-05) Previous editions are obsolete.   DESIGNATED FACILITY TO DESTINATION S

FIGURE 13-17 Example of the universal hazardous waste manifest issued by the U.S. EPA. *Courtesy of the Environmental*

References and Further Reading

National Fire Protection Association. (2008). *NFPA 472, Standard for Competence of Responders to Hazardous Materials/Weapons of Mass Destruction Incidents*. Quincy, MA: Author.

Occupational Safety and Health Administration. (1990). 29 CFR 1910.120, *Hazardous Waste Site Operations and Emergency Response (HAZWOPER)*. Washington, DC: U.S. Department of Labor.

Weber, Chris. (2007). *Pocket Reference for Hazardous Materials Response*. Upper Saddle River, NJ: Pearson/Brady.

Weber, Chris, and John Meyers. (2009). *Chemical Warfare Agent–Biological Warfare Agent Illicit Laboratory Response Course*. Longmont, CO: Dr. Hazmat, Inc.

LIST OF ACRONYMS

AAR	Association of American Railroads
ABCs	airway, breathing, circulation
AFFF	aqueous film-forming foam
AHJ	authority having jurisdiction
ALARA	as low as reasonably achievable
AMU	atomic mass unit
ANFO	ammonium nitrate–fuel oil
ANSI	American National Standards Institute
APR	air-purifying respirator
ARF	alcohol resistant foam
ATF	Bureau of Alcohol, Tobacco, Firearms and Explosives
ATR	attenuated total reflectance
BLEVE	boiling liquid expanding vapor explosion
BWA	biological warfare agent
CAA	Clean Air Act
CAMEO	Computer-Aided Management of Emergency Operations
CAS	Chemical Abstract Services
CBRN	chemical, biological, radiological, and nuclear
CBRNE	chemical, biological, radiological, nuclear, and explosive
CDC	Centers for Disease Control and Prevention
CERCLA	Comprehensive Environmental Response, Compensation, and Liability Act
CGI	combustible gas indicator
CHEMTREC	Chemical Transportation Emergency Center
CHRIS	Chemical Hazard Response Information System
CIH	certified industrial hygienist
CO	carbon monoxide
CPC	chemical protective clothing
CPR	cardiopulmonary resuscitation
CRC	contamination reduction corridor
CST	civil support team
CWA	chemical warfare agent
DEA	Drug Enforcement Administration
DEQ	Department of Environmental Quality
DHHS	Department of Health and Human Services
DHS	Department of Homeland Security
DOE	Department of Energy
DOT	Department of Transportation
EL	exposure limit
EMS	emergency medical services
EOD	explosive ordnance disposal

EPA	Environmental Protection Agency
EPCRA	Emergency Planning and Community Right-to-Know Act
ERG	*Emergency Response Guidebook*
ERPG	Emergency Response Planning Guideline
ESLI	end of service life indicator
eV	electron volt
FBI	Federal Bureau of Investigation
FEMA	Federal Emergency Management Agency
FFPE	firefighter protective ensemble
GC/MS	gas chromatography/mass spectrometry
HAZWOPER	Hazardous Waste Operations and Emergency Response (29 CFR 1910.120)
HD	distilled mustard agent (vesicant)
HEPA	high efficiency particulate air
HMIS	Hazardous Materials Identification System
HMTD	hexamethylene triperoxide diamine
HVAC	heating, ventilation, and air-conditioning
IAP	incident action plan
IBC	intermediate bulk container
ICS	incident command system
ID	identification
IDLH	immediately dangerous to life or health
IM	intermodal
IMO	International Maritime Organization
IMS	incident management system; ion mobility spectroscopy
IP	ionization potential
K-9	canine
LC_{50}	lethal concentration, 50% (by inhalation)
LD_{50}	lethal dose, 50% (by injection, skin contact, or ingestion)
LEPC	local emergency planning committee
LEL	lower explosive limit
LPG	liquefied petroleum gas
MAWP	maximum allowable working pressure
MC	motor carrier
MOPP	mission oriented protective posture
MSDS	material safety data sheet
MUC	maximum use concentration
MW	molecular weight
NFPA	National Fire Protection Association
NIMS	National Incident Management System
NIOSH	National Institute for Occupational Safety and Health
NLM	National Library of Medicine
NPL	National Priorities List

OJP	Office of Justice Programs	SLUDGEM	salivation, lacrimation, urination, defecation, gastrointestinal upset, emesis, and miosis
OSHA	Occupational Safety and Health Administration		
PAPR	powered air-purifying respirator	SOG	standard operating guideline
PASS	personal alert safety system	SOP	standard operating procedure
PCR	polymerase chain reaction	SSP	site safety plan
PEL	permissible exposure limit (OSHA)	START	simple triage and rapid treatment
PET	positron emission tomography	STCC	Standard Transportation Commodity Code
PID	photoionization detector	STP	standard temperature and pressure
PIO	public information officer	SWAT	special weapons and tactics
PPE	personal protective equipment	TATP	triacetone triperoxide
PSI	pounds per square inch	TDS	time, distance, shielding
PSIG	pounds per square inch gas	TI	transport index
RAIN	recognize, avoid, isolate, notify	TIC	toxic industrial chemical
RCRA	Resource Conservation and Recovery Act	TIH	toxic inhalation hazard
RDD	radiological dispersion device	TIM	toxic industrial material
REL	recommended exposure limit (NIOSH)	TLD	thermoluminescent detector
REM	Roentgen equivalent man	TSCA	Toxic Substances Control Act
RGasD	relative gas density	TSD	treatment, storage, and disposal
RQ	reportable quantity	TWA	time weighted average
SARA	Superfund Amendments and Reauthorization Act	UEL	upper explosive limit
		UN/NA	United Nations/North America
SAW	surface acoustic wave	UV	ultraviolet radiation
SBCCOM	U.S. Army Soldier and Biological Chemical Command	VRD	victim retrieval device
		WISER	Wireless Information System for Emergency Responders
SCBA	self-contained breathing apparatus		
SERC	State Emergency Response Commission	WMD	weapon of mass destruction

INDEX